VULNERABILITY

Alongside globalization, the sense of vulnerability among people and populations has increased. We feel vulnerable to disease as new infections spread rapidly across the globe, while disasters and climate change make health increasingly precarious. Moreover, clinical trials of new drugs often exploit vulnerable populations in developing countries that otherwise have no access to healthcare and new genetic technologies make people with disabilities vulnerable to discrimination. Therefore the concept of 'vulnerability' has contributed new ideas to the debates about the ethical dimensions of medicine and healthcare.

This book explains and elaborates the new concept of vulnerability in today's bioethics. Firstly, Henk ten Have argues that vulnerability cannot be fully understood within the framework of individual autonomy that dominates mainstream bioethics today: it is often not the individual person who is vulnerable, rather that his or her vulnerability is created through the social and economic conditions in which he or she lives. Contending that the language of vulnerability offers perspectives beyond the traditional autonomy model, this book offers a new approach which will enable bioethics to evolve into a global enterprise.

This groundbreaking book critically analyses the concept of vulnerability as a global phenomenon. It will appeal to scholars and students of ethics, bioethics, globalization, healthcare, medical science, medical research, culture, law, and politics.

Henk ten Have is Director of the Center for Healthcare Ethics at Duquesne University in Pittsburgh, Pennsylvania, USA. He studied medicine and philosophy in the Netherlands, and worked as a Professor in the Faculty of Medicine of the Universities of Maastricht and Nijmegen. From 2003 until 2010 he joined UNESCO in Paris as Director of the Division of Ethics of Science and Technology. His recent publications include *Handbook of Global Bioethics* (2014) and *Global Bioethics: An Introduction* (2016). He is currently working on the *Encyclopedia of Global Bioethics* (2016).

'With one eye on the recent bioethical discourse and the other on philosophical texts, Henk ten Have provides a comprehensive study of vulnerability understood in all its dimensions. His study includes medical and humanitarian concerns and issues of justice. However, ten Have does not simply oppose vulnerability to autonomy. Not only is the acceptance of vulnerability the condition for autonomous agency, but the insistence on the positive aspect of this notion also refreshes the way we frame many other important ethical and political categories. Informative and stimulating, this book is also well written and available to a large audience.'

Corine Pelluchon, Full Professor in Philosophy and Applied Ethics at the University of Franche-Comté, France

'Drawing on a rich fund of resources, including medical anthropology, Continental European philosophy, feminist bioethics, care ethics, and social and political philosophy, Henk ten Have has produced a masterly account of vulnerability. But – more than that – this work reframes bioethics itself, moving it from a narrowly individualistic account of autonomy to using vulnerability as the stimulus for a global and politically relevant account of ethics. This book must become essential reading for students of bioethics, challenging their preconceptions and introducing them to the full depth and complexity of the field.'

Alastair V. Campbell, Professor in Medical Ethics and Director, Centre for Biomedical Ethics at the National University of Singapore

VULNERABILITY

Challenging bioethics

Henk ten Have

Routledge
Taylor & Francis Group

LONDON AND NEW YORK

First published 2016
by Routledge
2 Park Square, Milton Park, Abingdon, Oxon OX14 4RN

and by Routledge
711 Third Avenue, New York, NY 10017

Routledge is an imprint of the Taylor & Francis Group, an informa business

© 2016 Henk ten Have

British Library Cataloguing-in-Publication Data
A catalogue record for this book is available from the British Library

Library of Congress Cataloging-in-Publication Data
Names: Have, H. ten, author.
Title: Vulnerability : challenging bioethics / Henk ten Have.
Description: Abingdon, Oxon ; New York, NY : Routledge, [2016] | Includes
bibliographical references.
Identifiers: LCCN 2015044013 | ISBN 9781138652668 (hardback) |
ISBN 9781138652675 (pbk.) | ISBN 9781315624068 (e-book)
Subjects: | MESH: Bioethical Issues. | Vulnerable Populations.
Classification: LCC R724 | NLM WB 60 | DDC 174.2—dc23
LC record available at http://lccn.loc.gov/2015044013

ISBN: 978-1-138-65266-8 (hbk)
ISBN: 978-1-138-65267-5 (pbk)
ISBN: 978-1-315-62406-8 (ebk)

Typeset in Bembo
by diacriTech, Chennai

For Mary Elodie Lemmens-Ponder and her five daughters who show everyday that vulnerability is a force that makes human existence livable and enjoyable.

CONTENTS

1

THE IDEA OF VULNERABILITY

Bioethical discourse nowadays often expresses concerns about vulnerability and vulnerable populations. People in countries where access to healthcare is limited can easily be seduced to participate in clinical trials of new drugs and thus become vulnerable to exploitation. People with disabilities can become targets of new genetic technologies, making them vulnerable to discrimination. Children with life-threatening diseases are vulnerable since they cannot take charge of their own decisions and depend on parents and proxy decision-makers. In these cases, special measures by other persons are required to assist and protect the vulnerable. Vulnerability functions as a warning sign that for some people special attention is needed. Ordinarily, persons can decide for themselves, identify their interests, and protect themselves. This is the general assumption in contemporary bioethics since it is dominated by the principle of respect for individual autonomy. Vulnerability is the flipside of this dominance. From the perspective of autonomy as the normative ideal, vulnerability is diminished autonomy. It is the exception to the general assumption, and therefore needs specific scrutiny.

The term 'vulnerability' is not often used in daily conversations. It sounds too sophisticated, perhaps even bombastic. Many people don't like the word because it conveys the impression that they are not in control of their lives or cannot cope with the conditions they are facing. The term also evokes images of miserable and dependent existence. At the same time most of us are aware that we are not always as strong as we like to be; we can catch a disease or have an accident and be wounded. In some conditions, the likelihood to be hurt is even higher – for example when we are suddenly confronted with a snowstorm or waiting in a busy hallway in the midst of the influenza season. We also know that some people are more prone to misfortune than others; they will be more easily hurt and will suffer injuries. Young children should not be allowed to walk out into a busy street because they are not even aware of the possible harms. Other people, such as the homeless, do not have a choice; for them everyday life is precarious but they have few means to protect themselves. In all these conditions people are vulnerable even if they are not labeled as such. Like bioethical discourse, ordinary language therefore seems to connect vulnerability with loss of control and power, with diminished individual autonomy. At the same time, it conveys the practical experience that all of us are vulnerable, at least in certain conditions and periods of life.

This book argues that the exceptionalist view of vulnerability is one-sided and biased. In fact, vulnerability is the general predicament of humans, while autonomy is the exception. Interpreting vulnerability as diminished individual autonomy, bioethics will not be able to fully understand and address issues of vulnerability. This is not merely a theoretical controversy about the relationship between autonomy and vulnerability. It has significant practical implications. It will be argued in this book that a new language of vulnerability has emerged in the wake of globalization. This language directs attention to the changing socio-economic conditions of human existence that often impair and reduce the decision-making capacity of individuals. Vulnerability therefore introduces a broader view of human beings within their existential settings across the world. Incorporating vulnerability as deficient autonomy within the dominant view of bioethics is not only inadequate for understanding and addressing vulnerability but also misses the opportunity for transforming bioethics into a global discourse.

In short, the purpose of this book is twofold. First, it will argue that vulnerability cannot be fully understood within the framework of individual autonomy. Second, the language of vulnerability offers perspectives beyond the traditional autonomy model that will support bioethics to evolve into a global enterprise. The argument will be developed in five steps. First, the meaning of 'vulnerability' will be explored, using cases of individuals, categories of persons, populations and countries to identify significant dimensions of the concept (Chapter 1). Second, it will examine how the notion is used in various scientific disciplines (Chapter 2) and how it is incorporated into the dominant discourse of bioethics in particular (Chapters 3 and 4). Third, a critical analysis of vulnerability will be provided from two perspectives. One is the philosophical perspective that vulnerability is a defining characteristic of the human species (Chapter 5); the other the political perspective that vulnerability is exacerbated by social, economic, and political circumstances (Chapter 6). The next step will be to show how vulnerability has emerged as a universal phenomenon due to processes of globalization (Chapter 7). These processes have articulated human vulnerability through changing the circumstances of human existence. Focusing on individual autonomy is not only misplaced but will not allow bioethics to scrutinize the sources of vulnerability in a global world. Finally, the theoretical and practical implications of vulnerability as a critical global discourse will be elaborated in Chapters 8 and 9.

What is 'vulnerability'?

The word 'vulnerability' is related to the Latin verb 'vulnerare' (wounding) and the noun 'vulnus' (wound). The shortest definition of 'vulnerable' is 'able to be easily hurt, influenced or attacked' [1]. Dictionaries distinguish three uses of the term 'vulnerable' [2, 3]:

1.   susceptible of receiving injuries
2.   open to attack or damage
3.   capable of being physically or emotionally wounded.

The first two uses correspond to medical and military interpretations. 'Vulnerability' is commonly used in medicine and, in particular, in pathology referring to the capability to be hurt or the susceptibility for injury or disease ('the lung is vulnerable to damage from these aerosols'). But it is also used in a military sense indicating a weakness in a defense system confronted with attempts to disrupt or harm the system ('a vulnerable part of the frontier'). Both uses of the term have a technical connotation. It is important to describe and analyze the possibility of harm in order to better understand the normal functioning of the body and its components. But it is also the other way around. On the basis of the physiology of the body, organs, tissues, and cells we can understand and explain the possibility of pathology. Before one can imagine a medical intervention, one first has to comprehend where and how vulnerabilities will occur. The military interpretation of vulnerability also is primarily technical. It requires a technical risk analysis of possible threats and weaknesses in order to determine whether a particular individual, a group of citizens, or a country as a whole is at risk of attack. It is, like the medical interpretation, first of all descriptive; the relevant determinants need to be identified and analyzed before practical consequences can be outlined.

The third use of the term is more general. It associates vulnerability to a web of other notions such as damage, harm, or being hurt. It can also be associated with fragility, precariousness, weakness, frailty, and finiteness. One does not need to be a medical or military expert to understand the nature of vulnerability. A wound is a violation of the integrity of the person: it hurts, it disturbs, and it impedes daily functioning. This use of the term indicates that vulnerability cannot merely be described. There is a need not just for a technical analysis but for intervention and action. Because there is the capability of being hurt, there is at the same time the possibility for prevention and protection. This use of the term elicits normative connotations exemplifying its role in ethical discourse. This last use of the term will be explored by examining four examples where the term is applied to the elderly, the homeless, research subjects, and countries.

Examples of vulnerability

Senior citizens

In his analysis of vulnerability Barry Hoffmaster [4] discusses the example of his 85-year-old father who has lived in nursing homes since his heart attack and stroke 40 years ago. He is partly paralyzed, cannot eat and drink, and is immobile. He is dependent on the care of others but he also risks contracting multiple infections and diseases. Due to the physical damage, his father is extremely vulnerable. On top of his bodily impairment, he has developed dementia. Hoffmaster furthermore presents the case of his 84-year-old mother. For the longest possible time, she has taken care of her husband. She is in good physical health but because she is now excluded from caring for her spouse, she feels guilty and powerless and has become emotionally vulnerable.

In discussions of vulnerability, the elderly are paradigmatic examples. Confronted with declining biological and cognitive functions, they are at the same time exposed to higher risks for diseases and injuries. From a medical perspective, all the weaknesses and deficiencies associated with ageing can be identified and described. The elderly can be regarded as a vulnerable category for a number of specific medical reasons. But older persons are also vulnerable in a wider, general sense. Nowadays, more and more persons who are in this medically vulnerable category function very well until reaching a higher age. Even if they are in medically good condition, they are more vulnerable to external stress than younger people. In 2003, France was affected by an extraordinary heat-wave, making it the hottest summer in 150 years. An estimated 15,000 people died in the first three weeks of August due to the elevated temperatures, the vast majority over the age of 75. It was the holiday season and in cities like Paris, older people were left alone in small apartments without air conditioning. Precisely because of the cultural phenomenon that in France almost everybody goes on holiday at the same time during August, the abandoned elderly became vulnerable to the heat-wave. Medical and political authorities responded much too late and many medical services were closed because of holidays [5]. Senior citizens are not only vulnerable to extreme weather events. In most countries, the elderly no longer have a source of income through labor and instead depend on retirement benefits. They usually have more expenses for healthcare. This may lead to a precarious balance in daily existence. In September 2012, Spanish Prime Minister Mariano Rajoy recognized this delicate situation. He declared that in his fight against the economic crisis, he would not reduce pensions because of the disastrous consequences this reduction had in Greece and in Ireland. He explained, 'Senior citizens are the nation's most vulnerable'. About one-fifth of the country's population collects pension but many pensions are below the minimal wage. At the same time, with an unemployment rate of 25 percent, younger generations are increasingly dependent on the elderly for financial support and housing [6]. The U.N. Population Fund recently made a similar call to protect the elderly [7]. One in nine people all over the world are 60 years or older. This figure will rise to one in five (two billion) by 2050. Contrary to what most people think, the majority of older people (two in three) live in developing countries. Also, in these countries, the elderly population exceeding 80 years of age is the fastest growing population. The growing number of older persons will create enormous challenges worldwide.

The previous examination concerning senior citizens clarifies some dimensions of the notion of vulnerability. First, it is clear that there is a difference between category and individual. Although the elderly as a category are vulnerable, the status of older individuals can be quite different. Depending on age, not everybody will be equally vulnerable. While there are good reasons to regard senior citizens as a vulnerable category, we cannot simply apply the label of vulnerability to everybody in this category. Doing so would be counterproductive since it will cause older persons to be singled out with the potential risk for discrimination and abuse. The prevalence of age discrimination in many countries has been a reason to connect ageing with vulnerability in the first place [8]. Second, vulnerability has various dimensions: physical, emotional or psychological, social, and economic. Being vulnerable can relate to internal conditions such as frailty, disease, and disability. Vulnerability can also relate to external conditions such as income insecurity, lack of access to quality health care, or an environment that is unfriendly for people with diminished mobility, and visual or hearing impairments. For many senior citizens, internal and external conditions are often combined and co-exist at the same time.

Homelessness

Another example of vulnerability concerns homeless people [9]. Homelessness is a persistent problem in almost all countries but it is hardly visible. The category of the homeless includes many different persons: migrants, unemployed, divorced, and abused. Most of them will be homeless for a period of time, not permanently. Some live in shelters, others on the street. It is difficult to estimate the number of homeless people. But it is not an uncommon condition. Approximately one percent of the U.S. population experiences homelessness each year (3.5 million people) [10]. More than one-third of this population includes children under the age of 18. The numbers indicate that over the past two decades, homelessness has increased dramatically. Furthermore, the expectation is that homelessness will significantly increase in the near future among the elderly [11]. As far as data indicates, in other countries the situation is not different. For example, in the Netherlands, there are 60,000 homeless persons according to estimations made by the Salvation Army based on the use of shelter facilities. The Dutch Ministry of Health has defined homelessness as, 'Vulnerable people who have left their homes or were forced to leave because of a combination of problems and who are unable to live independently' [12]. Vulnerability is particularly noticeable among homeless children. In India, an estimated 11 million children live on the streets. In Brazil, the exact number is not even known; estimates vary between 1 and 10 million children, many of whom are orphans who have lost their parents. In developing countries, there are almost no shelter services for the homeless. More than one million children are living on the streets in Bangladesh, deprived of basic social needs such as shelter, food and water, education, sanitation, and healthcare [13]. Homeless people in general are vulnerable to disease and crime. They may also have specific problems because they lack the usual protection that other people have; they are susceptible to thermal injuries, sleep deprivation, and foot and leg ulceration, to name a few [14]. Furthermore, social support is often disconnected; they are on their own.

Homelessness is a situation of vulnerability because it implies loss of power and loss of control. Homeless people lack the resources that enable self care and struggle to find shelter against external threats like changing weather. The homeless person is confronted with potential threats that are usually unmanageable owing to diminished capacities, like mental illness or addiction, or severe threats, like extreme weather or unsafe environments. The vulnerability is multidimensional. Threats like homelessness, ageing, and childhood may be combined. The example of homelessness illustrates that vulnerability is a relational rather than individualistic concept: one is always vulnerable *to* something or somebody. In ideal circumstances, one would not be exposed to severe weather, risk of infection, or criminal activity. Vulnerability emerges when there are external conditions that may negatively impact on the person's well-being and health. Furthermore, the example demonstrates that vulnerability is a gradual phenomenon. It is not an all or nothing condition. Vulnerability implies not only internal or external threats but it also relates to our abilities to cope with those threats. This interaction between threats and coping mechanisms allows for the development of models, policies, and approaches to assist vulnerable persons. Regarding homelessness, for example, different factors of vulnerability can be distinguished: structural determinants (economic processes, increasing immigration, poverty, and foreclosures), institutional factors (the availability or lack of services, and alternative housing), relational factors (the breakdown of relationships), and personal determinants (disability, illness, addiction, age, ethnic status, and employment) [15]. Such analysis can be used to develop and implement strategies and to diminish vulnerability for populations and for specific individuals.

Vulnerable populations in research

Amar Jesani is a physician with a long-time interest in medical ethics and human rights. He was the coordinator of two national bioethics conferences in India (in 2005 and 2007), and he is currently the editor of the *Indian Journal of Medical Ethics*. In 2005 he established the Centre for Studies in Ethics and Rights (CSER) in Mumbai. The centre initiates and funds research in ethics and human rights. One of the projects is the Clinical Trial Watch. This project's duties include collecting data on clinical trials, monitoring the research industry, and publishing critical information about medical research in India. Jesani regularly voices concerns about medical research in his country. The number of clinical trials is growing rapidly. According to worldwide estimates, one in four clinical trials is now conducted there [16]. The low costs of trials (60 percent lower than in the U.S.) are a major reason for so many trials being conducted in India. The available infrastructure also contributes to the growth. There are many hospitals and well-trained doctors (more than 660,000 in 2010) who speak English. The Indian government incentivizes research for the clinical trials industry. In 2005, it softened regulations and created tax exemptions for research services. Another reason for the growth of clinical trials is more cynical: the country has an enormous population with a wide range of diseases, while the majority lack access to any form of treatment. Poverty and the absence of affordable health care make it impossible for many people to obtain treatment. These 'treatment-naïve' subjects are unable to afford treatment, so they are an ideal group to recruit for testing new drugs. Government interventions over the last decade have reduced public support for health care services. In 2010, the per capita government expenditure on health in India was $15 (compared to $69 in China and $3,426 in the U.S.) [17]. Official policies encouraging drug research are therefore an indirect way to supplement under-resourced health facilities. For many people, the invitation to take part in a clinical trial offers an opportunity for free treatment. Under these conditions, this is an issue of vulnerability. Is participation in a research project an unbiased choice? The question arises in part because of the socio-economic circumstances and cultural conditions that make it attractive to participate. Many research subjects are poor, unable to read and write, unaware of their rights, and tend to obey the authority of physicians [18]. Different ethnic and tribal groups are regarded as particularly vulnerable. In 2010 public outcry focused on an international study on the Human Papilloma Virus (HPV) vaccine in the states of Andhra Pradesh and Gujarat. The vaccine was administered to 25,000 healthy girls between 10 and 14 years of age. Most of the children were of tribal origin, very poor, mostly illiterate, and unable to speak the local language. Many of them lived in hostels near a government school. Researchers obtained consent from wardens and teachers without informing parents. There was no infrastructure for reporting adverse events, no follow-up, and no insurance for participants. Seven girls died after the administration of the vaccine [19]. In September 2013, a parliamentary panel condemned the trial. Later that month, the Indian Supreme Court urged the government to tighten regulations for clinical trials. The chief justice said that foreign companies consider the country a heaven for clinical trials but it has become a 'hell for India' [20].

The notion of 'vulnerable populations' is often used in connection to international research in developing countries. It usually refers to lack of choice. The general assumption is that it is morally permissible to recruit subjects for a clinical trial on the basis of informed consent. If subjects (or in the case of children, the parents) do not receive adequate information and do not voluntarily consent, they should not be included in the trial. This rule of informed consent is a standard requirement in developed countries although it is often challenged in many developing countries. Persons are vulnerable if they are not able to consent because their decision-making capacity is diminished, as in the case of the Indian school girls. But vulnerability is more than diminished autonomy. Even if the parents were adequately informed and had given consent, the vulnerability is still there. What choice did they have, if otherwise their daughters would not receive medical care; when at the same time government officials were advocating the vaccination; and when the parents are acutely aware to belong to an indigenous minority considered as 'primitive' and marginalized? Vulnerability here refers to general conditions that make groups of people susceptible to exploitation and abuse. The term 'vulnerable population' therefore not only refers to categories of individuals such as children who are not able to give consent due to lack of individual autonomy but also to groups of people whose decision-making capacity is compromised due to social and economic circumstances. The parents of the school children might have been fully autonomous persons, but as individuals they were not able to change and improve the conditions in which they were living. They might even wish to improve the existence of their children by providing them opportunities such as better protection against diseases. This shows the other aspect of vulnerability; it does not only refer to the capacity to choose but also to the actual choices that can be made.

The example furthermore makes clear that vulnerability is a conditional notion. It expresses a potentiality. We may be harmed, we run the risk of being wounded but we do not actually need to be in a damaged situation (the actual 'vulneration'). There is a difference between the potentiality of harm and the harm itself. Vulnerability means that there is the possibility of harm, injury, exploitation or abuse but it does not imply that these negative effects are actually happening or have occurred. However, the probability of such effects can be decreased through taking appropriate measures. The notion of vulnerability has a warning function: it signals that for certain persons or populations additional assistance and protection should be provided. This is the normative appeal of the notion that may explain why it is often used in contemporary ethical discourse: using the label 'vulnerable' should make us more cautious, instigate us to take care, and to provide protection.

In the example of the HPV trial in India it seems that protective mechanisms are lacking. There are insufficient regulations, there is hardly any government oversight, and ethical review committees are not adequately functioning. A tradition of public reflection learning lessons from scandals is absent [21]. But the authorities in India are not the only ones to blame. The HPV study was carried out by an international NGO, funded by a U.S. foundation, and supported by government officials and the Indian Council of Medical Research. The research proposal was approved by ethics review boards in India and abroad. The government committee investigating the case concluded that the ethical violations were 'minor'; the report was not even published [22]. A group of scholars and institutions had warned in an early stage about the potential dangers and risks of the project but to no avail [23]. And there already was a long list of unethical research projects since the 1990s demonstrating the importance of vulnerability in clinical trials executed in India [24].

Vulnerable countries

A dark blue ocean, white sandy beaches lined with palm trees, singing and dancing people in traditional costumes – the short movie *Tuvalu: That sinking feeling* puts one of the smallest countries in the world into the spotlight. The atoll country of Tuvalu includes nine islands scattered over 500,000 square miles in the Pacific Ocean, midway between Hawaii and Australia. Having gained independence in 1978, its population now is approximately 11,000. Tuvalu is known as the paradigmatic case of vulnerability to climate change. As a small island developing state, it is the most vulnerable of all vulnerable countries. This is what Elizabeth Pollock, the filmmaker wants to show with her short feature [25]. Higher tides, inundation and flooding, rising groundwater levels, salt water intrusion damaging agriculture, shoreline erosion, and loss of islets document that the islands are disappearing. It seems that in the near future only one solution remains: migration to higher places. Indeed, in August 2014, New Zealand formally granted refugee status to a family from Tuvalu. It is too dangerous for them to return home. The expectation is that these first 'climate refugees' will be followed by many other victims of global warming [26].

Vulnerability to environmental degradation, in particular climate change, is only one factor. Small island states are vulnerable in several ways. They can be impacted by natural hazards such as extreme weather with cyclones and hurricanes, as well as rising sea-levels. They are also characterized by economic vulnerability. Small islands are by definition fragile because they have a small size, poor natural resources, unique and fragile ecosystems, and have problems with transport and communication. Basically, they are confronted with many forces outside their control.

Applying the term 'vulnerability' to countries has at first sight negative implications. The discourse of vulnerability presents Tuvaluans as 'helpless' and as 'victims' [27]. They have a choice between drowning and migrating. The focus on vulnerability stresses the weaknesses rather than the strengths and opportunities for the population. It marginalizes alternative discourses of adaptation to changing environmental conditions. On the other hand, islanders are known for resilience and resourcefulness. Traditionally, they have a history characterized by survival and self-reliance. Periodic migration between islands was an intermittent practice among Pacific Islanders. They consider the sea and not the land as their home [28]. They know how to benefit from sea resources. In the case of Tuvalu 'vulnerability' only considers the threats, not the coping mechanisms. The country may be vulnerable but for most Tuvaluans climate change is not a reason to migrate. They prefer making adaptations in their country, not leaving it [29].

The discourse of vulnerability in this example articulates primarily the physical dimension of vulnerability. Tuvalu is regarded as an inherently dangerous place that will become uninhabitable. In contrast, Australia is a secure place. But the physical vulnerability of Tuvalu is not synonymous with its social and cultural vulnerability. For the inhabitants migration is not an option since the existential threat is not merely the physical disappearance of the islands but rather the loss of national and cultural identity. Tuvaluans have always been self-sufficient until colonization in the beginning of the twentieth century, and they still rely on their own resources and institutions such as the extended family and the tradition of consensual decision-making in councils of elders. Like other Pacific countries Tuvalu has rich cultural resources and diversity, as well as traditional knowledge that is unique. The description of the country as 'vulnerable' is analogous to the medical and military interpretations mentioned earlier: the island is increasingly susceptible to harm, the damage can be identified and specified, and possible strategies of prevention and mitigation can be developed. This technical use of the term, however, cannot conceal that the real concern is the vulnerability of the people living on the island. They might be harmed, not only physically, but in a much broader sense of losing their culture and history.

The vulnerability of small island states finally demonstrates that the notion entails a wider context. Vulnerability is associated with global justice [30]. Tuvalu is sinking because the impact of climate change is disproportionally felt in small island states while they have minimally contributed to the global climate problem. Burdens and benefits are asymmetrically distributed. Tuvaluans are portrayed as victims but not as victims of the mass-consumption in the developed world that mainly contributes to climate change. They are vulnerable but not responsible for climate change. They are powerless to protect their way of life and the values they cherish – the real power and control is elsewhere. Vulnerability is putting the onus on them but does not address the people elsewhere who cause the damage. The potential harm that will damage Tuvaluans should be discussed in a wider, political context of increasing globalization.

Three dimensions of vulnerability

The examples of different uses of the term 'vulnerability' show that it is often applied in a wider sense than the traditional medical and military interpretations [31]. Subsequent chapters will further analyze vulnerability. But the examples illustrate that there are several dimensions to distinguish.

1. Vulnerability can apply to individuals, groups, communities and countries. However, individuals and collectives or categories are different. Identifying vulnerable populations does not imply that every individual in this population necessarily is vulnerable, or has the same vulnerability all the time. The distinction between category and individual is important since it is based on two characteristics of the notion of vulnerability itself: it is a gradual and relational concept. The example of homeless people indicates that vulnerability is dependent on the circumstances; it is not a static and permanent condition. Vulnerability is also associated with threats and risks that are faced, and in this way the notion is relational. It is therefore necessary to identify what exactly creates or produces vulnerability for individuals and groups of individuals. Even if vulnerability is primarily manifested in individual persons, it is not an exclusively individual feature. As will be discussed later, vulnerability is also used in a more fundamental, philosophical discourse in which it is not gradual and relational.
2. Various types of vulnerability can be distinguished. The examples show that there are physical, psychological, social, economic, and environmental vulnerabilities. Focusing on a medical perspective is often insufficient to understand the vulnerabilities at stake. The elderly may often be more at risk for harm resulting from deteriorating socio-economic circumstances than from deteriorating health, even if they are medically more vulnerable than younger persons. The threats that can possibly hurt people are therefore heterogeneous, and will require different methods of analysis and examination, as well as varying strategies and policies to reduce or mitigate possible harms. As the examples show, vulnerability not only implies exposure to possible harms but it also refers to the lack of coping and adaptive mechanisms to prevent or minimize the harms. Indigenous and illiterate poor people in rural areas in India have fewer possibilities to deal with vulnerability than well-educated persons with a job in cities like Mumbai.
3. Internal and external conditions explain why people are vulnerable. The physical constitution of senior citizens is frail compared to that of young people; for their flourishing they are more dependent on external circumstances such as pension income and weather. Often the external conditions make categories of people more vulnerable than others, like the citizens of Tuvalu are experiencing. But in the Tuvalu case, even if the country is vulnerable, what really concerns us is *human* vulnerability [32]. The physical

characteristics of the islands combined with the deteriorating environmental and economic circumstances produce the possibility of serious harm to its inhabitants. The likely disappearance of Tuvalu is a loss for humanity not because geography is changing but because human beings are hurt since their world and culture will be annihilated.

The dimensions of vulnerability identified on the basis of the discussed examples help to define vulnerability. Because the term is used in various disciplines ranging from philosophy, theology, and ethics to ecology, computer science, and physiology there is enormous diversity of formulations and interpretations. However, a general systems approach focusing on the conceptual components of the notion, regardless of the domains in which it is used and irrespective whether it is used for individuals, communities, or countries proposes a comprehensive perspective. Neil Adger, climate change researcher from the United Kingdom, defines vulnerability as 'the state of susceptibility to harm from exposure to stresses associated with environmental and social change and from the absence of capacity to adapt' [33]. This is a functional, not a content definition. It does not clarify the fundamental characteristics of vulnerability but shows how the notion functions and relates to other concepts. This approach is useful since it urges us to consider the conceptual elements that we need to take into account in understanding the notion. Vulnerability is regarded as a function of exposure, sensitivity, and adaptive capacity.

The first component is exposure. There must be external stresses or disturbances that produce potentially harmful threats. These threats are hard to avoid since human beings are continuously exposed to each other and to the social and natural environment.

The second component is sensitivity. This is susceptibility to harm or damage. In a general sense it is 'the degree to which the system is modified or affected by an internal or external disturbance or set of disturbances' [34]. From a medical perspective sensitivity is inherent in the body, organs, tissues, and cells: they can be affected for example by lack of oxygen. From a general perspective that focuses on the human person as a whole, sensitivity is inherent in the human predicament, existing prior to any external exposure.

The third component is the ability to adapt or the capacity of response. Sometimes a distinction is made between coping ability and adaptive capacity [35]. The first is the short-term capacity to overcome external stresses, the second the longer-term adjustments. Human beings are able to cope, adapt, and make adjustments; they can resist and overcome threats.

The functional definition underlines that vulnerability exists when all three components are involved. For example, when there is a threat of an infectious disease, the exposure is in principle the same for everyone, but the sensitivity is different: children and the elderly have more risks if they are affected. The adaptive capacity is better for persons who have access to medical care and medicines. The most vulnerable groups therefore are children and elderly with no or inadequate access to the healthcare system. Another example is that in severe winter conditions, the exposure is in principle the same for everyone, as is the sensitivity. But the adaptive capacity is insufficient for homeless persons. This is what makes them vulnerable to cold weather injuries.

The distinction between internal and external conditions of vulnerability complicates the functional definition. Sensitivity and adaptive capacity constitute internal conditions of vulnerability, while exposure constitutes the external one [36]. Systems in general and human beings in particular can be vulnerable to external stress but as long as they are not exposed to it there will not be any harm. If exposure is included in the definition of vulnerability there is no need to call a system or a human being vulnerable when the exposure to threats is absent or eliminated. Gallopín mentions the example of a person with immunodeficiency confined in a sterile environment where the individual is no longer vulnerable to infectious diseases [37]. Another example concerns the vulnerability of essential infrastructure to cyberattacks. As long as operating systems of banks and electric power companies are not connected to the Internet, they are not vulnerable [38]. Both examples consider exposure as a defining component of vulnerability. As indicated in the discussion of homelessness, and reiterated earlier in this paragraph, the notion is generally regarded as relational. It is the interaction between internal and external components that produces vulnerability. This view implies that human beings are not vulnerable as such; as long as there is no exposure to threats they are not. Another implication is that a range of policies and practices aimed at influencing exposure is available in order to reduce or eliminate vulnerability. Taking away exposure and removing threats will solve the issue of vulnerability.

As I will discuss later, this view is contested. The argument is that exposure should not be a defining component of vulnerability. Systems can be vulnerable prior to any disturbance or external stress. The same examples given by Gallopín can be used. Software is vulnerable even before the Internet came into existence; the immunologically challenged person is vulnerable whether or not he is exposed. The same is true for human beings. Particularly the philosophical view will argue that human beings are inherently vulnerable because of their fundamental sensitivity to threats. Such view of course will shape the possibilities to reduce vulnerability. Diminishing sensitivity and enhancing adaptive capacity will be more useful than protecting against exposure.

Vulnerability and bioethics

The field of bioethics also uses the functional definition of vulnerability. Schroeder and Gefenas recently proposed the following definition: 'To be vulnerable means to face a significant probability of incurring an identifiable harm while substantially lacking ability and/or means to protect oneself' [39]. Although the terminology is different, the internal and external components of the functional perspective can be recognized: the internal capacity of response and the external threat (although the term 'exposure' is not mentioned). The definition first of all applies to human beings who are weak and cannot protect themselves. What is not explicitly included in this definition is the component of sensitivity. This is remarkable since the authors in their publication discuss the fragility of the human condition (because this is what exposes us to the possibility of harm). It is also difficult to imagine how we would incur harm if sensitivity was lacking. What is furthermore not explicated in the definition is the normative dimension of vulnerability. This normativity is presumably what makes the notion appealing in bioethical discourse.

As will be shown in the next chapter, the use of 'vulnerability' in the scholarly literature is recent. In many cases it is just a descriptive or technical term. In the context of bioethics however it has an ethical connotation. Here, the term 'vulnerability' is not a neutral attribute of a particular person or group of persons. Making an observation or giving a description could be done by using words like 'exposed' or 'subject to' in order to indicate that a person is threatened or capable of being damaged. In the discourse of bioethics, and perhaps also in some other discourses, vulnerability has normative implications. This is expressed in the recent edition of the *Oxford English Dictionary* that in June 2012 has included a draft addition in relation to vulnerability. A new sense is added to previous descriptions of the entry 'vulnerable': 'Designating a person in need of special care, support, or protection (esp. provided as a social service) because of age, disability, risk of abuse or neglect, etc.' [40]. Vulnerability therefore evokes a response; it encourages other people to provide assistance; we cannot leave vulnerable persons to their fate. If we can reasonably and reliably prevent them being damaged or hurt we should take action.

Two aspects of the notion explain its normative force. First, vulnerability is conditional. A person is capable of being hurt but the damage has not yet occurred; this will probably happen unless appropriate measures will be taken. This conditionality generates a responsibility to take care and preventive action, if at least harm could be prevented. Second, vulnerability is associated with possible harm, not with positive outcomes. A person is not vulnerable to win the lottery or to be awarded a bioethics degree. Vulnerability indicates that something negative might happen. The consequence of both aspects is that unless some action is taken a vulnerable person will probably be harmed or damaged. Because vulnerability is a potentiality there is also room for intervention; perhaps the person can be assisted to protect himself or can be protected by others against harm, or made less vulnerable through various care arrangements; perhaps harm can be prevented from taking place, or the impact of harm can be mitigated. As will be argued later in this book, there are multiple ways of responding to vulnerability.

The discourse of vulnerability in bioethics is not uniform. Although vulnerability has commonly been interpreted within the normative framework of individual autonomy (as will be explained in Chapter 4) and has resulted in practical recommendations to protect persons with diminished autonomy in healthcare and medical research, bioethical discourse has problems in reconciling this framing of vulnerability with two fundamental views that question the reduction of vulnerability to impaired autonomy.

One is the philosophical view that argues that vulnerability is a basic and inherent characteristic of being human. It strongly emphasizes the component of sensitivity. Since vulnerability is essential for human beings it should be respected. In the end we have to accept that vulnerability is part of our human predicament and that the power of modern medicine and technology cannot overcome it. Of course, human beings have adaptive capacity and the ability to protect themselves against harmful exposures. But they can and should not change their basic human condition that makes them inherently sensitive to internal and external stress. This fundamental vulnerability becomes more pronounced if people grow older or become disabled. They can try, but they can never fully protect themselves so that harm and damage will be avoided. What this vulnerability primarily demands is care and assistance. In this view, vulnerability precedes individual autonomy; it is the condition that nurtures the growth and maturation of personal autonomy.

The other is the political view of vulnerability that considers it as a component of the social context. This perspective emphasizes exposure as essential to vulnerability. Human existence is necessarily precarious because we live together with other human beings within a life world that may easily deteriorate. Unforeseen negative things can always happen. But in certain social, economic, and political conditions, some human beings are more vulnerable than others. Even if human beings share the same inherent and common vulnerability as argued in the philosophical perspective, vulnerability can be exacerbated because of human interconnectedness and living conditions, making some of us more vulnerable. We certainly have a responsibility to remediate, reduce, or eliminate this type of vulnerability, even if we have to accept the inherent vulnerability underlined in the other perspective. According to the political perspective, the focus should be on the vulnerabilities that are created not on the ones that are given. The increasing awareness of vulnerability nowadays is related to social processes of globalization, as discussed later in this book. These processes have produced more risks and threats for most people worldwide and at the same time they are undermining social coping mechanisms. This context requires more than an individual response. What is needed is social and political action. In this view, impaired autonomy is not the crux of vulnerability. Since it is a relational condition, being vulnerable is the result of interaction between internal and external conditions. Focusing on the incapacity to choose or the loss of control and power as the hallmark of vulnerability, ignores that it is often the impossibility to make actual choices given the circumstances in which people live, that determines vulnerability.

The two views of vulnerability that will be elaborated in later chapters explain the dialectics in the approaches to vulnerability. Should human vulnerability be accepted and respected, as is argued in the philosophical view? If modern science and technologies continuously try to remove or overcome this essential characteristic of human beings through enhancing the human body, intervening against diseases and impairments, and extending the human lifespan, will they then endanger the humanness of human beings? Making persons invulnerable does not sound attractive in this view since the opposite of vulnerability suggests a being that is untouchable, unexposed, not open or sensitive, and also self-contained and not sociable. On the other hand, the political view of vulnerability points out that actions should be taken in order to reduce or eliminate exposure to harmful threats, precisely because the vulnerability results from our activities as social beings. Some persons are made more vulnerable than others, even if we share the same basic vulnerability as human beings, since the changing social context is exposing them to more threats. In this view, vulnerability is unacceptable since it is always the result of human action. As will be argued later, both views of vulnerability are related. Without the basic vulnerability that is inherent in being human, the political view could not be applied. We would not be able to recognize that vulnerability is socially produced. Since human beings are essentially vulnerable, they are all the same; nobody is invulnerable. It is the angry Apollo in Homer's *Iliad* inciting the Trojans to face their Greek foes, who said: 'They, like ourselves, are vulnerable flesh, Not adamant or steel' [41]. This shared vulnerability is the basis for the insight that we are fundamentally interdependent. There are always others, mostly anonymous and unknown, on whom we are depending; there is the possibility that they will hurt us. No person is self-sufficient and in full control of what is happening. But we can use this shared predicament to make the world more hospitable and to reduce as much as possible the vulnerabilities that are the result of our interactions. Vulnerability therefore is not only a generally shared possibility, the consequence of being human, but at the same time differently distributed and allocated depending on the social context of human existence. Rather than making people invulnerable, we can at least increase their resilience and make human existence more secure by influencing this context. Since we are all embedded in larger global processes our basic vulnerability can instigate us to imagine the 'possibility of community', creating possibilities for social and political action [42]. The experience that we are vulnerable to others is not merely negative, something that we should prevent, but it provides an opportunity for transformation.

Challenges to bioethics

Although the notion of vulnerability is often used in various discourses, it is in need of further clarification. Two perspectives can be used to clarify the content of the notion: the philosophical perspective that regards vulnerability as a characteristic of the human species, and the political one that considers it as the product of specific conditions. Both perspectives take into account the three functional components of vulnerability. The philosophical perspective articulates in particular the component of sensitivity while the political perspective puts more emphasis on the components of exposure and adaptive capacity. The interrelation of the two perspectives also clarifies why contemporary bioethics has a difficult time in understanding vulnerability. Both perspectives challenge the predominance of individual autonomy in contemporary bioethics and the resulting view that vulnerability essentially is an individual affair. They demonstrate, on the contrary, that the notion of vulnerability implies that human beings are social beings. The notion challenges the idea that individual persons are autonomous and in control. The social nature of vulnerability makes it a resource to create community and relationships; it refers to solidarity, the needs of groups and communities, and not just to those of individuals. As long as bioethics continues to associate vulnerability with diminished autonomy and impaired individual decision-making it will not fully understand the notion.

The idea of vulnerability therefore presents several challenges to contemporary bioethics. It not only suggests a broader view of human beings but it also defines a wider range of normative implications. Being vulnerable is not primarily regarded as a negative feature or a weakness but as a productive and positive human quality that promotes cooperation, solidarity, assistance, and care. Focussing on the conditions that produce or exacerbate vulnerability furthermore provides a better way to address the moral problems associated with processes of globalization. Finally, the notion of vulnerability is an incentive to enrich bioethics and to transform it into a global normative discourse.

Notes

1 American-English Dictionary. Cambridge Dictionaries Online, http://dictionary. cambridge.org/dictionary/american-english/vulnerable?q=vulnerability (accessed 4 April 2015).
2 Oxford English Dictionary, 'Vulnerable' 2012. Oxford University Press; online edition
3 Merriam-Webster Dictionary, http://www.merriam-webster.com/dictionary/ vulnerability. (Accessed 4 April 2015).
4 Hoffmaster, 'What does vulnerability mean?', 38–40.
5 The effect of the 'canicule 2003' was also severe in other European countries. In Europe 70,000 persons died as a result of the heat-wave, almost all older persons. See: Martine Perez. 'Canicule 2003: 70 000 morts en Europe', *Le Figaro*, 15 October 2007. http://www.lefigaro.fr/sciences/2007/03/23/01008-20070323ARTFIG90028-canicule_morts_en_europe.php. (Accessed 4 April 2015); see also INSERM, *Improving public health responses to extreme weather/heat waves,* 53–56.
6 Daley, 'Spain's jobless rely on family, a frail crutch', *The New York Times*, 28 July 2012, http://www.nytimes.com/2012/07/29/world/europe/spains-elders-bearing-burden-of-recession.html?pagewanted=all&_r=0. (Accessed 4 April 2015).
7 See UNFPA and HelpAge International, *Ageing in the twenty-first century. A celebration and a challenge.*
8 The UNFPA report points out that only 49 percent of the consulted older persons believe that they are treated with respect. See UNFPA and HelpAge International, 7.

9 See Sebastian, 'Homelessness: a state of vulnerability'.

10 National Coalition for the Homeless, *How many people experience homelessness?* 3.

11 Sermons and Henry, *Demographics of homelessness series: The rising elderly population*, 2.

12 Edgar, Doherty, and Meert, *Review of Statistics on Homelessness in Europe*, 5.

13 UNICEF, *Investing in vulnerable children*, 9

14 Sebastian, 'Homelessness: a state of vulnerability', 15 ff.

15 Edgar and Meert, *Fourth review of Statistics on homelessness in Europe*, 12.

16 According to Sengupta, 'Fatal trials: clinical trials are killing people', 118.

17 WHO, *World Health Statistics 2011*. Geneva. Total expenditure for healthcare in India in 2008 was 4.2 percent of the gross domestic product; in the US it was 15.2 percent.

18 Srinivasan, 'The clinical trial scenario in India', 31–32.

19 For the HPV vaccine trial see Srinivasan, 'HPV vaccine trials and sleeping watchdogs'; Buncombe and Lakhani, 'Without consent: how drugs companies exploit Indian "guinea pigs",' *The Independent*, 14 November 2011, http://www.independent.co.uk/news/world/asia/without-consent-how-drugs-companies-exploit-indian-guinea-pigs-6261919.html (Accessed 4 April 2015).

20 Buncombe, 'A heaven for clinical trials, a hell for India': Court orders government to regulate drugs testing by international pharmaceutical companies'.

21 Ramanathan and Jesani, 'The legacy of scandals and non-scandals in research and its lessons for bioethics in India', 4.

22 Srinivasan, 'HPV vaccine trials and sleeping watchdogs', 73.

23 See Memorandum, 'Concerns around the human papilloma virus (HPV) vaccine'.

24 Schipper and Weyzig, *Briefing paper on ethics in clinical trials; #1: Examples of unethical trials,* 7–10, 12–16.

25 Pollock, *Tuvalu: That sinking feeling*, http://www.pbs.org/frontlineworld/rough/2005/12/tuvalu_that_sin_1.html. See also Chambers and Chambers, 'Five takes on climate and cultural change in Tuvalu', 299–301; Patel, 'A sinking feeling'; Allen, 'Will Tuvalu disappear beneath the sea?'.

26 Aulakh, 'New Zealand decision created world's first climate refugees'. *The Star*, Sunday 17 August 2014, (http://www.thestar.com/news/world/2014/08/06/new_zealand_decision_created_worlds_first_climate_refugees.html) (Accessed 4 April 2015).

27 Farbotko, 'Tuvalu and climate change: Constructions of environmental displacement in the *Sydney Morning Herald*', 280, 286.

28 McCall, 'Clearing confusion in a disembedded world: The case for nissology', 77.

29 Mortreux and Barnett, 'Climate change, migration and adaptation in Funafuti, Tuvalu', 108; see also Paton and Fairbairn-Dunlop, 'Listening to local voices: Tuvaluans respond to climate change'.

30 Parks and Roberts, 'Globalization, vulnerability to climate change, and perceived injustice', 339.

31 Excellent recent analyses of vulnerability are: Kirby, *Vulnerability and violence;* Turner, *Vulnerability and human rights*. Maillard, *La Vulnérabilité. Une nouvelle catégorie morale?* Pelluchon, *Éléments pour une éthique de la vulnérabilité. Les hommes, les animaux, la nature.* Clark, *The vulnerable in international society*. Mackenzie, Rogers and Dodds, *Vulnerability. New essays in ethics and feminist philosophy.*

32 In this book, the emphasis will be on human vulnerability. Corine Pelluchon has developed a broader ethics of vulnerability that also includes animals and nature. Fragility is characteristic of all living organisms. They are all sensitive and therefore vulnerable. Pelluchon's ethics of vulnerability is not exclusively addressing vulnerable

persons or populations. It primarily assumes the fundamental responsibility of human beings within the 'communauté biothique' because they are vulnerable like other living beings and because they have the privilege of knowledge. Vulnerability, argues Pelluchon, invites us to rethink our relationships with other species, not only animals but also plants, with ecosystems and with the biosphere (Pelluchon, *Éléments pour une éthique de la vulnérabilité. Les hommes, les animaux, la nature,* 37, 40, 56; see also Pelluchon, *L'autonomie brisée. Bioéthique et philosophie*).

33 Adger, 'Vulnerability', 268.
34 Gallopín, 'Linkages between vulnerability, resilience, and adaptive capacity', 295.
35 Gallopín, 'Linkages between vulnerability, resilience, and adaptive capacity', 296.
36 In later chapters it will be argued that adaptive capacity itself depends on external conditions, so that it is not merely an internal determinant of vulnerability.
37 Gallopín, 'Linkages between vulnerability, resilience, and adaptive capacity', 295.
38 Brenner, *America the vulnerable. Inside the new threat matrix of digital espionage, crime, and warfare,* 93 ff.
39 Schroeder and Gefenas, 'Vulnerability: Too vague and too broad?', 117. The definition presented is very similar to the well-known definition of the United Nations Department of Economic and Social Affairs (*Report on the World Social Situation,* 14).
40 Oxford English Dictionary. Online edition 2012 (http://www.oed.com/view/Entry/224872#eid15169124). (Accessed 4 April 2015).
41 Homer, *The Iliad* 606–607, 103.
42 Butler, *Precarious life. The powers of mourning and violence,* 20.

2

GROWING INTEREST IN VULNERABILITY

The notion of vulnerability has become widely discussed in the scholarly literature. The first journal article using this notion was published in 1925. It describes the special vulnerability of the macular fibers in the eye [1]. There is no explanation or definition of 'vulnerability'. The author is using the term in a medical-technical sense as susceptibility to injury, lesions, or as liability to disease. Most of the scholarly literature has been using 'vulnerability' as a keyword in this technical sense. The number of publications referring to 'vulnerability' has increased significantly from a total of 11 journal articles in 1970 to 2,748 in 2013 [2]. The overwhelming majority of these publications have been published recently: 90 percent have appeared since 1990. The rate of journal publications using this keyword is also accelerating: 69 percent of all articles on the subject have been published since 2000. Similar trends can be discerned in bioethics publications. Particularly during the last decade, the notion has become widespread, although contested, in bioethics. Vulnerability is also no longer used in a narrow technical sense but has become endowed with broader meanings.

This chapter will examine the trend towards increasing and broadening use of the notion of vulnerability. It will address the question why this notion is so prevalent in contemporary discourses. Its popularity seems to reflect the growing perception and recognition that lately human existence itself is becoming more and more vulnerable.

Vulnerability in the scholarly literature

The majority of journal publications using the keyword 'vulnerability' do not relate this notion to ethics. Until the 1970s, nearly all relevant publications used the term in a technical sense, mostly for medical, less often for military readership. The notion captures concerns about anatomical structures that can be damaged or bodily functions that can be disturbed. The focus is for example on the vulnerability of the brain to ischemia, or vulnerability to psychiatric breakdown in old age. The term usually is not defined but is applied as a mere technical descriptor to indicate that an internal or external harmful influence can damage specific entities (cells, tissues, organs, persons, groups, and countries).

Since the 1970s the notion starts to obtain a larger scope. Its field of application widened beyond medical and military discourse. This trend was noticeable in the new and burgeoning field of nursing science. Olly McGilloway was the first to introduce vulnerability in 1976 in the nursing literature. She argued that the care activities of the new profession of nurse practitioners should be focused at the vulnerable situation of the patient which is characterized by helplessness, lack of competence, and emotional disturbance [3]. The notion of vulnerability is attractive for nursing because it centers upon the individual person and examines not only bodily structure or physical constitution but also the functioning of the person within a specific context. For example, some studies use the notion to explain that the interconnection between different areas of functioning (social, economic, mental, and physical) makes older persons increasingly vulnerable [4], that economic stress has differential adverse effects depending on the vulnerability of persons [5], that domestic violence affects particular persons, especially women and children, more frequently than others [6], and that fear of crime is more prevalent among certain populations such as the elderly [7]. Although the notion of vulnerability in these studies is focused on the human person as a whole, it is not elaborated conceptually. It is applied in a rather functional sense as exposure and reaction to stress [8]. Although the social context is taken into account and attention is not only focused on personal attributes that are important for coping and responding to stress, the general emphasis continues to be on the individual person; vulnerability explains why one person adapts while another being hurt in similar circumstances.

The shift to context is more obvious in social science literature. Here the argument is that socio-economic conditions such as lack of education and nutrition, of sanitation and of housing make individuals vulnerable [9]. Economic developments, particularly a competitive and unequal world market system, contribute to exploitation of vulnerable populations, especially women [10]. From a global perspective, particularly from the point of view of developing countries, the social and economic context is recognized as a relevant dimension in itself, not just as a determinant of individual vulnerability. The context creates vulnerable categories of people. Vulnerability therefore has a structural component that impacts individuals but is primarily important because it affects whole populations. The argument is that it is unfair to expect that individuals will be able to resist or reduce this harmful impact, so a policy or common approach is required [11].

In the 1970s and 1980s, vulnerability was also a core concept in the emerging field of disaster studies. Although at that time there was controversy over the question of whether or not there was an increase in the occurrence of disasters, it was emphasized that the damage caused by disasters could not be attributed merely to an extreme geophysical event. Rather, vulnerability was the result of the impact of this event on a human population: 'without people there is no disaster' [12]. It is vulnerability that is the real cause of disaster. This vulnerability is amplified, especially in poor and marginalized countries. It is irrelevant whether extreme events are occurring more often; it is the deteriorating social and economic conditions in these countries that make populations more vulnerable to these events that produce disasters [13].

In the 1990s, the use of the notion of vulnerability surged as a key concept in two new areas of scientific and policy interest: the HIV/AIDS pandemic and environmental degradation [14]. Approaches to HIV prevention moved in the early 1990s from individual to social vulnerability [15], identifying contextual factors that put individuals at risk for HIV/AIDS. At the global level, it was recognized that poor social conditions can lead to AIDS (described as 'the disease of poverty' [16]) but that also the disease itself often had severe social and economic consequences. Women in particular became the primary vulnerable group. Almost invisible at the beginning of the HIV/AIDS pandemic, women eventually were regarded as the paradigmatic vulnerable population, especially in developing countries, because arguably they are biologically more susceptible to this infection, but also because they have greater social vulnerability (due to gender inequalities, poverty, social disadvantages, or reduced sexual autonomy)[17]. The discourse of vulnerability exemplifies the transition in HIV prevention from individual to structural determinants such as poverty, inequality, powerlessness, discrimination, stigmatization, and marginalization [18].

The second area in which the concept of vulnerability during this period of time found a fruitful application was environmental science, particularly since the Rio Declaration on Environment and Development, adopted in 1992. This declaration introduced the idea of 'environmental vulnerability' as one of its principles [19]. It was argued that vulnerability requires a holistic approach, integrating technical interventions and engineering with social and economic activities [20]. The notion was used to explain how societies transformed in history because of climatic changes [21]. Many attempts have been made to clarify the concept of vulnerability in this environmental context. Arguably, a conceptual framework of vulnerability is necessary since climate change demands interdisciplinary research on the basis of clear understanding of the concept [22]. But at the same time, its application in this domain used to be very technical, for example in efforts to construct a vulnerability index or model frameworks in order to assess and rank countries like Tuvalu [23]. Though still used in a technical sense, the scope of the notion has significantly expanded.

Finally, since 2000 the number of publications discussing the notion of vulnerability further increased in analyses and discussions on bioterrorism and human security. For example, food and water supplies are vulnerable to deliberate sabotage with biological and chemical agents [24]. There is also the danger of violent conflict, organized crime, and terrorist attacks making communities and even whole countries vulnerable. The latest trend today calls attention to 'cybervulnerabilities'. The critical infrastructure of countries and companies that increasingly depend on electronic networks can be seriously damaged by incidental, and recently more coordinated, cyberattacks [25]. The perception of dangerous threats has also intensified in healthcare. Particularly in the field of public health, more emphasis is nowadays put on human security, recognizing a more generalized vulnerability due to threats of infectious diseases such as SARS, avian flu, or ebola as well as the global interconnections of disease, poor economic conditions, and social injustice [26].

This brief overview of the use of 'vulnerability' in different domains of the scholarly literature shows that the notion has increasingly been applied beyond its traditional medical and military scope. It is used in a wide variety of research contexts such as public health, ecology, disasters, poverty, development, climate change, and security. In most of these contexts, the notion is employed as a technical or scientific concept. It is specified in different ways and applied at various levels. But in general it has the three basic components mentioned in the previous chapter: exposure (to harmful threats and hazards), sensitivity, and adaptive capacity. These components are then further identified, analyzed, and specified, whether it concerns a human being, a community, social or natural system, country, or integrated human-environment systems. The increasing use of vulnerability in different domains has enriched the concept. New types of vulnerability have been distinguished: psychological, social, economic, environmental, and even existential and cultural. Recent discourses on vulnerability emphasize the global dimension of developments that produce or uncover vulnerability: poverty, disasters, climate change, and pandemics. These are all misfortunes beyond an individual's control. They emerge from structural conditions that are simply beyond the power of individual persons. They also make clear that vulnerability is more than a physical characteristic; it is a social phenomenon. The growing awareness that vulnerability is associated with global and social developments has led to a conceptual rethinking. While its scope has been widening, the notion itself has expanded. The next sections will illustrate this expansion in two different ways; first by examining its evolving use in nursing literature, and second by comparing 'vulnerability' with the concepts 'resilience' and 'security' that are usually regarded as its opposites.

Expanding vulnerability

Over time, the term vulnerability not only received a wider scope and became applied to an enlarging number of areas, but it also evolved from a narrowly interpreted, technical concept to a broader notion. One can observe this expansion in the nursing literature. The considerable attention given by nurses to vulnerability was initially concentrated on the ill person. The task of nurses is to do for the patient what he or she cannot do for himself or herself with the intent of promoting independence [27]. This makes vulnerability a crucial perspective for nursing. However, this focus has two shortcomings.

First, vulnerability is approached from an individualistic perspective. The individual condition of the patient is the primary concern. Nursing as a professional discipline needs to identify hazards that may threaten the health of the patient or impede recovery, and create a micro-environment that protects the patient so that he or she may soon overcome his or her state of dependency. The social, economic, and political context of the health problems is usually not taken into account. Nursing assistance is also centered on restoring the adaptive capacity of individuals; they have to accommodate, to make adjustments, in order to cope with the health hazards, even if the social or economic conditions are unacceptable. The individual is adapted to the context without any effort to change or transform the context itself. Laura Tomm-Bonde, who worked as a nurse in Mozambique, has called this the 'naïve view of vulnerability' [28]. Concentrating solely on the individual's adaptive capacities takes for granted the structures and contexts that render whole populations vulnerable.

The second shortcoming is that vulnerability continues to be regarded as a technical notion, like in the traditional medical perspective. In the early nursing literature, a distinction was made between risk and vulnerability; the first refers to hazards within the environment, the latter to characteristics of the individual [29]. Here, vulnerability is associated with internal determinants only (confirming the individualistic focus) but 'internal' is at the same time conceived as medically relevant factors. These factors can be constitutional (or inborn, e.g. the genetic predisposition or the neurophysiology of the organism) or acquired (for example, traumas, diseases, and perinatal complications). The potential harm that can occur is primarily medical; it can result in pathology and dysfunctioning. On the basis of these factors individuals can be classified from vulnerable to invulnerable. The analogy is made with three dolls: a glass, plastic, and steel one [30]. The glass doll is very fragile, the plastic one is much more flexible and thus able to deal with various types of exposure, and the steel doll is relatively invulnerable. It is the internal makeup of the individual that determines his or her vulnerability.

This restrictive view as used in the early nursing literature became increasingly criticized. It has been pointed out that some groups of people have a higher probability of being ill than others while they have more or less the same internal constitution [31]. Vulnerability is not located in the individual alone. Nurses need to take into account that the roots of vulnerability are often social [32]. A more fundamental critique is directed against the conception of 'internal' determinants as endogenous factors such as genetic disposition and bodily constitution. Without denying that these factors play an important role in what makes an individual vulnerable (people with genetic diseases, disabilities, or advanced age will be more vulnerable than others), there is a more fundamental sense in which vulnerability is 'internal' to human beings: it is also an inherent characteristic of being human in general [33]. Basically, the doll analogy is wrong: there are no steel humans invulnerable to external threats. Because vulnerability is an ontological characteristic 'humans are always vulnerable' [34].

In the nursing literature it is indeed more and more emphasized that a move is needed towards a broader, more holistic concept of vulnerability [35]. Nurses are more than problem solvers at the level of the individual. The social, economic, and cultural context should be taken into account, noticing that the vulnerable person is embedded in a wider environment that may produce and sustain vulnerability. Similar ideas are expressed in the vulnerability paradigm in regard to HIV/AIDS: the underlying social conditions create vulnerability [36]. Furthermore, vulnerability cannot be reduced to a technical sense. Because nursing is focused on caring, its concerns need to go beyond the framework of health hazards and exposure to harm, acknowledging that human frailty is shared. All human beings have the same fundamental sensitivity to be damaged. This common weakness explains the crucial need for care in human interactions [37]. The notion of vulnerability in the nursing literature has therefore moved from the initial focus on individual persons in two ways: emphasizing that vulnerability can be produced by external conditions, and recognizing that the internal components of vulnerability characterize the human species rather than individual specimens.

Opposites of vulnerability

Security

Another way to demonstrate the conceptual evolution of the notion 'vulnerability' is examining its antonyms. Following the traditional military perspective, vulnerability was initially regarded as similar to insecurity. In this view, security means that the territorial integrity of states is protected against military threats. Security implies that vulnerability is reduced or eliminated. The threat is relatively straightforward: it is the possibility of physical violence and aggression by specific enemies. There is an inverse relation between the two notions: security increases when vulnerability decreases. Being secure means being safe from violent conflict, protected from danger, no longer vulnerable. It is clear that the notion vulnerability here applies to the state; the assumption is that the citizens will be secure as long as the state is secure. The scope of the notion is also restricted; it is vulnerability to physical violence during conflict.

This concept of security from a military point of view was increasingly contested from the 1990s [38]. Three developments have especially contributed to broadening the concept. First, the awareness that many new threats are non-military: natural disasters, famines, pandemics and new infectious diseases, environmental degradation, and climate change. It also became clear that existing conditions such as poverty should be regarded as threats contributing to widespread, often chronic insecurity of citizens. Second, the threats to security are global. Due to processes of globalization, not only the security of states or its citizens is at stake but that of all people around the world. The present-day world is characterized by 'mutual vulnerability', and therefore all people have a shared interest in promoting security [39]. This global dimension means that security is a universal concern relevant to people everywhere as well as an interdependent phenomenon; global threats have an impact anywhere [40]. Third, the collapse of the Soviet Union made it possible to shift the focus from protection of national boundaries to promotion of human development. The United Nations Development Programme (UNDP) played an important role, issuing Human Development Reports since 1990, reframing security as *human* security. The 1994 UNDP report addressed this issue explicitly: 'Human security is not a concern with weapons – it is a concern with human life and dignity' [41]. Poverty rather than war should be the focus of security.

The broader concept of security promoted by these developments emphasized that the focus should shift from state to individual and also that basic human needs should be the central concern. Human security, as the UNDP report underlines, is centered on people; it is concerned with the daily life of human beings, protecting them from chronic threats such as hunger, disease, and violence [42]. It is obvious that this broader notion refers to a wider range of vulnerabilities (e.g. economic, environmental, medical, and political). It is less obvious that vulnerability in this view is no longer the opposite of security. Rather it has become a defining characteristic of security itself. The Norwegian political scientist Astri Suhrke has argued that 'the core of human insecurity can be seen as extreme vulnerability' [43]. She distinguishes three categories of extreme vulnerability: war, poverty, and natural disasters. People are not only vulnerable to physical violence. They can also be victims of 'structural violence', conditions of abject poverty or powerlessness beyond individual agency and control. The main purpose of the new idea of human security is to protect the vulnerable. Emphasizing human security therefore entails obligations. Connecting with vulnerability reinforces the normativity of the notion of human security. Those who have the capacity to provide security to vulnerable people have the obligation to do so [44]. This perspective of security is applied in healthcare, for example in the World Health Report 2007 [45]. Global public health security ('a safer future', as the report is entitled) can only be accomplished through minimizing the vulnerability of people around the world to threats to health that are more and more crossing borders.

This new notion of human security has two implications. Its objective first of all is to provide a 'vital core' for human existence or a 'minimal set of conditions of life' [46]. Security is a precondition for human development. Policies should therefore aim at reducing as much as possible existing vulnerabilities in order to safeguard basic human needs, not only in the domain of health but also in other areas such as income, education, and political freedom. The second implication is that policies should be more positive. They cannot be merely defensive or protective (as in the traditional view of security) but should be comprehensive and pro-active [47].

The broadening of the discourse on security has changed the interpretation of vulnerability. Initially regarded as synonymous to insecurity, it is now often regarded as a more basic phenomenon. Insecurity is the result of vulnerability. Only by addressing the latter, it is possible to promote human security.

Resilience

The concept of resilience is applied as the opposite of vulnerability, particularly in psychological and psychiatric literature [48]. In the early stage, resilience was considered as equivalent to 'invulnerability'. People are vulnerable because they lack resilience. The term was first used in a famous longitudinal study of Hawaiian children born into poverty on the island of Kauai [49]. Among the children who grew up in similar adverse environments, one third developed into competent, caring adults without serious problems. The question was what made the difference between vulnerability and invulnerability. Some children were successful because they were resilient. They had the ability to cope, to bounce back, and to adapt despite adversity and various threats, and to overcome their vulnerability [50]. Resilience was regarded as a feature of an individual who has specific competencies to reduce or neutralize the negative impact of the context in which he or she is developing and living. It was perceived as a quality that could make vulnerable individuals invulnerable. The challenge was to identify protective factors. Like the notion of vulnerability itself, this view soon came to be regarded as too restrictive and static. It was argued that resilience should be regarded as a dynamic process that needs to take into account the context in which the individual is living [51]. The significance of contextual determinants cannot be reduced to protective competencies of resilient individuals. On the contrary, it is important to identify 'underlying protective processes' [52]. Resilience is not merely a personal quality but is often related to external factors such as the family, community, and the social environment. In this broader view, the opposition between resilience and vulnerability disappears. Nobody is invulnerable, and everybody needs resilience. Those who successfully adapt to some threats will face new vulnerabilities when the circumstances of life are changing.

Similar permutations in the concept of resilience are noticeable in the fields of global environmental change, climate change, and disaster studies. Resilience, vulnerability, and adaptation are here closely connected core concepts [53]. On the one hand is the view that resilience is lost when a social-ecological system has become vulnerable [54]. There is no longer an ability to respond to disturbances. Vulnerability and resilience apparently are opposites. On the other hand, in studies of global environmental change the concept of vulnerability is generally composed of the three elements mentioned in the previous chapter: exposure, sensitivity, and adaptive capacity [55]. Resilience is considered as part of the third element. It helps the (natural and/or social) system to cope, adapt, or recover from the effects of hazards and threats. In this view, resilience is a component of vulnerability [56]. It determines the degree in which a system is vulnerable. The best adaptation results from enhancing the resilience and reducing the susceptibility [57]. It is also noted that vulnerability is not necessarily a negative property. Disturbances, stress, and hazards may create transformations that are beneficial. Gallopín mentions as examples the emergence of a social group from chronic poverty and the collapse of an oppressive regime [58]. The transformative work is done through resilience; this is not simply the capacity to absorb shock or to cope with threats but also 'the capacity for renewal, re-organization and development' [59]. British sociologist Anthony Giddens in his recent book on politics and climate change also presents a positive view of resilience [60]. It allows constructive responses to overcome vulnerability at different levels. The first level is the physical environment where resilience refers to the capability of the environment to withstand shocks and threats. The individual is the second level where resilience refers to the quality of character, the drive to make the best of adverse circumstances. The third level is the group or community where resilience refers to cooperative qualities and the capacity to act together.

The use of vulnerability in ethics and bioethics

The use of the notion of vulnerability in bioethics is recent. The first publication using the concept of vulnerability in the context of research ethics (in 1977) applies vulnerability to the investigator and the pharmaceutical company, arguing that ethical review not only protects the participating subjects but also reduces the vulnerability of the researchers [61]. The authors, however, do not use the notion of vulnerability as an ethical consideration in itself. This is different in the second oldest article that shows up in PubMed searches using the keywords 'vulnerability' and 'research ethics'. This article, published in 1979, can be regarded as the first journal publication that explicitly used vulnerability as an ethical notion [62]. The authors, Robert Levine and Karen Lebacqz, examine the ethical principles that should guide clinical trials. They introduce the notion of vulnerability in connection to considerations regarding the selection of research participants. Subjects may be vulnerable 'by virtue of their capacities or situations', and then need 'extra protection' [63]. The notion is applied to the question whether it is ethical to select subjects for a randomized clinical trial exclusively from Veterans Administration (VA) hospitals. The authors conclude that patients in VA hospitals are vulnerable because they have limited options compared to private patients. But they do not infer that such vulnerable patients should not be used as research subjects because it is possible to reduce their vulnerability.

The interest of various scientific disciplines in the concept of vulnerability explains why the quantity of journal articles with this keyword alone has substantially increased since the 1970s. However, its combination with the keywords 'ethics', 'bioethics', and 'research ethics' is much less frequent and more recent. In general more than 80 percent of these publications (combining the keywords) have appeared since 2000. In the previous decade 1990–2000 a rather small number of articles was published combining these keywords [64]. In other words, bioethical discourse discovered the notion of vulnerability rather late, compared to other disciplines. At the same time, the focus of bioethics seemed to be more specific.

As a keyword in connection to ethics, research ethics and bioethics, the term 'vulnerable populations' is more often used than the notion of vulnerability itself. The reason probably is bibliometrical. The notion of 'vulnerable populations' was first introduced as a keyword in the *Bioethics Thesaurus* in 1997 [65]. It became a MeSH term (Medical Subject Heading) in 2003. However, 'vulnerability' is neither a bioethics keyword nor a medical subject heading. The same holds for the *Thesaurus Ethics in the Life Sciences* launched by the German Reference Centre for Ethics in the Life Sciences in Bonn: whereas 'vulnerable populations' is a keyword in the subject area Society, Politics, Economics, Law, Education, Media, 'vulnerability' is not.

In order to qualify as a keyword, a notion should already be in current use. A PubMed search of all journal articles combining the keywords 'vulnerable populations' and 'ethics' shows that 55 percent has been published since 2000. However, in the previous decade there was already a substantial production: 40 percent was published in the decade 1990–2000. The same is true for articles combining the keywords 'vulnerable populations' and 'research ethics': 65 percent were published since 2000, while 31 percent appeared in print in the decade 1990–2000. Finally, 54 percent of all articles combining 'vulnerable populations' and 'bioethics' were published since 2000, and 44 percent in the preceding ten years [66].

The conclusion on the basis of this analysis is that the majority of journal publications combining the notion of vulnerability with ethics, bioethics, and research ethics is recent (since 2000). The notion of 'vulnerable populations' in the context of ethics, however, has a somewhat older tradition, being substantially used since 1990. Compared to other disciplines, bioethics and in particular research ethics have been rather late in exploring and applying the notion of vulnerability. How the use of this notion has evolved in the context of bioethics is the subject of the next chapter. Before attending to this, it is necessary to briefly question why vulnerability is so prevalent in a wide range of contemporary discourses.

The attractiveness of vulnerability discourse

Vulnerability has become an important concept in nursing science, public health, and social science literature. It is also increasingly used in relatively new fields of study concerning disasters, HIV/AIDS, environmental degradation, climate change, bioterrorism, and human security. As discussed in this chapter, not only the scope of the notion has widened over the last few decades but also the concept itself has substantially evolved. The question is how to explain the rising status of this notion.

One of the answers is provided by the Director-General of the World Health Organization, Dr Margaret Chan when she says: 'Vulnerability is universal' [67]. Though her concern is global health and particularly the spread of infectious diseases, the same sense of widespread vulnerability can be noticed in many other fields. The fact that the world has become increasingly interconnected and interdependent has created a sense of mutual vulnerability [68]. It is not simply the case that human existence has become more fragile but everybody is facing similar threats. A pandemic disease in one country can quickly travel to countries on the other side of the globe. Ecological changes in one part of the world are impacting other regions. A major disaster in a country has repercussions far outside its borders. Of course not all countries and all persons are equally vulnerable, but the global nature of contemporary threats means that in fact everybody can be hurt. It also seems that the number and variety of threats has increased, creating various types of vulnerability. It is remarkable that many discourses explicitly focus on sources of vulnerability that are beyond the control of individual persons. Because individual lives are often impacted by more encompassing processes (the economy, the healthcare system, the environment, or the social security system) people feel that their power to influence the conditions in which they live is rather limited or even non-existent. People can therefore be vulnerable because these conditions – the social, economic, ecological, or political environment – make them vulnerable. Although vulnerability is manifested at the individual level, its roots are often in structural conditions and circumstances.

Another answer to explain the popularity of vulnerability discourse is that global processes necessitate international cooperation. The emphasis on vulnerability is in fact an effective moral appeal to global solidarity and international responsibility. If humanity is vulnerable, we all have an obligation to protect the most vulnerable, the more so since we ourselves are vulnerable. It is impossible to provide security for all if vulnerable populations are not better protected. Individual states, populations, groups, persons will not be able to secure themselves, to reduce or limit their vulnerability by their own efforts alone. They need to coordinate and cooperate with others. Traditional mechanisms of protection are furthermore insufficient because of the global dimension of threats and possible harms. Emphasizing vulnerability thus not only creates a sense of shared predicament, but it also encourages initiating common activities for the benefit of humanity. It is not without reason that vulnerability discourse is particularly employed by international organizations.

During the World Food Summit in 1996 member states of the Food and Agriculture Organization (FAO) pledged a commitment to reaching the vulnerable and to developing a systematic approach of vulnerability analysis and mapping, targeting especially women and small children as populations most vulnerable to hunger [69]. In following years FAO issued a series of reports on food insecurity and vulnerability with the aim to establish Food Insecurity and Vulnerability Information and Mapping Systems (FIVIMS) at national, regional, and international levels. These systems assist in identifying who vulnerable people are, where they are, and why they are vulnerable [70]. These efforts are connected with the Vulnerability Analysis and Mapping of the World Food Programme (WFP), monitoring vulnerable situations at national levels [71]. As discussed earlier, the United Nations Development Programme (UNDP) launched the Human Development Reports in 1990. The main purpose of the reports was to show that development is focused on people rather than income and wealth. The attention of policy-makers should be redirected from the economy and economic growth to human well-being. Already in the first report, the notion of vulnerability has been used [72]. One can observe a similar move here as in the reframing of the concept of security as human security. Vulnerability is a useful conceptual tool to focus on the basic needs of human beings instead of the prerequisites of military or economic systems. This reorientation, however, does not imply an individualistic perspective. In fact, persons everywhere are powerless in the face of global processes that engulf and threaten them. Vulnerability can only be properly addressed by targeting its sources in the social, economic, and political context. Another significant contributor to the debate is the United Nations Environment Programme (UNEP). It issued in 2003 a report on human vulnerability in connection to environmental change. The report advocates a broad concept of vulnerability considering it as a social phenomenon [73]. Arguing that vulnerability will only increase in the future, and specifically that developing countries are vulnerable, UNEP points out the need for international policies. A prior condition for planning and decision-making is that vulnerability is measured. Assessment of human vulnerability should therefore include a range of issues, from damage to health, and poverty to loss of natural heritage and extreme weather events [74]. The emphasis on the social context is reiterated in another influential international document: the Report on the World Social Situation published in 2003 by the Department of Economic and Social Affairs of the United Nations. This report is entirely devoted to social vulnerability. The report focuses attention on the vulnerable groups of 'the forgotten, the invisible and the ignored billions of people' who are disenfranchised, powerless, and voiceless [75].

Finally, the focus on vulnerability in contemporary discourse is attractive since it opens up possibilities for a new approach to policy-making. In the first place, vulnerability necessarily requires forward-looking, pro-active policies. We need to act before threats will have any impact, before the expected damage occurs. As human beings we can foresee harmful effects and we have to undertake pre-emptive actions. That is why there is a continuous need to analyze, assess, and respond to vulnerabilities [76]. Secondly, processes of globalization are not only creating new threats. They also demonstrate that the traditional protection mechanisms (social security and welfare systems, family support systems) are increasingly inadequate to cope with these threats. Furthermore, these processes are eroding the abilities of individuals and communities to address vulnerability. Vulnerability as a shared condition of humanity demands therefore the implementation of transnational norms to guide policy-making. Such a global normative framework is available in international human rights discourse where it is argued that policies to reduce vulnerability should be based on promoting and implementing human rights. Addressing human vulnerability is only feasible through respecting human rights [77].

Conclusion

The notion of 'vulnerability' is used in an increasing number of publications in various fields of interest. In many cases it is used as a technical concept with analyses focused on the components of exposure, sensitivity, and adaptive capacity. Over time, however, the notion has been widened. Vulnerability no longer is regarded as characteristic of individual persons but rather as social phenomenon. This expanding scope of the notion is observable in the nursing literature as well as in the new field of security studies. Being vulnerable is more and more regarded as the result of a range of social, economic, cultural and political conditions, and therefore beyond the power and control of individuals. This widening of the notion is associated with the discourse of globalization. Processes of globalization are associated with a growing sense of vulnerability. Global interconnectedness has resulted in a world that is more vulnerable to various, often relatively new threats. The interdependence of contemporary existence is projecting and reinforcing the idea that we live in an environment that is increasingly beyond our control. Vulnerability manifests itself in individuals but its sources are somewhere else. Rather than adapting individual subjects to conditions and processes of globalization, it is therefore important to critically examine globalization itself and to reorient it towards the needs of human beings and the qualities of human existence. This is the agenda of 'globalization with a human face' [78]. Within a context that is global but also abstract and universal, vulnerability is the expression of an emerging new humanitarianism that centers on human persons, not as isolated individuals but as embedded in encompassing processes that continuously produce vulnerability. The question is what is it exactly in these globalizing processes that articulates, produces, or exacerbates human vulnerability? The connection between vulnerability and globalization will be further analyzed in a later chapter. But first the evolution of vulnerability in the context of biomedical ethics must be explored.

Notes

1　See Traquair, 'The special vulnerability of the macular fibres and "sparing of the macula"'.
2　According to a PubMed search in August 2014 using the keyword 'vulnerability'.
3　McGilloway, 'Dependency and vulnerability in the nurse/patient situation', 235.
4　See Fillenbaum, 'An examination of the vulnerability hypothesis'.
5　See Aldwin and Revenson, 'Vulnerability to economic stress'.
6　See Shainess, 'Vulnerability to violence: Masochism as process'.
7　See Cutler, 'Safety on the streets: Cohort changes in fear'.
8　Aldwin and Revenson, 'Vulnerability to economic stress', 162.
9　See Zaidi, 'Poverty and disease: Need for structural change'.
10　Steady, 'African women, industrialization, and another development. A global perspective', 52, 57, 61.
11　See Chambers, 'Editorial introduction: vulnerability, coping, and policy'.
12　See O'Keefe, Westgate, and Wisner, 'Taking the naturalness out of natural disasters'.
13　Snarr and Brown studied permanent post-disaster housing in Honduras after Hurricane Fifi in 1974. The victims were mainly poor and marginal people. Most of the new buildings were not constructed to prevent or mitigate future disasters, so that vulnerability will remain. See Snarr and Brown, 'Permanent post-disaster housing in Honduras: Aspects of vulnerability to future disasters'. Lewis, in his study of natural disasters in the Pacific country Tonga, shows how vulnerability is related to loss of traditional methods of potable water storage and building construction techniques. Global interconnections have increased the degree of dependency, and reduced the capacity for self-reliance. See Lewis, 'Some perspectives on natural disaster vulnerability in Tonga', 152, 154, 158.
14　A bibliometric study of the use of the keyword 'vulnerability' in the field of human dimensions of global environmental change shows that the number of papers with this keyword has increased rapidly since the early 1990s, see Janssen, Schoon, Ke, and Börner, 'Scholarly networks on resilience, vulnerability, and adaptation within the human dimensions of global environmental change'.
15　Whelan, 'Human rights approaches to an expanded response to address women's vulnerability to HIV/AIDS', 21.
16　Ankrah, 'AIDS and the social side of health', 968.
17　Higgins, Hoffman, and Dworkin, 'Rethinking gender, heterosexual men, and women's vulnerability to HIV/AIDS', 436. See also Wadhwa, Ghosh, and Kalipeni, 'Factors affecting the vulnerability of female slum youth to HIV/AIDS in Delhi and Hyderabad, India'.
18　See Parker, 'Sexuality, culture, and power in HIV/AIDS research'.
19　Principle 6: 'The special situation and needs of developing countries, particularly the least developed and those most environmentally vulnerable, shall be given special priority. International actions in the field of environment and development should also address the interests and needs of all countries'. See UN Conference on Environment and Development, *Rio Declaration on Environment and Development*.
20　Lewis, 'The vulnerability of small island states to sea level rise: The need for holistic strategies', 244; see also Epstein, 'Framework for an integrated assessment of health, climate change, and ecosystem vulnerability'.
21　See Brown, 'Climate change and human history, some indications from Europe, AD 400–1400'.

22 Füssel, 'Vulnerability: A generally applicable conceptual framework for climate change research', 156. See also Adger. 'Vulnerability'.

23 See, for example, Briguglio, 'Small island developing states and their economic vulnerabilities'; Mimura, 'Vulnerability of island countries in the South Pacific to sea level rise and climate change'; Turner, Kasperson, Matson *et al.* 'A framework for vulnerability analysis in sustainability science'.

24 Khan, Swerdlow and Juranek, 'Precautions against biological and chemical terrorism directed at food and water supplies', 4.

25 Brenner, *America the vulnerable. Inside the new threat matrix of digital espionage, crime, and warfare*, 129.

26 See Gutlove and Thompson, 'Human security: Expanding the scope of public health'.

27 Following a famous definition of nursing: doing for the patient what he cannot do for himself with the intent of promoting independence; see Henderson, V. *The nature of nursing*, 21 ff.

28 Tomm-Bonde, 'The naïve nurse; revisiting vulnerability for nursing', 2.

29 Rose and Killien, 'Risk and vulnerability: a case for differentiation', 61.

30 Anthony, 'A risk-vulnerability intervention model for children of psychotic parents', 100.

31 Rogers, 'Vulnerability, health and health care,' 66.

32 See Aday, *At risk in America: The health and health care needs of vulnerable populations in the United States*; Nichiata, Bertolozzi, Takahashi and Fracolli, 'The use of the "vulnerability" concept in the nursing area'.

33 See Daniel, 'Vulnerability as a key to authenticity'; Johnstone, 'Ethics and human vulnerability'.

34 Daniel, 'Vulnerability as a key to authenticity', 191.

35 Tomm-Bonde, 'The naïve nurse; revisiting vulnerability for nursing', 9; Sellman, 'Towards an understanding of nursing as a response to human vulnerability', 4. See also: Basnyat, 'Beyond biomedicine: health through social and cultural understanding'.

36 Mann and Tarantola, 'From vulnerability to human rights', 463.

37 Stenbock-Hult and Sarvimäki, 'The meaning of vulnerability to nurses caring for older people', 32.

38 See Rothschild, 'What is security?'; Suhrke, 'Human security and the interests of states'; Newman, 'Human security and constructivism'; King and Murray, 'Rethinking human security'; Acharya, 'Human security: East versus West'.

39 Gutlove and Thompson, 'Human security: expanding the scope of public health', 17.

40 UNDP, *Human Development Report*, 22.

41 UNDP, *Human Development Report*, 22.

42 UNDP, *Human Development Report*, 23.

43 Suhrke, 'Human security and the interests of states', 272.

44 Newman, 'Human security and constructivism', 240.

45 See WHO, *The world health report 2007 – A safer future: global public health security in the twenty-first century*.

46 Gutlove and Thompson, 'Human security: expanding the scope of public health', 5.

47 In the prosaic language of the UNDP Report the implications of the concept of human security are described as follows: 'It acknowledges the universalism of life claims ... It is embedded in a notion of solidarity among people. It cannot be brought about through force ... It can happen only if we agree that development must involve all people'. UNDP, *Human Development Report, 24*. See also *UN, Human security in theory and practice. Application of the human security concept and the United Nations Trust Fund for Human Security*, 12–25.

48   See Dyer and McGuinness, 'Resilience: Analysis of the concept'; Luthar, Cicchetti, and Becker, 'The construct of resilience: a critical evaluation and guidelines for future work'; Johnson and Wiechelt, 'Introduction to the special issue on resilience'; Tusaie and Dyer, 'Resilience: a historical review of the construct'; Earvolino-Ramirez, 'Resilience: a concept analysis'.

49   See Werner, 'Risk, resilience, and recovery: Perspectives from the Kauai longitudinal study'; Dyer and McGuinness, 'Resilience: Analysis of the concept', 277; Johnson and Wiechelt, 'Introduction to the special issue on resilience', 658-9.

50   Resilience is defined as 'the ability to bounce back from adversity', see Dyer and McGuinness. 'Resilience: Analysis of the concept', 276.

51   Tusaie and Dyer. 'Resilience: a historical review of the construct', 6-7; Johnson and Wiechelt, 'Introduction to the special issue on resilience', 662-3.

52   Luthar, Cicchetti and Becker, 'The construct of resilience: a critical evaluation and guidelines for future work', 545.

53   See Janssen and Ostrom, 'Resilience, vulnerability, and adaptation: a cross-cutting theme of the International Human Dimensions Programme on Global Environmental Change'.

54   Folke, 'Resilience: the emergence of a perspective for social-ecological systems analyses', 262.

55   Smit and Wandel, 'Adaptation, adaptive capacity and vulnerability', 286.

56   Gallopín, 'Linkages between vulnerability, resilience, and adaptive capacity', 301.

57   Mimura, 'Vulnerability of island countries in the South Pacific to sea level rise and climate change', 141.

58   Gallopín, 'Linkages between vulnerability, resilience, and adaptive capacity', 295.

59   Folke, 'Resilience: the emergence of a perspective for social-ecological systems analyses', 253.

60   Giddens, *The politics of climate change*, 163.

61   See Ramsay, Tidd, Butler, and Venning, 'Ethical review in the pharmaceutical industry'.

62   See Levine and Lebacqz, 'Some ethical considerations in clinical trials'.

63   Levine and Lebacqz, 'Some ethical considerations in clinical trials', 732.

64   In the decade 1990–2000 the number of journal articles using 'vulnerability' as a keyword was 5,358 (20.8%). Combining this keyword with 'ethics' were 78 articles out of a total of 517 (15.1%), with 'bioethics' 15 out of a total of 89 (16.9%), and with 'research ethics' 30 out of a total of 243 (12.3%).

65   According to Flanigan, 'Vulnerability and the bioethics movement', 13.

66   The total number of journal articles combining keywords 'vulnerable populations' and 'ethics' amounted to 1,698 since 1973. Since 1990 (until 2012) 1,600 articles have been published (94.7% of the total number). In the decade 1990–2000 a total of 678 publications have appeared (40.1%). Combining 'vulnerable populations' and 'research ethics' were 798 articles since 1973. Since 1990 768 articles have been published (96.2%). In the decade 1990–2000 a total of 248 (31.1%) have been published. Journal articles combining 'vulnerable populations' and 'bioethics' totaled 261 since 1984; 98.1% of these articles were published since 1990 (total 256). In the decade 1990–2000 114 articles have been published (43.7%).

67   WHO, *The world health report 2007 – A safer future: global public health security in the twenty-first century*, 2.

68   Aginam, 'Between isolationism and mutual vulnerability: A South-North perspective on global governance of epidemics in an age of globalization', 303–304. See also Brown,

'"Vulnerability is universal": Considering the place of "security" and "vulnerability" within contemporary global health discourse'.

69 FAO, *World Food Summit. Rome Declaration on World Food Security*.

70 FAO, *Report on the development of food insecurity and vulnerability information and mapping systems (FIVIMS)*.

71 WFP, *Sampling guidelines for vulnerability analysis*.

72 UNDP, *Human Development Report 1990. Concept and measurement of human development*, 82.

73 UNEP, *Assessing human vulnerability to environmental change. Concepts, issues, methods, and case studies*, 3.

74 UNEP, *Assessing human vulnerability to environmental change. Concepts, issues, methods, and case studies*, 27 ff.

75 United Nations Department of Economic and Social Affairs, *Report on the World Social Situation, 2003. Social vulnerability: Sources and Challenges*, 8.

76 Giddens, *The politics of climate change*, 163 ff. See also Kaldor who argues that it is the primacy of human rights that distinguishes the human security approach from the more traditional approaches based on the security of states; Kaldor, *Human security*, 185-6.

77 Whelan, 'Human rights approaches to an expanded response to address women's vulnerability to HIV/AIDS', 23; Mann and Tarantola, 'From vulnerability to human rights', 465; Newman, 'Human security and constructivism', 242.

78 This is the title of the Human Development Report of 1999 (see: UNDP. *Human Development Report 1999. Globalization with a human face*. It is the fundamental policy choice, expressed by Kofi Annan, United Nations Secretary General in an address to the World Economic Forum in 1999: 'We have to choose between a global market driven only by calculations of short-term profit, and one which has a human face. Between a world which condemns a quarter of the human race to starvation and squalor, and one which offers everyone at least a chance of prosperity, in a healthy environment'. See Annan, Kofi. 'Secretary-General proposes global compact on human rights, labour, environment, in address to world economic forum in Davos', UN Press Release, 1 February 1999, (http://www.un.org/News/Press/docs/1999/19990201 .sgsm6881.html) (Accessed 4 April 2015).

3

VULNERABILITY IN THE CONTEXT OF HEALTH CARE AND BIOETHICS

'Vulnerability' is a relatively new concept in contemporary bioethics. This chapter will examine the emergence of vulnerability in authoritative bioethical documents. Its story begins with the publication of the Belmont Report in 1979. The notion first surfaced as a special consideration in the application of the general principles of respect for persons, beneficence, and justice in the context of medical research. The Belmont Report became influential in canonizing the normative framework for all areas of contemporary bioethics. The research context continued for some time to be the exclusive domain of vulnerability. The Council for International Organizations of Medical Sciences (CIOMS) referred to vulnerability in its first guidelines for international research in 1982. The Declaration of Helsinki of the World Medical Association (WMA) only referred to vulnerability for the first time in its fifth revision of 2000. In both documents, this first use is brief and only presented as a consideration to be taken into account. The notion of vulnerability itself is not explained; also, the criteria for the identification of individuals or groups as vulnerable are not explicitly presented. The status of vulnerability changed in the 1990s. The new CIOMS Guidelines in 1991 specifically mention vulnerability as a fundamental principle and present it as included in the principle of respect for persons. This position is not reiterated in later guidelines that affirm vulnerability as a special application of the principle of respect for persons and the principle of justice. Vulnerability, however, was now upgraded as a guideline for research itself. The emerging status of vulnerability culminated in the UNESCO Universal Declaration on Bioethics and Human Rights (2005). This is the first international document that articulates respect for vulnerability as an ethical principle. At the same time, the scope of vulnerability is broadened: it is no longer only relevant for medical research but for healthcare as such.

The increasing number of references to vulnerability in global bioethics documents does not imply that it is clear what vulnerability really is and how it should be applied in bioethical discourse. The notion also has a rather isolated biography in bioethics literature. There rarely are references to other disciplinary domains in which the notion is used. Conceptual reflection and analysis to clarify the notion are not often presented. And lessons from other practical settings applying it, such as discussed in the previous chapter, are difficult to find. A more extensive analysis of bioethical debates concerning vulnerability will take place in the next chapter. This chapter will examine how the notion has emerged in the history of bioethics, following some landmark documents, and how it is presented in the dominant framework of bioethics. A review of these documents will also show that major controversies exist concerning the status of the notion, its content and scope, as well as its consequences for healthcare practice.

The birth certificate of vulnerability

The first reference in the bioethics literature to the notion of vulnerability is in the Belmont Report in 1979 [1]. There an analytical framework of three principles is presented: respect for persons, beneficence, and justice, as well as three applications to the conduct of research. Vulnerability is mentioned under each of the three applications.

The Belmont Report was produced by the National Commission for the Protection of Human Subjects of Biomedical and Behavioral Research in the United States. This commission was created in 1974 as part of the National Research Act because of concerns with new medical technologies but also after public scandals in medical research such as the Tuskegee Syphilis Study [2]. The law mandated the commission to 'identify the basic ethical principles which should underlie the conduct of biomedical and behavioral research involving human subjects' [3]. Bioethicist Albert Jonsen, one of its members, narrates how the commission started to study the basic principles during 1974–75. In February 1976 the commission had a retreat in the Belmont Conference Center in Maryland. The final report was drafted with the help of philosophers Stephen Toulmin and Tom Beauchamp. It was approved by the commissioners in June 1978 and published in the *Federal Register* in April 1979, several months before the mandate of the commission ended [4].

The Belmont Report has become a landmark in the development of bioethics. According to Jonsen it was 'the classic principlist statement' formulating the basic principles for bioethics in general and not just for research ethics [5]. Its influence was strong because it was produced in the context of a legal and political mandate (even when this was the only commission report without recommendations). It created a clear framework of principles that could guide normative ethics, turning reflection into action, and that could be used in academic as well as public policy-making deliberations. The impact of the principles approach of the report was reinforced because in the same year *Principles of Biomedical Ethics,* one of the most influential books in the recent history of bioethics, was published [6]. One of the authors, Tom Beauchamp, philosopher at the Kennedy Institute of Ethics at Georgetown University since 1977, worked as a staff member of the commission that had been charged with writing the final version of the Belmont Report. Concomitantly he wrote with theologian James Childress the now famous textbook in which the three principles of the Belmont Report transmuted into four principles (making a distinction between beneficence and nonmaleficence). The publication of the report as well as the textbook provided a powerful theoretical framework for the new discipline of bioethics.

Informed consent is the application of the first general principle identified by the commission, i.e. respect for persons. Informed consent has three elements: disclosure, comprehension, and voluntariness. The report argues that the element of voluntariness of the consent process can be compromised through undue influence. This may occur for two reasons: either the offer to participate in research is difficult to resist because the inducement is excessive or because the subject is vulnerable (' ...inducements that would ordinarily be acceptable may become undue influences if the subject is especially vulnerable') [7]. The second reference to vulnerability is in the section on assessment of risks and benefits which is the application of the second general principle, beneficence. Here it is argued that there are five considerations to be taken into account in the assessment of the justifiability of research. One of these is the involvement of vulnerable populations: 'When vulnerable populations are involved in research, the appropriateness of involving them should itself be demonstrated' [8]. Although the condition of the particular population involved should be considered in the judgment, no further explanation is offered. A somewhat more elaborated reference is in the section on selecting subjects which involves the application of the general principle of justice. The report argues that injustice may appear even if individual subjects are selected fairly. Burdens and benefits of research may still not be distributed fairly at the social level. It continues: 'One special instance of injustice results from the involvement of vulnerable subjects' [9]. Four examples are provided: racial minorities, the economically disadvantaged, the very sick, and the institutionalized. The report furthermore explains very briefly what the problem with these groups is. These subjects are readily available for research but at the same time should be protected because of 'their dependent status and their frequently compromised capacity for free consent'. It is also mentioned that as a result of their illness or socio-economic condition, these subjects are 'easy to manipulate' [10].

Vulnerability therefore is mentioned in connection to all three basic ethical principles proposed in the Belmont Report. The references, however, are very brief; they serve to call attention to specific conditions and considerations for the application of the principles. No description, definition, or explanation of the notion is given. It should be noticed that the terminology used is 'vulnerable', 'vulnerable populations', and 'vulnerable subjects'. The noun 'vulnerability' is not used. The reference is to an attribute of a subject or population, not to a relevant quality or characteristic as such that has the same conceptual status as the three ethical principles. Notwithstanding its summary mention, the notion of vulnerability made its way in bioethical discourse since the publication of the report. At the same time, it is obvious that its actual birth in bioethics was earlier. The notion is used in the background papers commissioned by the National Commission and published in 1978. The papers introducing the notion were written as early as 1975, so that the idea of vulnerability must have circulated in the conference at Belmont House in early 1976 [11]. That the idea was in the air is not surprising. The hearings of the U.S. Senate on human experimentation in 1973 demonstrated that disadvantaged groups in society are most often involved in research. The chair of the subcommittee, Senator Edward Kennedy, referred in his opening statement of the hearings of 7 March 1973 to what later would be called 'vulnerable populations': the risk of being involved in research ' …is taken most often by the poor, the minority groups and the institutionalized. Often they are not even aware of those risks. Often…that risk is taken in order to escape from intolerable living conditions' [12]. The Guidelines of the Department of Health, Education, and Welfare on protection of human subjects, issued in May 1974, defines 'subject at risk' as 'any individual who may be exposed to the possibility of injury, including physical, psychological, or social injury, as a subject in any research…'[13]. Although this is not a circumscription of vulnerability it at least expresses the idea that certain persons are capable of being wounded in research.

The background expert reports

In order to examine the basic ethical principles for research involving human beings, the National Commission had invited a range of experts to prepare essays on relevant aspects of research involving human subjects. These materials were used in the deliberations in the retreat at Belmont House [14]. The background essays were published in a two-volume appendix in 1978 prior to the Belmont Report itself [15]. The papers on ethical principles, written by ethics experts Kurt Baier, Tom Beauchamp, James Childress, Tristram Engelhardt, Alasdair MacIntyre, and LeRoy Walters, do not mention vulnerability. The only person elaborating the notion of vulnerability is the medical advisor of the Commission, Dr Robert Levine, Professor of Medicine and Lecturer in pharmacology at Yale University School of Medicine. Hired as a special consultant to the commission in 1974, he prepared four substantial 'preliminary' papers (July 1974 to February 1976) on various dimensions of research involving human beings: the boundaries between research and medical practice, risk benefit assessment, informed consent, and selection of human subjects for research [16]. Levine is the first scholar to use the term 'vulnerability' in publications in the field of bioethics [17].

In his paper on the role of assessment of risk-benefit criteria (October 1975), Levine uses the term 'vulnerable' to refer to individuals who are most susceptible to harm. In the same section he refers to subjects from populations that are especially vulnerable [18]. There are three categories in which the assessment of harms and benefits is compromised. The first is classes of persons with limited capacities to consent (children, fetuses, prisoners, the institutionalized mentally infirm). The second is 'the unconscious and the inebriated' (by alcohol, narcotics, etc.) [19]. In both cases, the capacity to consent is limited. The third category is persons who are vulnerable 'as a consequence of their life situations' [20]. One example is persons legally capable but in practice incapable to consent because comprehension is inadequate (for instance adult persons with mental disability who are not legally incompetent). Another example is poor persons and persons with prolonged chronic illness because they are 'especially motivated to take risks' in return for possible economic or therapeutic benefits [21]. Like other people in specific situations (infertile, obese, or depressed persons, or people who believe they are imminently dying) they are desperately seeking a cure and are therefore motivated to take more risks. In fact, the examples illustrate that Levine distinguishes between two different categories of vulnerable persons. Vulnerability may exist because of compromised consent (due to insufficient comprehension) or compromised assessment of risks (because of life situations). However, if we look more carefully in light of the ethical principles stated in the Belmont Report but not elaborated by Levine at this point in time, apparently both categories can be explained in relation to the diminished or restricted ability to make autonomous decisions.

In the paper on the nature and definition of informed consent (December 1975) Levine elaborates the connection between vulnerability and autonomy. The normative requirement to obtain informed consent presupposes an assessment of a person's autonomy. Limits to individual autonomy may occur in three cases: the capacity to consent is limited, the individuals have subordinate relationships to the investigator or his institution, or the individuals have special vulnerability 'by virtue of other aspects of their life situations' [22]. Levine refers to the classification in his previous essay but he provides another example of a vulnerable research population: patients recently admitted to a coronary care unit with known or suspected myocardial infarction. At least some of these patients are vulnerable: they might perceive themselves as being in the process of dying, and their abilities to make rational judgments are impaired because they have received a narcotic [23]. Vulnerability in this essay is clearly a case of 'limited autonomy' but not simply because the capacity of consent is limited or because the subject is dependent. Rather the person is vulnerable because the circumstances have reduced his freedom to make choices. In this case the person should be regarded as an autonomous individual, but he is not able to exercise his autonomy due to impediments in his life situation.

The most extensive discussion of vulnerability is in Levine's paper on the selection of human subjects for participation in research (February 1976). The focus in this paper is slightly different. Limited autonomy here is interpreted as the lack of capacity to protect one's rights and welfare. This is exactly the wording of the May 1974 Guidelines of the Department of Health, Education, and Welfare on protection of human subjects [24]. Levine identifies three classes of subjects incapable of protecting their own rights and welfare [25]:

a.   The uncomprehending (unable to understand the information necessary to give valid consent),
b.   The vulnerable (defined very generally as 'those who are either capable of being wounded or defenseless against injury')[26].
c.   The dependent (defined as 'unable to exist, sustain himself, or act suitably or normally without the assistance or direction of another or others')[27].

Each of these categories consists of a spectrum 'ranging from slightly to absolutely incapable' [28]. What should be of concern are capabilities that are so reduced that different treatment is required. But in practice the three categories often overlap.

Levine further distinguishes vulnerable individuals in three groups depending on the origin of the vulnerability [29].

The first group is the sick. Characteristic of this group is that the sick role is providing barriers to autonomy. For example, persons with prolonged chronic illness are willing to take any risk for possible relief when they are difficult to treat with standard therapeutic modalities. Also certain conditions of disease and illness such as depression, obesity, infertility, and imminent death compromise individual autonomy when people are desperate for cures and assistance. These are similar examples as in his earlier papers. Now in this paper, Levine gives most attention to dying persons.

The second vulnerable group is the impoverished. This condition is defined as: 'a condition in which a person considers it necessary to take extraordinary risks to secure money or other economic benefits that will enable him to purchase what he considers to be the necessities of life' [30]. This does not imply that such persons cannot partake in research. But they are vulnerable in the sense that their circumstances may motivate them to take extraordinary risks. They may, for example, be too poor to purchase medical care. If they have the possibility to participate in research, they will assume that the costs of care will be underwritten by the sponsor of the research. This group will therefore be easy to recruit in research trials so that special precautions need to be taken. For example, they should not be paid to assume risks (only compensated for injury) and economic inducements should not be based on estimated risks but on the amount of inconvenience.

The third vulnerable group is minority groups (determined by race, gender, or ethnicity). This group is not vulnerable a priori but they have the potential to be vulnerable. They can be the object of discriminatory societal customs that will make them vulnerable in practice. The best protection, according to Levine, is to have this group represented on the Institutional Review Boards.

Characteristic for all three groups is that vulnerability is situational: a person is vulnerable because his ability to make autonomous choices is reduced in certain social and existential conditions. As a consequence he is no longer able to protect his own rights and welfare, and his assessment of the risks of participating in research is compromised. However, as Levine argued, autonomy can also be limited due to lack of comprehension; this may result in compromised consent or compromised assessment of risks. But in the background papers it is not considered as vulnerability. The same goes for the condition of dependency that is also a limitation of autonomy. The problem with this approach is that if the new notion of vulnerability is inscribed in the logic of the ethical principles framework of bioethics and associated conceptually with diminished autonomy, it is not clear why it only covers some external, situational limiting factors and not others such as dependency, and why it does not cover internal factors such as lack of comprehension and undue influences. The Belmont Report itself links all factors when it discusses vulnerable subjects: dependent status, compromised capacity for free consent, and ease of manipulation due to illness or socio-economic condition [31]. As will be shown in the next paragraph, the notion of vulnerability would soon develop into the broader category, subsuming compromised consent, weakness due to social and existential conditions, and dependency. If vulnerability indicates diminished autonomy, whatever the precise reason, the normative implication will be similar: those persons should be offered protection.

The CIOMS Guidelines

The Council for International Organizations of Medical Sciences (CIOMS) was established in 1949 jointly by WHO and UNESCO. Its membership includes academies of science, research councils and specialist organizations, all in the field of biomedicine. The main purpose of the organization is to promote international activities in the biomedical sciences. Since the 1980s CIOMS has been active in issuing international ethical guidelines for biomedical research and epidemiology. In fact, the story begins much earlier [32]. In 1967 CIOMS organized a conference in Paris, France, on biomedical science and the dilemma of human rights. This conference was specifically focused on medical experimentation. It would be the first of a long series of annual round tables on the ethical aspects of medicine. In the same year the Director-General of WHO established a Secretariat Committee on Research Involving Human Subjects. This committee was charged with developing a set of guiding principles for research initiated or sponsored by the World Health Organization. Shortly afterwards, in December 1968, the General Assembly of the United Nations adopted a resolution concerning human rights and scientific and technological developments. The assembly was concerned that recent discoveries and advances might endanger the rights and freedoms of individuals and peoples. It therefore requested a study of possible problems, emphasizing among others the following standpoint: 'Protection of the human personality and its physical and intellectual integrity, in the light of advances in biology, medicine and biochemistry' [33]. The result was a WHO report, years later, summarizing the major ethical problems of contemporary biomedicine [34]. But, more importantly, Dr Zbigniew Bankowski, the newly appointed Secretary-General of CIOMS took the initiative to direct the focus of the organization that until then had been primarily concerned with the merely scientific aspects of biomedicine to the social and ethical implications of scientific and technological progress [35]. In 1976, CIOMS appointed a standing Advisory Committee on Bioethics. Bankowski furthermore centered the annual international conferences explicitly on ethics and medical experimentation. Round tables on this issue took place in 1973 (in Switzerland), 1978 (in Portugal), and 1981 (in the Philippines) bringing together researchers, ethics experts, and policy-makers from around the world [36]. Through intensive dialogue consensus gradually emerged to develop guidelines that could be internationally recognized. In 1979 WHO and CIOMS launched a joint project aimed at elaborating the principles of the Helsinki Declaration into a set of practical guidelines with a global scope. Bankowski recalls how this effort was particularly supported by U.S. authorities who were keen to have guidelines for international research efforts [37].

In 1982 CIOMS issued *Proposed International Ethical Guidelines for Biomedical Research Involving Human Subjects* [38]. The purpose of these guidelines was two-fold. The first relates to the question of how the principles of the Declaration of Helsinki (with the first revision adopted in 1975 by the World Medical Association, member of CIOMS) can be put into practice. These principles have 'universal validity', but their mode of application will vary in particular circum-stances [39]. Different contexts require different applications. An international, independent organization like CIOMS is in the best position to promote ethical principles worldwide. The experts drafting its guidelines are not merely medical professionals (as in the WMA) but they represent a variety of disciplines. Differing in social, cultural, and political background, they are also sensitive that application of ethical principles requires understanding culture and traditions in other parts of the world [40].

The second purpose is related to the organization's mission: the document focuses on research from the perspective of developing countries. In order to frame the guidelines in a way that could be applied in these countries, CIOMS undertook an extensive consultation process involving experts from 60 develop-ing countries from all WHO regions. At that time biomedical research was only conducted on a limited scale in the developing world, and most countries did not have a national framework for protecting research subjects against abuse. Being closely associated with the United Nations system, the approach of CIOMS was characterized by its emphasis on human rights and public health. The main con-cern was how the principles of research ethics can be applied in countries with prevailing conditions of vulnerability [41]. The working group preparing the draft had already discussed the problem of vulnerable groups in its first meeting in July 1980 [42].

The 1982 Proposed Guidelines was the first international document that explicitly refers to vulnerability in a heading in the text and that does more than simply mention it. Taking consent of subjects as the basic starting point (as in the Declaration of Helsinki), the document states that there are many individuals for whom consent is compromised. Subsequent sections discuss children, 'pregnant and nursing women', and the 'mentally ill and mentally defective persons', followed by the special section on 'other vulnerable social groups'. The last category includes 'members of a hierarchically-structured group' [43]. Examples are medical and nursing students, employees of the pharmaceutical industry, and members of the armed forces. Their participation in research may be unduly influenced by expectations of benefit. Vulnerability is here regarded in a rather restricted sense: primarily as impaired capability or incapability to give adequate consent due to either internal conditions (compromised consent) or one particular external circumstance (subordinate position).

Bankowski points out how the HIV/AIDS pandemic impacted the ethics strategy of CIOMS [44]. First recognized as a new disease in 1981, HIV/AIDS led to the rapid growth of multinational research. Epidemiological studies analyzed the dissemination of the disease, while clinical studies tested experimental vaccines and possible medication. At the same time there were increasing concerns about the involvement of vulnerable population groups, reinforced by fears for possible stigmatization and discrimination of HIV patients. CIOMS therefore faced the challenge of providing clear guidance to protect vulnerable communities and individuals.

The response came several years later. In 1991, the *International Guidelines for Ethical Review of Epidemiological Studies* were published [45]. These guidelines articulate vulnerability in two ways. For the first time in an international document, vulnerability is specifically mentioned as a fundamental principle included in the general principle of respect for persons. This general ethical principle in fact incorporates two fundamental principles. One principle is autonomy; the other is 'protection of persons with impaired or diminished autonomy, which requires that those who are dependent or vulnerable be afforded security against harm or abuse' [46]. The second articulation of vulnerability specifies that it is an important item to consider in the ethical review procedure. Here the text enumerates several vulnerable populations, linking vulnerable and dependent groups in the same article 43: 'Ethical review committees should be particularly vigilant in the case of proposals involving populations primarily of children, pregnant and nursing women, persons with mental illness or handicap, members of communities unfamiliar with medical concepts, and persons with restricted freedom to make truly independent choices, such as prisoners and medical students' [47]. Similar populations as in 1982 are mentioned but prisoners are added as a new group. It is obvious that the main criterion for vulnerability is compromised autonomy.

The rather limited interpretation of vulnerability in this document is surprising. It is not coherent with its upgraded status as an ethical principle. During the international conference in 1990 to finalize these guidelines voices were raised arguing for an ethical perspective beyond the level of individual autonomy. The Director-General of WHO, addressing the conference, pointed out that usually ethics is focused on the rights and interests of individuals but that the purpose of the meeting was to extend this focus in order to include communities and populations. He criticized the tendency of research to be orientated on the values of developed countries, and advocated that ethics takes into account 'the voices of the millions of uneducated and deprived people in urban slums and developing countries' [48]. Articulating a broader ethical approach was a recurrent theme during the conference. It was argued that epidemiology requires a focus on communities and populations rather than individuals. Gostin, for example, proposed macro-ethical principles for population-based research. One of these principles should be 'protection of vulnerable populations' [49]. On the other hand, addressing the needs of developing countries requires not only the acknowledgement that cultures are different and that individual autonomy is often not a primary value in non-Western cultures [50]. It also requires the recognition that vulnerability is particularly widespread in these countries with extensive poverty, under-development, failure of the public system, and pervasive social injustices [51]. The text of the guidelines perhaps reflects a typical compromise. Vulnerability and autonomy are both regarded as fundamental principles but at the same time subordinated under the even more fundamental principle of respect for persons. But considering vulnerability as a principle comes with a prize because of its mirroring to the principle of autonomy. Being vulnerable demands protection since it is a deficiency of autonomy. In this frame, autonomy is always the primary and positive principle, while vulnerability is secondary and negative.

Only a few years later, in 1993, CIOMS published new guidelines for biomedical research in general, superseding the 1982 Proposed Guidelines [52]. The *International Ethical Guidelines for Biomedical Research Involving Human Subjects* are basically a statement of general ethical principles (resuming the three principles of the Belmont Report) and 15 guidelines with extensive commentaries. Vulnerability is frequently mentioned. However, it is no longer identified as a fundamental principle. The notion is introduced in one of the first paragraphs of the document in connection to the principle of respect for persons with exactly the same wording as in 1991, however, this time as a 'fundamental ethical consideration', not a principle. Vulnerability is also associated with the principle of justice. It is identified as a morally relevant distinction between persons that justifies a difference in distribution of burdens and benefits. For the first time, the text provides a description of what vulnerability is: 'a substantial incapacity to protect one's own interests owing to such impediments as lack of capability to give informed consent, lack of alternative means of obtaining medical care or other expensive necessities, or being a junior or subordinate member of a hierarchical group' [53].

Vulnerability is repeatedly mentioned in the commentaries to several guidelines, specifying categories of vulnerable persons, namely children and persons unfamiliar with modern medical concepts (Guideline 1), persons with mental and behavioral disorders (Guideline 6), and prisoners (Guidelines 7). These are the same categories mentioned in previous guidelines, except that women are no longer associated with vulnerability. Guideline 10 now makes a distinction between 'traditional' classes of vulnerable individuals (mentioned in Guidelines 1, 6, and 7) and other vulnerable social groups. The first is characterized by limited capacity or freedom to consent, the second by subordinate social position. For this second group the wording is more or less the same as that in the 1982 Proposed Guidelines and similar examples are mentioned. However, beyond these paradigmatic and previously listed cases, the text expands the number of vulnerable groups considerably, based on the same rationale of dependency. Additional groups now labeled as vulnerable are: 'residents of nursing homes, people receiving welfare benefits or social assistance and other poor people and the unemployed, patients in emergency rooms, some ethnic and racial minority groups, homeless persons, nomads, refugees, and patients with incurable disease' [54]. They all may have attributes resembling those of the other vulnerable classes, and therefore they are in need of special protection. In these cases, there is no internal limitation of autonomy (as in the 'traditional' class of vulnerable persons) but vulnerability is associated with external conditions that impact on autonomous decision-making (weakness due to social and existential conditions but also dependency).

The most recent revision of the CIOMS Guidelines in 2002 has the same structure as previous documents but more guidelines (21 instead of 15) [55]. It also puts stronger emphasis on human rights. In the introduction, it is stated that human rights law in the application of ethical principles to research concerns primarily two principles: respect for autonomy and protection of dependent or vulnerable persons and populations [56]. This statement seems to reiterate the position of the 1991 Guidelines that protection of vulnerability is a principle. But in the paragraph on general ethical principles itself nothing has changed. The text repeats that vulnerability is an ethical consideration when applying the principle of respect for persons. The text in relation to the principle of justice has also remained exactly the same. There is one new dimension. Vulnerability is not only relevant because it requires special protection, but it also demands special responsiveness: 'Justice … requires that the research be responsive to the health conditions or needs of vulnerable subjects' [57]. This positive approach to the consequences of vulnerability is reflected in the commentary to Guideline 12 where it is stated that members of such groups have the same entitlements to access potential therapeutic benefits of research as non-vulnerable persons.

The major change in the 2002 Guidelines, however, is that vulnerability now is upgraded to a guideline itself. Guideline 13 explicitly states: 'Special justification is required for inviting vulnerable individuals to serve as research subjects and, if they are selected, the means of protecting their rights and welfare must be strictly applied' [58]. There are also two changes that considerably enlarge the scope of the notion. One is that a broader definition is provided: 'Vulnerable persons are those who are relatively (or absolutely) incapable of protecting their own interests'[59]. Vulnerability no longer is a 'substantial incapacity' but a relative quality that may result from a wider range of insufficiencies. It is not only the result of diminished or absent autonomy. But it can be the consequence of other factors that make people susceptible to being damaged: ' ...insufficient power, intelligence, education, resources, strength, or other needed attributes to protect their own interests' [60]. The broader definition leads to a second change compared to previous guidelines: the number of vulnerable groups is growing. First, there is the group with limited capacity or freedom to consent (now called the 'conventional' group). Then, there are the groups that are vulnerable due to their social position of dependency. Third, there are simply 'other vulnerable groups' (the text of the commentary does not make a connection here with the social context). New groups concern elderly persons, displaced persons, individuals who are politically powerless, but also 'countries or communities in which resources are limited to the extent that they are, or may be, vulnerable to exploitation...'[61]. Finally, at least some women are back in the vulnerability framework: ' ...in some parts of the world women are vulnerable to neglect or harm in research...'[62].

The Declaration of Helsinki

The story about the authoritative statements of the World Medical Association (first adopted in 1964) can be much shorter since the Declaration of Helsinki did not mention the notion of vulnerability until its fifth revision in 2000. It can be argued that the idea was present in the background all the time though the term was not expressly used. When the notion was finally introduced, it was briefly mentioned in paragraph 8 of the introduction to the basic principles. The declaration points out that some research populations are vulnerable and in need of special protection. Five groups are listed although it is not clear that they are all considered as vulnerable: 'the economically and medically disadvantaged'; 'those who cannot give or refuse consent for themselves'; 'those who may be subject to giving consent under duress'; 'those who will not benefit personally from the research'; and 'those for whom research is combined with care' [63]. Especially this part of the declaration was criticized because it has made the category of vulnerability too broad, essentially making every person vulnerable, so that the requirement of special protection has become pointless [64].

The 2008 revision presents a reduced and changed statement regarding vulnerability in the introduction. The text no longer lists vulnerable populations. It summarily refers to two criteria, i.e. inability or refusal to give consent, and being vulnerable to coercion and undue influence [65]. But at the same time, vulnerability is now mentioned under principles. Provision 17 stipulates the consequences of involving a vulnerable population or community in medical research. In this case research is only justified if it is responsive to the health needs and priorities of the population or community, and if there are potential benefits for them from the results of the research [66].

The latest revision, adopted in October 2013, maintains the emphasis on vulnerable populations; this issue is now mentioned as a separate section and no longer under principles. It also proposes a description of vulnerability as 'increased likelihood of being wronged or of incurring additional harm', therefore emphasizing particularly the element of exposure. The text provides more criteria for justifying inclusion of vulnerable groups in medical research; besides responsiveness and sharing of benefits, research is only justified if it cannot be carried out in a non-vulnerable group. Furthermore, consideration should be given to ensuring that a fair level of additional benefits will be provided to this group [67]. Most commentators applaud the attention given to vulnerable groups and the efforts to provide more protection as well as the strengthening of the requirement for post-trial provisions [68].

The Universal Declaration on Bioethics and Human Rights

In 2005 member states of UNESCO unanimously adopted the *Universal Declaration on Bioethics and Human Rights*. This first statement of bioethical principles adopted by governments is part of international human rights law, although not legally binding. The declaration states 15 ethical principles for global bioethics, including the three principles mentioned in the Belmont Report, although in slightly different formulation: autonomy and individual responsibility; benefit and harm; equality, justice, and equity. Particularly at the instigation of developing countries, the declaration aimed to present an ethical framework for bioethics that included perspectives and values from various countries, cultures, and traditions. The dominant focus on individual decision-making that is prevalent in Western bioethics was therefore broadened. For this reason, the member states agreed to include ethical principles that referred to the social and cultural context, such as 'solidarity and cooperation', 'social responsibility and health', and 'respect for cultural diversity and pluralism'. They also reached consensus on principles that recognize that human beings are dependent on other living beings and their environment ('protecting future generations'; 'protection of the environment, the biosphere and biodiversity') [69].

In a relatively late stage of the drafting and finalization of the text, the governmental experts concurred to include 'respect for human vulnerability and personal integrity' as a principle in the text of the declaration [70]. Article 8 formulating the principle reads as follows: 'In applying and advancing scientific knowledge, medical practice and associated technologies, human vulnerability should be taken into account. Individuals and groups of special vulnerability should be protected and the personal integrity of such individuals respected' [71].

The text tries to combine two perspectives: vulnerability is an attribute of certain individuals or populations but also a basic characteristic of human beings in general. The first sentence emphasizes that human vulnerability is intrinsic in the human condition; it should be taken into account. The second sentence indicates that in particular circumstances this basic vulnerability can be exacerbated. In these cases protection is required while in the first case respect is more appropriate [72]. The interpretation of the balance between these two perspectives, however, can be different. In a later report on the principle, the UNESCO International Bioethics Committee argues that the specific focus of the article is to address the special vulnerabilities that occur due to specific circumstances; the general vulnerability of human beings should be regarded as constituting a kind of background consideration [73]. This difference in interpretation is related to the contrast between philosophical and political views of vulnerability that will be elaborated in subsequent chapters.

Vulnerability is not explicitly defined in the declaration but somewhat elaborated in various other articles. In the preamble, the declaration refers to the special needs of 'vulnerable populations'. Article 24 (international cooperation) emphasizes that not only individuals may be rendered vulnerable, but also families, groups, and communities. The article furthermore refers to certain circumstances, not listing specific vulnerable categories but explicitly mentioning conditions that may produce vulnerability: disease, disability, 'other personal, societal and environmental conditions', and limited resources [74].

It is not the first time that UNESCO documents refer to vulnerability. In the *Universal Declaration on the Human Genome and Human Rights* (adopted in 1997) vulnerability is mentioned in the context of solidarity and international cooperation, emphasizing that states should respect and promote the practice of solidarity 'towards individuals, families and population groups who are particularly vulnerable to or affected by disease or disability of a genetic character' [75]. Vulnerable groups are explicitly mentioned in Article 24 as one of the parties concerned in the implementation of the declaration; appropriate consultations with them should be organized [76]. Obviously, the scope of this earlier declaration is limited to genetic diseases and genetic interventions. The *International Declaration on Human Genetic Data*, adopted by UNESCO member states in 2003, also refers to vulnerability but in the even more restricted context of collecting, processing, using, and storing human genetic data. In its preamble it explicitly refers to 'the special needs and vulnerabilities of developing countries' [77]. The 2005 Declaration could built on these earlier formulations and give them a wider scope beyond the field of genetics.

It is important to note that in the *Universal Declaration on Bioethics and Human Rights* respect for human vulnerability is promoted as one of the fundamental ethical principles for global bioethics. The scope of the declaration is wider than health research though a stringent conceptual definition of bioethics is not given. The document merely declares that it is addressing ethical issues in the field of medicine, life sciences, and associated technologies, covering therefore healthcare, science, and technology. As a principle, vulnerability is relevant not merely for research. The principle is furthermore applied to individuals but also to families, groups, communities, and populations. The declaration points out that there are groups of special vulnerability; no list of such groups is provided but specific conditions 'rendering' vulnerability are mentioned. Finally, explicit reference is made to *human* vulnerability. The focus therefore is on an anthropocentric interpretation and application of vulnerability. The point of reference is human life, not life in general. This is consistent with the scope of the declaration that applies to human beings.

Controversial dimensions of vulnerability

The above review of official and authoritative bioethics policy documents shows how the notion of vulnerability has emerged in national and international discourses. But this brief history also highlights that views and interpretations differ. Since its introduction into bioethics, the notion has been the subject of intense scholarly debate. Different dimensions of the notion continue to produce controversies: its status, content, scope, and implications.

Status

The status of the notion of vulnerability has significantly evolved over time. In most of the documents it was first mentioned as a special consideration that needs to be taken into account in the application of general principles in research. Following the lead of the Belmont Report, the CIOMS guidelines of 1982 as well as the Declaration of Helsinki in 2000 introduced vulnerability as a special consideration when applying general principles. However, the status of the concept of vulnerability was enhanced significantly in the CIOMS guidelines of 1991 that were the first to refer to vulnerability as a fundamental principle. However, the statement is ambiguous since it does articulate on the one hand that it concerns a principle; on the other it argues that it is incorporated in the principle of respect for persons. The ambiguity continues to exist in subsequent CIOMS guidelines. In 1993, vulnerability no longer is a principle but rather a special application of principles. In 2002, vulnerability is mentioned at the same time as a principle in human rights law, an ethical consideration when applying the principle of respect for persons, as well as a guideline itself. Finally, the UNESCO Declaration of 2005 clearly promotes vulnerability at the level of fundamental bioethical principles not only in research but also in health care. The status of vulnerability has therefore been strengthened over the last decade. It is no longer a consideration within the application of ethical principles, or subsumed under another principle, but itself is an ethical principle that needs to be balanced against other ethical principles such as respect for persons, beneficence, and justice.

The reasons for this rising conceptual status are not clear, or at least not mentioned in the documents themselves. It is presumably related to the globalization of bioethics, given the international status of the documents that have contributed to it. The CIOMS and UNESCO documents not only present a global perspective but also articulate the human rights perspective. Awareness of abuse and harms of vulnerable populations (e.g. HIV/AIDS patients) may have contributed to the growing significance of vulnerability. Another possible reason is that bioethics has long been dominated by the discourse of individual autonomy as characteristic for the Western context in which bioethics has emerged. As long as vulnerability has been framed as limitation of autonomy, it is considered as a special consideration or special application of the fundamental principle of respect for autonomy. Vulnerability could only be conceptualized as a separate principle if its contents were emancipated from autonomy discourse.

Content

The second issue is the content of the notion of vulnerability. In many instances, the documents do not define vulnerability. Individuals and groups are labeled as vulnerable, but it is not always clear what the criteria are for doing so. The Belmont Report relates vulnerability to dependency and compromised capacity to consent. In the background papers to the report, Robert Levine pays attention to the notion of vulnerability. He identifies vulnerable persons as a subcategory of persons not capable to protect their own rights and welfare. Vulnerability therefore is a special class of limited or impaired autonomy. But it is not the same as lack of comprehension or dependent status that may also compromise autonomy and autonomous decision-making. Levine seems to imply that vulnerability emerges because of the special condition of the individual or group of individuals. His examples, namely the sick and the impoverished, do not refer to dependency or lack of comprehension but to the circumstances that might impede really free decision-making. Vulnerability therefore is limited autonomy but not simply because of lack of autonomy but because of situational impediments to exercise autonomy.

The ambiguity about the criteria and thus the content of vulnerability is reflected in the 1982 CIOMS guidelines. Two criteria are used here: compromised consent and dependency (specifically persons who are in a subordinate position in a hierarchy). We have seen how the 1993 CIOMS guidelines for the first time present a description of the notion of vulnerability ('a substantial incapacity to protect one's own interests'). They also distinguish three subcategories. The first (lack of capacity to give informed consent) identifies what has become known as the traditional or conventional class of vulnerable persons. The second (lack of alternative means to obtain medical care) refers in fact to the social position that renders persons vulnerable, while the third (subordination in a hierarchical group) emphasizes the criterion of dependency. This conceptual approach is reminiscent of the earlier analysis of Levine but restructures it. Vulnerability is now the name of the broad category and it includes the subcategories of lack of comprehension and dependency. At the same time, the notion is moving beyond the perspective of limited autonomy. The above description of vulnerability is literally repeated almost ten years later in the 2002 CIOMS Guidelines. But in another section of the same document it is slightly reformulated, not as 'substantial incapacity' but as 'relatively (or absolutely) incapable' [78]. As shown earlier in this chapter, possible explanations of vulnerability are provided: insufficient power, intelligence, education, resources, and strength may make people unable to protect their own interests. This formulation is similar to the one used by Levine in the first edition of his textbook on research ethics [79]. Levine has been intimately connected with the drafting of the CIOMS guidelines, as co-chair of the drafting committee for the 1993 Guidelines and as chairperson of the steering committee writing the 2002 Guidelines. It is obvious that his ideas on vulnerability have been broadly accepted. The implication of the reformulation is that the scope of vulnerability is enlarged; it is a relative quality resulting from a range of insufficiencies, both internal and external, and not limited to impaired autonomy.

Scope

The third controversy concerns the scope of vulnerability. The notion is used to qualify individuals, but also families, groups, communities, populations, and even countries. It is often assumed that vulnerability is transferable from one category to another. What has started as mere examples of vulnerable groups has resulted in growing lists of vulnerable populations. Gradually, consensus has emerged on some criteria and categories, for example, limited capacity or freedom to consent (and thus, children, persons with mental or behavioral disorders, and prisoners) and social position of dependency (thus, subordinate members in a hierarchy, the institutionalized, and the poor). However, the list of groups included in the second category has steadily grown. The identification of a third category ('other vulnerable groups') has opened the door to the labeling of even more groups as vulnerable, especially the elderly and women. This broadening scope has made the notion of vulnerability not only increasingly difficult to apply in practical settings but has undermined its relevancy. If almost every person is a member of one or more vulnerable groups what will be the moral significance of identifying lists of vulnerable categories? The 2002 CIOMS Guidelines present the most encompassing classification of vulnerable groups. Since then, a different approach can be discerned. The Declaration of Helsinki in 2008 only uses vulnerability for populations and communities. In distinction to the 2000 fifth revision which lists five categories of vulnerable populations, the sixth revised text only briefly mentions, by way of example, two groups. The latest revision of 2013 does not mention any example of vulnerable groups at all. Likewise, the UNESCO Declaration does not list vulnerable categories. However, it identifies conditions that may produce vulnerability. It recognizes that vulnerability may be manifested at various levels (individual, family, group, etc.) and that 'groups of special vulnerability' can be distinguished, but a simple categorization is impossible. Vulnerability can be the result of internal conditions (impaired autonomy) or external conditions (dependency and weakness) but more often of interactions between these conditions, posing threats to some individuals or groups of individuals rather than whole categories of persons. The identification of vulnerability therefore should be subtle and nuanced. Threats can best be identified at the level of the individual (as part of a group or community) since the declaration associates vulnerability with the personal integrity of individuals. This more subtle approach of identifying conditions of vulnerability rather than classes of vulnerable individuals is also accommodating the critique that classification of vulnerability may result in discrimination and stigmatization (as will be elaborated in the next chapter).

Implications

Finally, it is debatable what exactly the moral implications of vulnerability are. This is the dimension of the practical consequences of the notion of vulnerability. At least three different consequences are pointed out in the documents. First, there is a need of extra justification for involving vulnerable persons and groups. This implication is emphasized in the Belmont Report. The assumption is that vulnerable populations are easy to recruit for research; the appropriateness of involving them should be specifically demonstrated and justified. The 1991 CIOMS Guidelines take a similar approach. They require that ethical review committees are particularly vigilant when vulnerable populations are involved. A second consequence is that vulnerability demands special protection. Because vulnerable persons cannot protect their own interests, others will be responsible for protecting their interests. This is the implication articulated in the 1993 CIOMS Guidelines and the 2000 Declaration of Helsinki. The third consequence formulates a more positive approach. The 2002 CIOMS Guidelines point out that vulnerability demands that research should be responsive to the needs of vulnerable persons. This responsiveness as well as the probability of specific benefits of research with vulnerable populations is even more stringently articulated in the latest revision of the Declaration of Helsinki, adding as a further condition that such research is only justified if it cannot be carried out in a non-vulnerable group. The UNESCO Declaration also relates vulnerability to a positive action. States should promote solidarity and international cooperation in order to answer to the special needs of vulnerable populations.

Although the three consequences are different, they cannot simply be characterized as pointing towards negative or positive duties. The underlying assumption is that vulnerable populations can be involved in research but that extra caution is necessary. Researchers are not asked to refrain from including vulnerable subjects but they should provide special justification and special protection.

To a certain extent restraint is required (negative duty) but at the same time, if vulnerable persons are included, additional activities are requested (positive duty). The most recent documents put more emphasis on the positive demands. Vulnerability is not merely a condition to take into account, but it is regarded as a moral reason to help people cope with their vulnerable condition. This emphasis is related to the application of the notion to health care in general and not just to the area of health research. Respect for human vulnerability formulates a normative prescription to take care for those threatened by biological and social circumstances as well as by the power of medicine itself. It stipulates that bioethics involves more than respecting individual choices and personal autonomy: it aims at safeguarding care for persons even if their situations are extremely different and highly unequal.

An indispensable concept for bioethics

Being introduced in the bioethical discourse via the Belmont Report in 1979, vulnerability has played an increasingly important role in international policy documents, culminating in 2005 in its formulation as a separate bioethics principle. The notion has been criticized as being too vague, too broad, too narrow, and futile since it does not provide clear moral guidance. Nevertheless, in the context of global bioethics, the language of vulnerability has become indispensable; contemporary bioethics needs a concept of vulnerability [80].

Despite its increasing use and emerging status, the notion continues to be controversial. One reason is the tension within the concept between a general and specific interpretation. On the one hand, vulnerability is regarded as a general characteristic of the human condition. It expresses the fragility and finitude of human existence, and is therefore a feature shared by all human beings. On the other hand, vulnerability refers to the fact that some people are more vulnerable than others, due to natural or social conditions such as disease or poverty. Both interpretations are based on different perspectives of vulnerability (examined in Chapters 5 and 6). They present not only a contrasting conceptualization of vulnerability but also diverging normative conclusions. The philosophical perspective implies respect. All human beings demand care and solidarity, but one has to accept that the power of medicine and technology is limited. Moreover, respect requests reflexive action: our acts should not foreseeably threaten and endanger other people. The political perspective requires action, primarily protection, but also assistance in overcoming and ameliorating the conditions that produce vulnerability. It expresses the normative requirement that some vulnerable fellow human beings need special care. More is needed than respect and non-interference; vulnerable people should receive assistance that will enable them to realize their potential as human beings. This tension between the two conceptual interpretations of vulnerability is often eliminated through denying one of the competing perspectives.

The question is how the tension can be made more productive in better understanding vulnerability, and how the two perspectives can be reconciled.

Before this question can be addressed in the following chapters, it is necessary to examine how contemporary bioethics has analyzed the notion of vulnerability, particularly within the dominant theoretical framework of principles that strongly emphasizes the significance of individual autonomy. This chapter has demonstrated that vulnerability is often incorporated and framed in this theoretical mould, rather than considered as a notion that opens up a new and different perspective. From the point of view of the functional definition described in the previous chapters, vulnerability can easily be explained in the moral vocabulary of autonomy. Some persons may be more sensitive to harm since their autonomy is limited, reduced, or absent. Normally, if people are autonomous they will also have the adaptive capacity to deal with harm. The response to vulnerability therefore is to enhance autonomy in order to increase adaptive capacity or to reduce exposure by minimizing harm. However such explanation does not take into account that vulnerability is often not an individual affair. The terminology of 'vulnerable populations' for example indicates that it is associated with common and shared conditions beyond the individual situation. This raises the question: how is vulnerability related to autonomy? Such basic questions notwithstanding, bioethical analyses have resulted in some agreement concerning at least some dimensions of the notion of vulnerability. This will be the subject of the next chapter.

Notes

1 National Commission for the Protection of Human Subjects of Biomedical and Behavioral Research, 'The Belmont Report'. For this historical overview, see also Ten Have, 'The principle of vulnerability in the UNESCO Declaration on Bioethics and Human Rights'.
2 Rothman, *Strangers at the bedside,* 168 ff.
3 Jonsen, *The birth of bioethics,* 100; Jonsen, 'Foreword', xiv.
4 See Jonsen, *The birth of bioethics,* 104. There is confusion about data: Jonsen later mentions that the report was first published for comments in the *Federal Register* in 1976 and officially promulgated in 1978; see Jonsen, 'Foreword', xv. But first publication was in April 1979, inviting comments until July 1979.
5 Jonsen, *The birth of bioethics,* 104; Jonsen, 'Foreword', xv.
6 See Beauchamp and Childress, *Principles of Biomedical Ethics.* The preface to the first edition is dated December 1978; the book was therefore completed before the Belmont Report was published in April 1979 but after the draft of the report was approved by the commission in June 1978.
7 National Commission for the Protection of Human Subjects of Biomedical and Behavioral Research, 'The Belmont Report', 23195.
8 National Commission for the Protection of Human Subjects of Biomedical and Behavioral Research, 'The Belmont Report', 23196.
9 National Commission for the Protection of Human Subjects of Biomedical and Behavioral Research, 'The Belmont Report', 23197.

10 National Commission for the Protection of Human Subjects of Biomedical and Behavioral Research, 'The Belmont Report', 23197.

11 Jonsen, (*The birth of bioethics*, 103) mentions that initially seven ethical principles were selected. Philosopher and special consultant Stephen Toulmin suggested that 'protect the weak and powerless' should be added to this list.

12 United States Senate, *Quality of health care – human experimentation, 1973*, 793.

13 DHEW Guidelines, 'Protection of Human Subjects', 18917.

14 See Jonsen, *The birth of bioethics*, 102.

15 National Commission for the Protection of Human Subjects of Biomedical and Behavioral Research. *The Belmont Report. Appendix Volume I & II.*

16 According to Jonsen (*The birth of bioethics*, 152) Levine's papers are about 'the nature of research'. But they are clearly more than medical papers since they extensively discuss ethical considerations in the practice of research. It is easy to notice how Levine's papers have contributed to the structure of the final Belmont Report, especially paragraph A (Boundaries between practice and research) and paragraph C (Applications). The three subsections of this last paragraph follow three of Levine's papers.

17 Dr Levine recalls that the idea of using the term originated during his reading of John Rawls', *A theory of justice* (Levine: personal communication). Rawls uses the adjective 'vulnerable' in two places. On page 126–127 he discusses the circumstances of justice as the conditions under which human cooperation is possible and necessary. The objective circumstances refer to the coexistence of many individuals with more or less similar powers: 'They are vulnerable to attack…'(p. 127). The second reference is in the final part of the book where he compares justice with love; being a just person may be hazardous just as loving another person: 'Once we love we are vulnerable' (page 573). Although Rawls frequently mentions 'the least advantaged' or 'the less fortunate' members of society, due to the harmful influence of natural fortune or social contingencies, he does not connect the idea of vulnerability with such groups. There are also references to 'vulnerability' or 'vulnerabilities' in other bioethical publications in the 1970s but the concept is not further elaborated or problematized. An example is Outka, 'Social justice and equal access to health care', 17.

18 National Commission for the Protection of Human Subjects of Biomedical and Behavioral Research. *The Belmont Report. Appendix Volume I & II*, 2-52.

19 National Commission for the Protection of Human Subjects of Biomedical and Behavioral Research. *The Belmont Report. Appendix Volume I & II*, 2-52.

20 National Commission for the Protection of Human Subjects of Biomedical and Behavioral Research. *The Belmont Report. Appendix Volume I & II*, 2-53.

21 National Commission for the Protection of Human Subjects of Biomedical and Behavioral Research. *The Belmont Report. Appendix Volume I & II*, 2-53.

22 National Commission for the Protection of Human Subjects of Biomedical and Behavioral Research. *The Belmont Report. Appendix Volume I & II*, 3-38; This is the first time that Levine is using the noun 'vulnerability'.

23 National Commission for the Protection of Human Subjects of Biomedical and Behavioral Research. *The Belmont Report. Appendix Volume I & II*, 3-50.

24 DHEW Guidelines, 'Protection of Human Subjects', 18917. The ethics review committee should assess the risks for subjects and should determine whether 'the rights and welfare of any such subjects will be adequately protected'.

25 National Commission for the Protection of Human Subjects of Biomedical and Behavioral Research. *The Belmont Report. Appendix Volume I & II,* 4-2; 4-30.

26 National Commission for the Protection of Human Subjects of Biomedical and Behavioral Research. *The Belmont Report. Appendix Volume I & II,* 4-85.

27 National Commission for the Protection of Human Subjects of Biomedical and Behavioral Research. *The Belmont Report. Appendix Volume I & II,* 4-90.

28 National Commission for the Protection of Human Subjects of Biomedical and Behavioral Research. *The Belmont Report. Appendix Volume I & II,* 4-30.

29 National Commission for the Protection of Human Subjects of Biomedical and Behavioral Research. *The Belmont Report. Appendix Volume I & II,* 4-34 ff.

30 National Commission for the Protection of Human Subjects of Biomedical and Behavioral Research. *The Belmont Report. Appendix Volume I & II,* 4-42.

31 National Commission for the Protection of Human Subjects of Biomedical and Behavioral Research, 'The Belmont Report', 23197.

32 See Bankowski and Dunne, 'History of the WHO/CIOMS project for the development of guidelines for the establishment of ethical review procedures for research involving human subjects', 441–442.

33 United Nations General Assembly, 'Human rights and scientific and technological developments', 54.

34 See WHO, *Health aspects of human rights: with special reference to developments in biology and medicine.*

35 Zbigniew Bankowski, medical doctor and experimental pathologist, was born in Poland. He served as WHO staff member from 1965 to 1975 in Switzerland and Tunisia. He was elected Secretary-General of CIOMS in 1975 and retired in 1999. He died in April 2010. His contributions to international bioethics have been significant but are not well known. His colleagues and friends have established the Zbigniew Bankowski Lecture Fund to support an annual lecture at CIOMS on ethical aspects of health policy (see: Announcements, 357).

36 See for the 1978 conference: Howard-Jones, 'Human experimentation in historical and ethical perspectives', 1441, and Rudowski, 'World Health Organisation biomedical guidelines and the conduct of clinical trials'.

37 Bankowski, 'Ethics and Health', 122. The joint WHO/CIOMS project was in fact approved by both organizations in 1978 but it took a year to secure the financial support (from Canadian, German, Swedish, and US funds) enabling the start of the project in June 1979. See Bankowski and Dunne, 'History of the WHO/CIOMS project for the development of guidelines for the establishment of ethical review procedures for research involving human subjects', 443.

38 CIOMS, *Proposed International Guidelines for Biomedical Research Involving Human Subjects.* The text of the 1982 Guidelines is also published in Howard-Jones, 'Human experimentation in historical and ethical perspectives', 1445-1448; as well as in Scholle Connor and Fuenzalida-Puelm, *Bioethics. Issues and Perspectives,* 220–226.

39 CIOMS, *Proposed International Guidelines for Biomedical Research Involving Human Subjects,* 222.

40 Levine, 'International codes and guidelines for research ethics: A critical appraisal', 236, 243; see also: Christiakis and Panner, 'Existing international ethical guidelines for human subjects research: Some open questions'; Macrae, 'The Council for International Organizations and Medical Sciences (CIOMS) Guidelines on ethics of clinical trials'.

41 CIOMS, *Proposed International Guidelines for Biomedical Research Involving Human Subjects,* 412.
42 See Bankowski and Dunne, 'History of the WHO/CIOMS project for the development of guidelines for the establishment of ethical review procedures for research involving human subjects', 451.
43 CIOMS, *Proposed International Guidelines for Biomedical Research Involving Human Subjects,* 415.
44 Bankowski, 'Ethics and Health', 122.
45 CIOMS, *International Guidelines for Ethical Review of Epidemiological Studies.*
46 CIOMS, *International Guidelines for Ethical Review of Epidemiological Studies,* see under General Ethical Principles.
47 CIOMS, *International Guidelines for Ethical Review of Epidemiological Studies.* See under Article 43 with the heading 'Vulnerable and dependent groups'.
48 For the speech of Hiroshi Nakajima who was Director-General of WHO from 1988 to 1998, see Bankowski, Bryant and Last, *Ethics and Epidemiology: International Guidelines,* 4. The CIOMS Conference in November 1990 was hosted in the headquarters of WHO in Geneva.
49 Gostin, 'Macro-ethical principles for the conduct of research on human subjects: Population-based research and ethics', 31, 35-38. Almost the same paper is reprinted as Gostin, 'Ethical principles for the conduct of human subject research: Population-based research and ethics'.
50 Levine, 'Informed consent: Some challenges to the universal validity of the Western model', 50 ff.
51 See in particular Khan, 'Epidemiology and ethics: The perspective of the Third World', 73.
52 CIOMS, *International Ethical Guidelines for Biomedical Research Involving Human Subjects.* 1993.
53 CIOMS, *International Ethical Guidelines for Biomedical Research Involving Human Subjects,* 1993, 10.
54 CIOMS, *International Ethical Guidelines for Biomedical Research Involving Human Subjects,* 1993, 30.
55 CIOMS, *International Ethical Guidelines for Biomedical Research Involving Human Subjects.* 2002.
56 CIOMS, *International Ethical Guidelines for Biomedical Research Involving Human Subjects,* 2002, 11.
57 CIOMS, *International Ethical Guidelines for Biomedical Research Involving Human Subjects,* 2002, 18.
58 CIOMS, *International Ethical Guidelines for Biomedical Research Involving Human Subjects,* 2002, 64.
59 CIOMS, *International Ethical Guidelines for Biomedical Research Involving Human Subjects,* 2002, 64.
60 CIOMS, *International Ethical Guidelines for Biomedical Research Involving Human Subjects,* 2002, 64.
61 CIOMS, *International Ethical Guidelines for Biomedical Research Involving Human Subjects,* 2002, 51.
62 CIOMS, *International Ethical Guidelines for Biomedical Research Involving Human Subjects,* 2002, 58.
63 WMA, *Declaration of Helsinki, fifth revision,* specific provision 8.

64 Forster, Emanuel and Grady, 'The 2000 revision of the Declaration of Helsinki: a step forward or more confusion?', 1451.

65 WMA, *Declaration of Helsinki, sixth revision,* provision 9.

66 'Medical research involving a disadvantaged or vulnerable population or community is only justified if the research is responsive to the health needs and priorities of this population or community and if there is a reasonable likelihood that this population or community stands to benefit from the results of the research' (WMA, *Declaration of Helsinki, sixth revision,* provision 17). The most recent revision has a more succinct formulation: 'Medical research with a vulnerable group is only justified if the research is responsive to the health needs or priorities of this group and the research cannot be carried out in a non-vulnerable group. In addition, this group should stand to benefit from the knowledge, practices and interventions that result from the research'. (WMA, *Declaration of Helsinki,* 2013, provision 20).

67 See, WMA, *Declaration of Helsinki,* October 2013.

68 Eggertsen, 'Helsinki doctrine under review'; Kompanje, 'Strong words, but still a step back for researchers in emergency and critical care research? The proposed revision of the Declaration of Helsinki'; Morris, 'Revising the Declaration of Helsinki'; Nathanson, 'Revising the Declaration of Helsinki. Your chance to influence research governance'; Wilson, 'An updated Declaration of Helsinki will provide more protection'.

69 Ten Have and Jean, *The UNESCO Universal Declaration on Bioethics and Human Rights. Background, principles and application,* 39 ff.

70 Patrão Neves, 'Article 8: Respect for human vulnerability and personal integrity', 155. Maria Patrão Neves is professor of ethics at the University of the Azores in Portugal and at that time special advisor to the President of the Portuguese Republic. She took the initiative to introduce and formulate the principle during the finalization stage of the declaration.

71 UNESCO, *Universal Declaration on Bioethics and Human Right,* Article 8.

72 See Patrão Neves, 'Article 8: Respect for human vulnerability and personal integrity', 159.

73 IBC, *Report of the IBC on the principle of respect for human vulnerability and personal integrity,* 1. See also: Evans, 'Commentary on the UNESCO IBC report on respect for vulnerability and personal integrity'.

74 UNESCO, *Universal Declaration on Bioethics and Human Right,* Article 24.

75 UNESCO, *Universal Declaration on the Human Genome and Human Rights,* Article 17.

76 UNESCO, *Universal Declaration on the Human Genome and Human Rights,* Article 24.

77 UNESCO, *International Declaration on Human Genetic Data,* preamble.

78 See CIOMS, *International Ethical Guidelines for Biomedical Research Involving Human Subjects,* 2002, 64.

79 Levine, *Ethics and regulation of clinical research,* 54. Levine explains: 'In general, we identify as vulnerable those who are relatively (or absolutely) incapable of protecting their own interests'.

80 As argued in Solbakk, 'Vulnerability: A futile or useful principle in healthcare ethics?'; see also Rogers, Mackenzie, and Dodds, 'Why bioethics needs a concept of vulnerability'.

4

THE BIOETHICAL DISCOURSE
OF VULNERABILITY

'All Americans have been touched by and have profited from the products of bio-medical research'. This statement of Senator Kennedy at the start of the Senate hearing on 7 March 1973 concerning the use of prisoners as subjects of medical research epitomizes the first basic premise of research ethics as it came into existence in the documents examined in the previous chapter, and this is not limited to American citizens [1]. Without scientific research health cannot be improved and diseases cannot be reduced or eliminated. The CIOMS Guidelines and the Declaration of Helsinki both underline that progress in healthcare is based on experiments involving human beings [2]. The UNESCO Declaration recognizes that scientific and technological developments have been greatly beneficial to humankind [3]. The implication is that it would be unethical not to continue searching for new medication and treatment modalities, and not to rigorously test them before they are used by human beings. But the second premise is that research with human beings may lead to abuses. This was often the immediate cause for drafting the declarations and guidelines. The conclusion from the two premises is that human subjects should be protected. How can human beings be protected from harm while at the same time the good of medical research is safeguarded? In response, the Belmont Report introduced a line of reasoning that would set the tone for subsequent bioethical analyses of vulnerability. The answer to the above question is self-determination. The first defence against possible harms of participating in medical research is to enable subjects to protect themselves, hence the importance of informed consent. The other side of this answer is that certain individuals will not be able to protect themselves because they cannot give consent or because their consent is not valid. They are therefore vulnerable to harm and should be protected by others. This approach to vulnerability assumes that it is essentially characterized by a deficiency of individual autonomy.

This chapter will analyze how vulnerability is conceptualized and interpreted within the dominant framework of bioethics. While it was, at least initially, recognized that the notion was related to the principle of autonomy as well as justice, it has nonetheless primarily been framed as the deficit of autonomy. The chapter begins with a discussion of how vulnerability is conceived as a normative rather than functional notion. The next section makes the argument that this normative framing was primarily driven by the primacy of the principle of respect for autonomy and that the association between vulnerability and the ethical principle of justice was considered less significant. One of the reasons for this narrow framing was increasing critique of the notion in bioethical discourse. As discussed in the subsequent section, as long as vulnerability is a vague and broad concept it will not be useful in practice. Assimilating the concept with the framework of autonomy will therefore enhance its usefulness. The other reason for the narrow framing relates to the initial setting in which the notion of vulnerability has been explored. In the context of research ethics, the focus is on protection of vulnerable populations and individuals against abuse and exploitation. Reinforcing and restoring individual autonomy is regarded as the best approach to protection. However, when the notion of vulnerability is used in other areas of healthcare, as discussed in the two following sections, it is often applied in a broader sense, emphasizing the conditions in which individual autonomy can or cannot be exercised. In ethical debates on healthcare, public health, and health policy vulnerability is employed as a heuristic tool that can open up new perspectives that often escape the more narrow focus of bioethical debate, particularly since these debates re-engage the connection between vulnerability and justice. The final section provides an assessment of the current status of vulnerability in bioethics, as a result of its normative framing. First, it is commonly accepted that there are different types of vulnerability. But at the same time, it is difficult to accept within mainstream bioethics that vulnerability is a general or ontological condition of being human. Such a philosophical perspective is hard to reconcile with the view that autonomous individuals determine the course of their lives. Second, it is commonly accepted that categorizing individuals in vulnerable groups and populations is not productive and perhaps even harmful. However, refocusing vulnerability on individual characteristics would not give credit to the reasons why the language of vulnerability has centered on populations in the first place. Populations are often the target of discrimination and stigmatization. Instead of deleting population language an analysis of the conditions that produce vulnerability should be encouraged. Third, it is commonly assumed that the practical implications of vulnerability are primarily concerned with the need for special or additional protection. This common assumption resulting from the normative framing of vulnerability has produced a specific conceptual understanding and a limited, circumscribed role of the notion of vulnerability in bioethical discourse. However, in producing such understanding and role, the moral force of the concept of vulnerability as well as its fundamental critique of the dominant framework of bioethics has been neglected and effectively neutralized.

Vulnerability as bioethical notion

Scientific disciplines commonly use 'vulnerability' as a functional notion. They define it as a function of exposure, sensitivity, and adaptive capacity (as discussed in Chapters 1 and 2). It is the interplay of external and internal factors that determines whether an entity or a human being will in fact be damaged. In the context of bioethics a different approach is used. The definition of vulnerability in the later CIOMS Guidelines as well as in Levine's textbook is a normative one: the incapacity to protect one's own interests. The characterization of a person as 'vulnerable' is not a mere observation or description but it has normative force.

What exactly makes vulnerability into a bioethical concept? The first aspect is that it refers to a value. Saying that a person is 'vulnerable' implies a value judgment. The concerned person may be damaged; he or she runs the risk of being harmed. The application of an ethical notion implies a valuation of persons or groups of persons. Someone's good is at risk and we should take precautions that he or she will not be hurt. This aspect of course needs further clarification that is not very often provided in bioethics literature. What are the values and disvalues involved? One question concerns the good that is threatened in vulnerable persons. In a medical context the good is health, so that vulnerability is considered as susceptibility to poor health [4]. In a policy perspective the good at risk is human security; developments such as climate change, bioterrorism, pandemics, and disasters make human life vulnerable to insecurity. The good articulated in the above bioethical documents is one's own interest. This focus on interest is not self-evident. The good that is at stake in vulnerability could be different – for example human flourishing, human needs, or human dignity. Rendtorff and Kemp argue that it is human integrity that requires protection, and that vulnerability is closely linked to integrity, just like the Universal Declaration on Bioethics and Human Rights did by formulating the principle of respect for human vulnerability and personal integrity [5]. Emphasizing lack of ability to protect one's interests as the core of vulnerability is consistent with liberal and utilitarian traditions arguing that each individual is in the best position to judge what is in his or her own interest. It is also consistent with the doctrine of informed consent and the important role of self-determination. Well-informed and free individuals will follow what is in their interests when they consent. If they lack decisional capacity others can only decide for them if they also follow the person's interests or best interests. However, as will be elaborated in the next section, the focus on one's interest already presupposes one particular moral perspective, namely the primacy of the principle of respect for autonomy.

The other question concerning values and disvalues regards the harms that may occur. Earlier, various types of vulnerability have been examined, corresponding to different types of harm: physical, psychological, social, economic, and environmental. In a certain way these harms do occur in all circumstances for everybody; they are not specific for bioethical concerns. We are all vulnerable to these types of harm. In the context of medical research however, specific kinds of harm are associated with vulnerability: coercion and exploitation. In healthcare practice the relevant harms can be discrimination, lack of access to care, or denial of rights [6]. It has been pointed out that the concept of harm may be too narrow; one can also be vulnerable to wrongs such as lack of respect [7]. The point, however, is that the potential harms or wrongs are disvalues because they do not acknowledge the significance of self-determination.

The second aspect of vulnerability as a bioethical concept is that an ethical concept implies an appeal to do something; it makes a claim on us, it tells us what to do [8]. In the words of Bernard Williams, such a concept provides a reason for action [9]. If somebody is labeled as 'vulnerable' we ought to protect him, we need to create a certain practice in which he will not be hurt or damaged. For example, it will be our obligation to ensure that vulnerable groups in developing countries will not be the primary targets of recruitment for clinical trials if similar research can be done with other groups. Or, we should provide special scrutiny of research protocols involving children as vulnerable individuals. Of course, what kind of action is most appropriate, and who will be the subject of that action, will depend on the specific circumstances. Furthermore, the action-guiding provided by the concept of vulnerability needs to be balanced against reasons for action provided by other relevant ethical concepts.

As a bioethical concept, 'vulnerability' is very much like the concept of human dignity. This notion is also often used in international documents, commonly without any definition. It is frequently regarded as a normative notion [10]. Human dignity refers to the value or intrinsic worth of persons. And it implies that the person ought to be respected; his or her dignity cannot be violated. In a similar way the question can be raised: what is violated if vulnerability is not respected or protected?

Bioethical framing of vulnerability

What is the value that is embedded in the notion of vulnerability? What is the good that is at risk? The common definition of vulnerability already seems to presuppose the primacy of the moral principle of respect for autonomy. In this regard, the influence of the Belmont Report has been transformative. It introduced vulnerability as an ethical concept in the bioethical discourse by associating it with two fundamental ethical principles: respect for persons and justice [11]. The argument used is that it can be wrong to involve vulnerable subjects in research because their capacity to decide may be compromised. But it can also be wrong because it is not just to recruit them due to the situation in which they live and because they run the risks or have the burdens of participating in research without receiving the benefits. Unfortunately, the text of the Belmont Report interprets the second condition in terms of the first. While vulnerability is associated with justice, it is primarily regarded as lack of autonomy. The injustice of involving vulnerable groups is considered as the result of their dependent status and their compromised capacity for free consent [12]. The principle of respect for persons therefore overrides the principle of justice in the interpretation of the notion of vulnerability. This approach has been reiterated in subsequent interpretations in other documents, reports, and regulatory frameworks. The fact that the notion was first articulated in the context of biomedical research has probably reinforced this specific interpretation. The practical significance of informed consent as one of the basic rules in research ethics was clearly reflected in efforts to interpret vulnerability. Crucial for this notion ultimately is limited capacity to consent. Richard Nicholson for example argues that vulnerability has always been implicit in research ethics 'since the earliest attempts to regulate medical research' [13]. For him, the approach is simple: free and informed consent eliminates the vulnerability of a potential research subject. For others it can be more complicated since limitations can have various sources. Furthermore, it is not informed consent per se but the basic principle of respect for autonomy that defines who is vulnerable. As discussed in the previous chapter, the capacity to consent can be present, but there may be circumstances in which this capacity cannot be adequately exercised. Therefore, whether or not internal or external factors are at stake, in both cases the person's capacity to make autonomous decisions is impaired or reduced. The more this autonomy is decreased, the more vulnerability will increase [14].

Prioritizing the principle of autonomy

Interpreting vulnerability from the perspective of the ethical principle of respect for persons interestingly allows a reassessment of the components of the functional definition. Normally, persons will be able to protect their own interests, and they are therefore not vulnerable. In the framework of self-determination only the internal components of sensitivity and adaptive capacity are relevant. The inability to protect oneself can be the consequence of lack of autonomy as such, or of impeded exercise of autonomy. In the first case, the person has no capacity to make free choices because he is not yet an autonomous being (a child for example) or because he has temporarily or permanently lost autonomy (e.g. a person who is unconscious or in a state of severe dementia). Usually, he has no adaptive capacity while his sensitivity to threats is increased. In the second case, the person is an autonomous being in principle but she cannot exercise her autonomy because of certain impediments in her environment (e.g. prisoners or institutionalized persons). In principle, nothing is wrong with her sensitivity but the adaptive capacity is reduced or eliminated. However, in this framing of the functional definition, exposure is not considered as an essential feature of vulnerability; it is merely accidental. The main point in this perspective is that some human beings can be vulnerable even if there are no threats at all and exposure is absent. In such circumstances, protective efforts will be warranted not because of possible exposure but because the persons are unable to self-protect. On the other hand, in case of an autonomous person, if there is exposure, whether it is harmful or not is normally up to the assessment of the person concerned. As human beings we always live in a material and social environment that may possibly harm us. But as autonomous persons we ourselves normally decide whether some risks are worth taking while pursuing the goods that we value. The focus on 'interests' is crucial here. It emphasizes that we are the only ones who can assess what our interests are. Perhaps somebody else will take a completely different decision in order to avoid any threats. But as long as it is our choice to be exposed to some possible threats, our vulnerability should not be an ethical concern or a reason for others to act.

The consequence of this framing of vulnerability in terms of deficient autonomy is twofold. One is that the recognition of vulnerability does not imply that subjects cannot participate in medical research or health care activities. It is important to remove the ethical concerns raised by the language of vulnerability. This is feasible since the concerns are associated with individual autonomy. On the other hand, if persons lack autonomy, their interests should be scrutinized by others and special protection should be provided. If they are not able to exercise their autonomy, special precautions should be taken to safeguard their interests from being harmed. These actions are essentially not different from what is normally done: treating individuals as autonomous agents. They all express the same fundamental ethical principle of respect for persons.

Neutralizing the principle of justice

What is lost is highlighted in the consequences of this particular framing. The exposure to (potential) harm as an external component of vulnerability is no longer relevant for the ethical debate. The social and economic conditions that may produce vulnerability should be taken into account only if the exercise of individual autonomy of subjects can be restored. It is not the role of bioethics to change or ameliorate the conditions that expose vulnerable persons to (potential) harm. Rather, bioethical discourse should empower individuals so that they are no longer in a weak or defenseless position, and will be able to protect their own interests. If that is not possible, they should receive special protection. The idea that it is unjust to involve persons in biomedical research who cannot exercise their autonomy due to disadvantageous, unfortunate, or repressive circumstances is no longer worth considering as soon as serious efforts are made to restore their autonomy. In this perspective, the association between vulnerability and justice is not a primordial concern. The fact that contemporary bioethical discourse has framed vulnerability primarily in terms of personal autonomy and its deficiencies will be critically analyzed later. But first we have to examine in more detail how the notion of vulnerability is assessed, applied, and clarified in bioethical discourse.

Diverging assessments of vulnerability

The extensive literature provides multiple and widely different appreciations of the notion of vulnerability [15]. On the one hand, it is regarded as a useful and even essential contribution to contemporary bioethics. Vulnerability is welcomed as a relevant ethical principle that should guide the practice of clinical research [16]. Others stress its significance beyond the area of research ethics since it is 'at the heart of bioethics' [17]. The notion is also regarded in a wider perspective: as the source of our special moral responsibilities to certain persons [18], as a necessary foundation for a medicine that is sustainable, i.e. economically afford-able, and equitably distributed [19], and as a basic ethical principle 'essential to policy-making in the modern welfare state' [20]. Warren Reich has applauded vulnerability as 'the single most important idea that will shape ... the ... development of bioethics' [21].

On the other hand, the notion of vulnerability is severely criticized. One issue is that it is too vague and ambiguous. There is no commonly accepted definition, although certain criteria (notably age, gender, ethnicity, health status, and liberty status) are often used to identify the vulnerable [22]. The concept is also diffi-cult to define since it leaves open many possibilities for interpretation. National and international regulations therefore do not demonstrate a coherent approach to vulnerability [23]. This type of critique is common in other scientific areas. Environmental scientists for example complain that the term is too broad and that it is more a 'conceptual cluster' since one single conceptualization is absent [24].

The second criticism is that as long as there is no agreement on the meaning of the concept its utility in healthcare practice and research is limited, the more so since it does not provide clear moral guidance [25]. Ever more groups and popu-lations have been labeled as 'vulnerable' so that the implications of the notion are becoming less clear. If everyone is vulnerable, the 'elastic' concept of vulnerability will be 'too nebulous to be meaningful' [26]. Since it is no longer clear what kind of additional protection will be required and how such protection can be pro-vided beyond the usual level, the normative force of the concept is disappearing.

The third criticism addresses the adverse consequences of using the notion. Some consider it a rhetorical device, often used as a 'conversation-stopper'[27]. or an 'artificial portmanteau' [28]. Who can really be opposed to protecting the vulnerable? But the use of the term is not innocent. It can divert attention away from real problems in the provision of health care. Bruce Vladeck calls it a 'euphemism' in order not to discuss deficiencies in health insurance, access to ser-vices and quality of care but rather to focus on characteristics of individuals and groups [29]. A similar critique is voiced by Carol Levine and colleagues: focusing on the characteristics that make people vulnerable does not take into account that harm may result from other sources such as the social and economic context [30]. Another possible consequence is that labeling persons and populations as 'vulnerable' can be stigmatizing and stereotyping [31]. Not all individuals within a group or population will be vulnerable because they are simply members of a category. A more subtle and differentiated approach will be necessary. At the same time, the term itself has negative connotations so that characterizing individuals as vulnerable can be 'patronizing' [32]. They are labeled as weak, deficient, and pitiful, so that they are regarded as tragic cases or victims in need of assistance and protection. Such labeling itself can provide a justification for a biopolitics of social control and legal intervention in the life of a 'vulnerable' person [33]. The paradox then is that the notion of vulnerability has been introduced in the bioethical discourse to provide special protection to certain individuals but can itself be a major cause for damaging those individuals [34].

Notwithstanding these criticisms, it is striking that nearly all scholars, critical or not, concur that the notion of vulnerability should not be abandoned, perhaps willy-nilly since it is already so commonly used in bioethical discourse. It is unlikely, as Carl Coleman concludes 'that vulnerability language will disappear any time soon' [35]. As will be further explored in this chapter, instead of discarding the concept many efforts were undertaken to clarify and rethink it so that it will be more nuanced and robust [36]. For example, the notion has been defined (for example in the 2002 CIOMS Guidelines). New, more precise definitions have been proposed [37]. Specific criteria for vulnerability have been developed [38]. Recently, new theoretical frameworks have been proposed to clarify and justify the notion in order to better understand its use in bioethics discourse and to specify its practical implications [39]. Analytic proposals for systematic interpretation of vulnerability will be examined in the next chapter. But first it will be helpful to consider how and why the concept has been introduced in other areas of health care besides biomedical research. This will demonstrate that more conceptual progress has been made than commonly admitted by critics.

Need for vulnerability as critical concept

The concept of vulnerability has originated within the ethical setting of biomedical research and against the background of specific historical experiences. In this setting it has been framed in a particular way as impaired capacity of autonomous decision-making. The concerns are practical: can vulnerable individuals or groups be included in the research, and if so, with what kinds of protection? However, as noticed in the previous chapter, the concept soon became used in other health care settings and bioethical debates. Here it is often applied in a different, wider sense. The notion may be used to identify people who are able to make autonomous decisions but nonetheless at risk of being harmed or wronged because they lack power and resources. This is a reminder that concerns with vulnerability have become prominent in contemporary bioethics for particular reasons. We should not forget the underlying worry that has accompanied the growth of medical science and technology: the possibility that people might be abused or exploited. Particularly in biomedical research there is always the (conscious but most often unconscious) temptation to take advantage of individuals or groups of people who are already living in disadvantaged and poor conditions. Rejecting the notion of vulnerability will therefore be dangerous since some people might no longer be protected [40]. Particularly in clinical trials, protection of vulnerable populations is necessary as a moral safeguard [41]. But what initially was recognized as potential harm in the specific area of research has now become a general worry in regard to biomedicine itself. New vulnerabilities are generated by the therapeutic, economic and social powers of biomedicine [42]. Talking about vulnerability is a way to express concerns about the present-day powers of medical science, biotechnology, and health care. Because of these new powers, the current framework of ethical principles and specifically the emphasis on autonomy, liberty, and individual rationality is no longer sufficient to protect persons against the possible threats of biomedicine itself. This is the main reason why we need a principle of vulnerability.

An example is the debate on global organ trafficking. The growing awareness that especially vulnerable populations in developing countries are being used as sources for organs has instigated the international transplant community to take action. The global summit of medical and scientific professionals, government officials, social scientists, and ethicists in Istanbul in 2008 condemned the commercial targeting of vulnerable populations as organ vendors [43]. The notion of vulnerability apparently played two innovative roles in the debate. First, it directed attention to the living donor or vendor rather than the organ recipient. Anthropological and ethnographic studies had shown for some time that people who sell their kidney were often the most disadvantaged members of society [44]. People did this because they were poor or in debt. Most of them did not benefit from the sale; their well-being, health, and social status had frequently deteriorated after the organ had been removed [45]. By identifying them as 'vulnerable' their plight became at least more visible; it also underlined the need to protect them against exploitation.

Second, emphasizing vulnerability carried the debate from the level of individual decision-making to that of injustice and inequity. Transplant commercialism often induces vulnerable individuals and groups to sell their organs. The Istanbul Declaration explicitly refers to 'illiterate and impoverished persons, undocumented immigrants, prisoners, and political and economic refugees' [46]. In regard to these individuals and groups the language of rational choice and autonomous decision-making is inappropriate. Because of poverty, debt, paternalistic structures, and marginalization they have no real choice. The notion of vulnerability effectively undermines the argument that selling an organ is a transaction between two autonomous individuals [47].

Vulnerability as heuristic tool

The debate on organ trade illustrates that vulnerability can be used as a concept that opens up new insights, that discovers relevant dimensions beyond the usual scope of bioethical discourse. The notion can in fact be employed as a 'heuristic device', following the suggestion made by Phil Bielby in his study of vulnerability in biomedical research. It is not simply applied but is employed as a tool for tentatively exploring phenomena and suggesting new perspectives [48]. Furthermore, in the process of its employment the notion is itself tested and reinterpreted. In analyzing what is wrong with organ trade the point is not that organ vendors are lacking decisional capacity. Although some of them might be coerced, many are perfectly able to voluntarily engage in transactions and are conscious of what they are doing. It is therefore not difficult to remediate vulnerability if it is interpreted as limited autonomy and to argue that vendors are morally respected as human agents involved in a transaction. However, this initial interpretation of vulnerability is not sufficient to explain why vendors are nonetheless vulnerable to exploitation and harm. A broader interpretation of the notion needs to be engaged in order to ethically assess the basic problem. Looking through the conceptual frame of vulnerability, the presentation of organ sales as free individual transactions must be deconstructed since it is not the violation of the principle of respect for autonomy that makes the vendor vulnerable but disrespect for the principle of justice. In this analysis, the notion of vulnerability makes us aware that the susceptibility to be damaged is not determined by the characteristics or capacities of individuals or groups but by the circumstances in which they happen to live. The tragic choice between poverty and selling an organ is imposed on people because of the social, economic, and political living conditions. Constructing this choice as autonomous decision-making is concealing the lack of options that people are facing due to structural determinants which make any choice unfair. The notion of vulnerability is used on the one hand as a critical remedy against this limited focus on individual capacity. On the other hand, it opens up a new direction in the debate, concentrating on the context of ethical decision-making as well as on issues of inequality and injustice. Earlier it was argued that the connection between vulnerability and justice is present from the beginning, although it has not been further explored in research ethics. This connection may also explain the uneasiness with the broadening of the concept of vulnerability. Carol Levine and colleagues at least suggest that 'broader concerns about inequalities of power and resources' have facilitated the multiplication of vulnerable groups [49]. This has broadened the notion of vulnerability in their view to such extent that it has become pointless.

Other areas where the concept of vulnerability is increasingly used are public health and health policy. One example is pandemic preparedness planning. Ethical guidelines generally refer to vulnerable populations. One reason is that such populations are often disproportionately affected; existing inequalities can furthermore be exacerbated in a pandemic. The qualification 'vulnerable' is often ambiguous. It refers to internal determinants that characterize some categories of people such as the elderly, children, and pregnant women. But it often also refers to external conditions so that economically and socially disadvantaged groups are primarily regarded as vulnerable. However in a public health perspective, the internal sources are closely connected to external ones, so that the individual members of a category are not the target of policies. Vulnerability is the result of interactions between individual and contextual risks. Typically, individual persons will only be partially able to influence these contextual hazards. The conclusion from this perspective is that vulnerability is not an individual deficiency [50]. Identifying groups as vulnerable implies that planning for pandemics is a matter of social justice [51].

Vulnerable populations are furthermore frequently mentioned in pandemic planning because they can increase the risk of infection for all other persons. The National Strategy for Pandemic Planning in the US argues that there is a need to 'pre-identify vulnerable populations' in order to build 'community resilience' [52]. Public health professionals should detect the sources of vulnerability and develop profiles of vulnerable groups so that policy-makers will be able to protect the entire population. Labeling groups as vulnerable transforms them into targets of pre-emptive action. This may be the result of the paradigm shift in public health interventions described by Canadian scholars Frohlich and Potvin [53]. The notion of vulnerable populations has moved public health according to these authors beyond the usual population-at-risk approach. All individuals with a higher risk for some disease are categorized as a population at risk (for example all men with high blood pressure); interventions aim to reduce risk exposure by monitoring and surveying each person in this group. A vulnerable population however is different: it includes individuals who share particular social characteristics; these characteristics generate exposure to risks. This population is vulnerable because it is 'at risk of risks' [54]. However, not every individual in the population is vulnerable. Frohlich and Potvin refer to people of aboriginal descent who constitute a vulnerable population because the mean distribution of risk exposure for them is higher than for other Canadians [55]. Social conditions may cause diseases and create health disparities which may increase exposure to other risks. It does not imply that every aboriginal person throughout his or her life course will be equally exposed. Targeting a vulnerable population will primarily aim at ameliorating social conditions and reducing health disparities.

Employing the notion of vulnerability in the field of public health has done more than directing attention away from the exclusive focus on individual autonomy and introducing a strong emphasis on contextual factors that contribute to rendering groups vulnerable. The nature of public health is to examine health at the level of communities and populations, and that has made the notion attractive in this area of health care in the first place. But vulnerability has also transformed the concept of population within public health interventions. It has encouraged a more nuanced view of populations, not based on characteristics of individuals according to various risk profiles, or on the population as a whole but on examining and modifying fundamental underlying mechanisms that make certain groups of people more susceptible to possible harm than others. Again, such view brings together vulnerability and justice.

Conceptual progress

The argument so far has been that within mainstream bioethics vulnerability is conceptualized as deficient autonomy. This framing makes the notion more useful in research practices. It has also produced a consensus about how the notion should be interpreted and applied. This section will analyze this consensus. Two decades of critical bioethical examination have resulted in at least basic agreement concerning three issues: (1) not all vulnerability is the same but there are two basic types of vulnerability, (2) categorization of groups and populations as vulnerable lacks subtlety and a more refined approach is required, and (3) implications of vulnerability should be better specified in order to go beyond the requirement of protection. However, the section will also show how the consensus has been accomplished at the price of narrowing and reducing the concept of vulnerability. It will argue that the mainstream conceptualization effectively disregards the broader scope of the notion and therefore does not take into account the reasons why the language of vulnerability has emerged as an influential discourse in the first place.

Basic types of vulnerability

Onora O'Neill has distinguished persistent vulnerability as an essential feature of being human from variable or selective vulnerability due to specific circumstances [56]. What she has in mind here is that there is a vulnerability that all human beings share because it is inherent in being human. This type of vulnerability cannot be removed; it will always accompany human beings as long as they exist. Martha Fineman has characterized it as 'a universal, inevitable, enduring aspect of the human condition' [57]. The other type of vulnerability emerges from the environment in which human beings are living; some people are more vulnerable than others since they are more exposed to threats, or more sensitive, or have less adaptive capacity. This basic distinction between a universal or general vulnerability and an accidental or special vulnerability has been reiterated by many scholars, though in different terminology [58]. In 1994, the point was made in a CIOMS Conference in Mexico: there are two types of vulnerable people, i.e. those who are inherently vulnerable and those who are rendered vulnerable by their society [59].

Sometimes confusion arises since the same terms are used to make another distinction within the last type of special vulnerability. Schroeder and Gefenas for example discuss the fragility of the human condition (as the first type of vulnerability), but then distinguish between intrinsic and contingent vulnerability because some people (for example, children) are not able to protect themselves since they lack autonomy while other people (e.g. illiterate and socially disadvantaged persons) cannot exercise their autonomy due to the circumstances [60]. The first group is intrinsically vulnerable; the second group is contingently. Bielby also differentiates between intrinsic and extrinsic vulnerabilities depending on whether the source of vulnerability, namely the limitation of the person's decisional capacity, is within or outside the agent's being [61]. Children are cognitively vulnerable (intrinsic) while prisoners are circumstantially vulnerable (extrinsic). Another example provided by Dunn and colleagues contrasts inherent and situational vulnerability. The first refers to characteristics of the individual person such as age or disability, the second to circumstances within which individuals are living [62]. Although these distinctions might be useful to emphasize various sources of special vulnerability, they differ from the distinction made by O'Neill and others. This last typology is more fundamental for two reasons. It is more generic since it relates vulnerability to the human species rather than to characteristics of specific exemplars. Moreover it does not presuppose a particular conceptualization of vulnerability. The distinction between intrinsically and extrinsically vulnerable individuals is based on the question of whether their autonomy is limited due to absent decision-making capacity or compromised exercise of this capacity. The more fundamental distinction between general and special vulnerability does not frame the notion in terms of limitations of personal autonomy.

The use of the terms 'internal' or 'intrinsic' versus 'external' or 'extrinsic' can be confusing. For that reason, the more fundamental distinction will be clarified in this book by differentiating 'ontological' (general, basic or primary) from 'special' (or secondary) vulnerability [63]. Subsequently, a second distinction is made within the last type of vulnerability, depending on individual or contextual determinants. Doing so, European scholars such as Jan-Helge Solbakk, Peter Kemp, and Jacob Rendtorff will be followed [64]. For them, vulnerability is primarily ontological. It expresses the fragility that unites all human beings, that brings moral strangers together, and that instigates them to care for each other. It is for this reason, according to Rendtorff, that vulnerability is 'the essential foundation of the treatment of human beings in the hospital and in the legal system' [65]. It therefore is more basic than other ethical principles such as respect for autonomy. This conception of vulnerability is articulated in what earlier has been called the philosophical perspective (elaborated in the next chapter). First of all, it is important to understand the fundamental role of vulnerability in and for ethics. Only then one can elaborate what its implications are in various domains of health care.

Scholars from developing countries, although making the same fundamental distinction, primarily focus on special vulnerabilities. A philosophical perspective that characterizes the human condition as fragile and weak implies that sources of vulnerability are hard to change, if they can be changed at all. For this reason, preference is given to a political perspective, emphasizing that social and economic conditions make some people more exposed to threats and to the possibility of harm than others, especially in developing countries. Michael Kottow wants to separate the philosophical view from bioethical discourse [66]. He argues that the term 'vulnerability' should only be used for the ontological condition of human frailty. We should not use this term for external, specific conditions that make people vulnerable; it is better to speak here of 'susceptibility'. In fact, Kottow suggests that in using the same term 'vulnerability' for two different situations the political discourse that could address harmful conditions in bioethics is at risk of being eliminated by the philosophical one. If everybody is vulnerable, there is no need for additional protection for special categories of the vulnerable (the 'susceptible') since the human condition is unalterable. The same danger of inaction is pointed out by Florencia Luna [67]. She opposes 'naturalizing' vulnerability. If it is a characteristic of human nature to be vulnerable, then vulnerability cannot be reduced or remedied (unless human nature can be enhanced). She argues that instead of distinguishing between ontological and special vulnerability, we should focus primarily on situations that render a person vulnerable. In this perspective, the political discourse should have priority over philosophical interpretation.

In mainstream bioethics, the fundamental distinction between two types of vulnerability is often made but there seems to be a hesitation to explore it. Perhaps this is due to a narrow conception of bioethics considering philosophical explorations of the concept as too distant from bioethics discussions and therefore as irrelevant [68]. But it also seems that the idea of ontological vulnerability is not easily compatible with the dominant ethical principle of respect for autonomy. As such, two strategies are pursued for dealing with the typology in much of the bioethics literature. One is rejection. Hurst for example simply dismisses a broad definition relating vulnerability to the human condition since it cannot provide a reason for special protection [69]. The fundamental distinction between ontological and special vulnerability is simply considered irrelevant. The other strategy is to define vulnerability in terms of a common denominator. The CIOMS Guidelines in 1993 describe vulnerable persons for the first time as being incapable of protecting their own interests [70]. Other definitions of vulnerability also want to avoid reference to the fragility of the human condition, but maintain the inability to protect one's own interests as its core constituent [71]. This approach presupposes the primacy of autonomous decision-making that may be compromised by certain internal and external conditions. Defining vulnerability in the language of individual autonomy is bridging the gap between general and special vulnerability but 'general vulnerability' does no longer refer to ontological vulnerability. It refers to inherent or intrinsic features of individuals rather than to the inherent or ontological frailty of humankind. As a consequence, the philosophical as well as political perspectives of vulnerability flow together into the discourse of bioethics. In fact, there no longer is a difference between the two distinctions made earlier since the first one (between ontological and special vulnerability) is merged into the second one (between intrinsic and contingent vulnerability). Defining vulnerability in terms of lack of autonomy as the common denominator for different types of vulnerability however has two consequences. First, it eliminates what is typical for ontological vulnerability. Second, it makes it difficult to grasp the meaning of special vulnerability. Social conditions that make people vulnerable are now only understandable as long as these conditions are diminishing autonomy. In this strategy therefore, the political discourse of vulnerability is neutralized through the language of contemporary bioethics while the philosophical discourse is exorcised.

Categorizing groups and populations

As pointed out in Chapter 2, the term 'vulnerable populations' has been more often used as a keyword in the bioethics literature than 'vulnerability'. Since the first mention in the Belmont Report an ever wider range of categories and sub-categories of vulnerable persons has been identified resulting in seemingly exhaustive lists of vulnerable populations. Depending on the criteria used, more categories of people came to be labeled as 'vulnerable'. Compromised capacity for consent was used to identify what we have called earlier (in Chapter 3) the conventional class of vulnerable persons (for example, children). But on the basis of disadvantaged social and economic positions, a second population was identified (for example, the poor and racial minorities) while dependency was the demarcating criterion for a third one (for example, institutionalized persons). The 2002 CIOMS Guidelines present the most encompassing classification of vulnerable groups [72]. Since then, the approach to categorization has been more reticent. The Declaration of Helsinki lists, in 2000, five groups that are vulnerable, but in 2013 no specific groups are mentioned [73]. The UNESCO Declaration on Bioethics and Human Rights does not list vulnerable categories at all [74].

The dwindling enthusiasm to identify vulnerable populations may reflect the growing criticism in the literature. Many scholars argue that categorization is too broad [75]. The idea that persons are vulnerable simply because they belong to specific groups or populations is inadequate. Talking about groups or populations is not very helpful in determining who needs protection. David Resnik gives the examples of a South African white, male, urban businessman and an Ethiopian black, female villager to illustrate this point [76]. If research subjects in developing countries are considered as vulnerable populations, both individuals within such populations should receive the same protection. This would be counter-intuitive. Rather, in order to give the notion of vulnerability practical leverage it should be differentiated; we should be able to argue that one person needs protection, not the other, even if both belong to vulnerable populations. The example shows that a more refined analysis is required to determine which individuals within a category are vulnerable and which are not. It is also argued that the use of categories is counterproductive [77]. Bringing together individuals in the same population, regarding them as 'vulnerable' and in need of protection can be stereotyping, discriminatory, patronizing, and paternalistic. It is clear that there are too many differences among individuals within the same category. But including them in a vulnerable population turns everybody into victims; that is, powerless and weak persons who are not in control of their life.

Children are consistently mentioned as a vulnerable population. This example illustrates that within this category there are significant differences. Not only decision-making capacity can be variable, but it also depends on the type of decision to be made and the maturity of the child whether or not it can participate in the decision-making process. Classifying women as a vulnerable population has probably raised the most controversies. In the 1982 CIOMS Guidelines pregnant women were mentioned as one of the vulnerable populations, in the same category as children and mentally ill persons because consent would be compromised [78]. This classification implied that they were excluded from research projects, with little possibility to improve treatment options during pregnancy. This point of view has been strongly criticized because women, pregnant or not, are not generally lacking decision-making capacity and are most of the time not incapable of consent. Denying them participation in research for this reason is therefore paternalistic [79].

The criticism of categorization has resulted in a growing consensus that a more subtle approach is needed. It is not useful to compile an ever growing list of vulnerable populations. Alternatively, one should examine why and how people are vulnerable. This approach is advocated by the National Bioethics Advisory Commission (NBAC) that argues that situations should be identified and classified in which individuals might be considered vulnerable [80]. Many scholars have supported this proposal. Vulnerability should not be regarded as a general characteristic of persons because they are members of a category or population, but one should focus on situations of vulnerability that make individuals vulnerable [81]. However, in elaborating a situational approach, it is tempting to (again) concentrate on incapacities, deficiencies, or restraints in the process of individual decision-making. The types of vulnerability in research distinguished for example by the NBAC are almost all explained by compromised autonomous decisional capacity [82]. Also in studies elaborating the question whether vulnerable populations such as the homeless can be involved in research, the discussion is primarily concerned with the question whether the individual members of these populations are sufficiently free to make autonomous choices [83].

This return to interpreting vulnerability primarily in terms of individual characteristics is a curious phenomenon. The language of vulnerable populations had been intentionally introduced to direct attention towards generic factors that produce vulnerability for groups of people. For one thing, one might argue that particular populations are and have been attractive targets for research. The field of research ethics has at least in part been built on cases of exploitation of vulnerable populations: poor and uneducated African-Americans in the Tuskegee project, intellectually disabled children in the Willowbrook experiment, and more recently, poor, untreated, and critically ill children in the Nigerian Trovan trial [84]. Researchers should therefore be alerted if they intend to include such populations in biomedical research. Risk of exploitation, unfair treatment, or harm does indeed exist not only for individuals but for some categories of people. Also, it can be argued that the purpose of population language is to emphasize special vulnerability, not vulnerability in general. It elaborates only one part of the fundamental distinction discussed in the previous paragraph. The population language avoids the philosophical perspective of ontological vulnerability and the difficulties it creates for bioethical analysis. Whether or not everybody is vulnerable as a human being, it expresses the view that some populations of human beings are vulnerable, or more vulnerable than others. They need extra protection or additional safeguards beyond what is normally provided.

Furthermore, it has been highlighted, especially by researchers of vulnerable populations that the concept of 'population' is more nuanced, as suggested by the criticism [85]. There is a difference between an at-risk population and a vulnerable population. Every member of the first population has the particular risk characteristic (for example, hypertension). But not every individual member of the last one is vulnerable [86]. The vulnerability of the specific population is determined by social, economic, and cultural determinants such as differences in social status, access to resources, social support, education, employment, income, housing, and characteristics such as age, gender, race, and ethnicity. There are particular conditions that produce vulnerability. Identifying vulnerable populations requires what Aday calls 'the language of community' [87]. Different groups in society have different opportunities and resources to enjoy a healthy existence. At the same time, some individuals will be more harmed by threats and negative events than others. Therefore, the community perspective needs to be combined with an individual perspective. The consequence is that the above criticism of categorization must not result in discarding the focus on populations but should combine the aggregate and individual levels of vulnerability. Additionally, it is argued that vulnerability is multidimensional [88]. One determinant is often compounded with others. Racial or ethnic minorities frequently have a lower socio-economic status, are more likely uninsured, and have a higher burden of disease. Poverty is frequently associated with low social status, lack of access to resources, lack of education, and reduced social networks. Since vulnerability is the result of the interaction of determinants that are usually beyond the control of individuals, it does require the articulation of communal, social, and governmental responsibilities rather than solely individual responsibility. To assist and protect vulnerable populations, a policy approach is necessary that does more than individual empowerment and that addresses the common characteristics and determinants of vulnerability in populations. Finally, the focus on populations implies that vulnerability is not a personal deficiency [89]. The critique that vulnerability language is victimizing or stigmatizing is not correct because it is too general. Identifying populations as vulnerable is not negative but has the intention to prevent (further) discrimination and stigmatization. When research shows that the health of certain populations is threatened by conditions that are not under their control, the appropriate response is to initiate policies to ameliorate these conditions. The point of identifying a vulnerable population is precisely to call attention to their plight and to call for action. An example is the role of women's vulnerability in the HIV epidemic [90]. Initially in the earliest phases of the epidemic, women were invisible since the focus was not on heterosexual transmission. In the early 1990s the situation was changed since women were increasingly recognized as a vulnerable group. This change not only directed attention towards prevention of women's risks but also to women's social and cultural disadvantages such as poverty and gender inequality. Public health efforts could take account of structural influences rather than individual prevention. Another example is the vulnerability of indigenous communities. They can be vulnerable for example because their traditional knowledge is taken away and used by others, as happened with the San Peoples in Southern Africa [91]. Already very poor, marginalized, and often discriminated, they were exploited by medical science and business since their traditional knowledge was appropriated without fair benefit sharing. They were only able to conclude an agreement with the help of outsiders. Such communities share a collective vulnerability which is obviously multidimensional. Entire communities have disappeared in the process of modernization because their way of life as expressed in a particular culture has ended. The use of the term 'vulnerable' to identify such communities is not attributing negative or stereotyping qualifications; on the contrary, it is criticizing processes of cultural devastation, intolerance, and cruelty [92]. Focusing on populations directs attention to the conditions that might be harmful to groups and categories of individuals, and that require remedial action beyond the level of individual agency. It does not blame individuals within such groups and categories. The argument that such focus carries the risk of stigmatization and stereotyping seems to presuppose the primacy of the ethical framework of deficient autonomy. Individuals within vulnerable populations are stigmatized as victims because it is assumed that they lack the power to protect themselves. They will then suffer the negative social consequences of their weakness and powerlessness. However, the purpose of identifying populations is different: it aims to critique the social context within which individuals are embedded.

The language of vulnerable populations is therefore useful to identify contextual determinants that affect groups of people. It draws attention to conditions in which individual persons are living but that also affect entire groups and populations. This language is the expression of a political perspective on vulnerability that emphasizes the role of structural determinants such as poverty or gender inequality that are not under the control of individuals but that require a political response. Recruiting these populations in research projects is not fair if they have the burdens or risks of participating while not receiving the benefits from research [93]. Being vulnerable in this perspective is not a matter of individual capacities (at least not in many cases) but is the consequence of social ills and threats that affect large groups of people at the same time. Protecting these populations is a matter of justice. Explaining vulnerable populations in terms of lack of autonomous decision-making therefore does not recognize the political perspective that is specifically critical of the significance of personal autonomy. On the other hand, it also explains the dissatisfaction with population language in mainstream bioethical discourse: it introduces a perspective that is not associated with the moral view that respect for persons is the primary bioethical principle.

The criticism that categorization is too broad has contributed to a more subtle interpretation of the notion of vulnerable populations. It now refers to what Shi and Stevens have called 'a comprehensive model' of vulnerability that includes both individual and contextual factors [94]. Identifying populations does not imply that there is no wide variety among its members. The focus of analysis should be on determinants that make the entire population vulnerable but also on determinants that make individual members more vulnerable or more resilient. In other words, this is the concern with situations of vulnerability, advocated by the NBAC. However, the relevant determinants in these situations are of two kinds. One kind is 'intrinsic': decision-making capacity is absent, diminished, or constrained. On this basis several populations are distinguished, such as children, intellectually disabled, and demented elderly. Another kind is 'situational' or 'contextual': groups of people are marginalized, discriminated, or in unequal relationships with others in society (populations such as the homeless, the poor, and the uninsured). Both kinds can be exploited in medical research, harmed in healthcare and health policy, but on different grounds. Examining situations of vulnerability also requires a more refined analysis of why persons within a particular population are vulnerable. Being a member of a vulnerable population indicates, as Resnik has suggested, that the likelihood of being damaged or harmed is higher [95]. It does not imply that every member is equally vulnerable. This more refined approach can prevent the stereotyping of entire populations of individuals.

Finally, the language of vulnerable populations has been criticized because it has multiplied and now includes almost anybody. Nicholson ironically observed that only 'a Western, white, well-educated male' is not vulnerable, and even this population is shrinking [96]. This broadening of the concept implies for some that it has become 'almost meaningless' [97]. Vulnerability can no longer be an argument for special protection if everyone is vulnerable [98]. This proliferation of special vulnerabilities has in practice the same effect as ontological vulnerability. The scope of application for the ethical principle of respect for autonomy is gradually contracting since there are fewer and fewer people who are able to protect their own interests. But this is precisely the point of vulnerability population studies; there are so many conditions and risk factors that influence health that everybody is vulnerable within some social circumstances or during some moments in the course of life. Vulnerability is not exceptional but rather the rule in human existence. This conclusion is important for health policy. It shows that the notion of vulnerability, as pointed out by Danis and Patrick, is not 'fixed and immutable' [99]. Each one of us will at some time in his or her life be a member of some or several vulnerable populations. Concluding this is not the same as arguing for ontological vulnerability. Even if we are not convinced that human beings as species are basically vulnerable, at least we have to admit that we all have special vulnerability in certain periods of our life: when we were children, when we are suffering serious diseases or disabilities, when we face difficulties with jobs and income, and when we grow old.

Implications of vulnerability

Finally, the scholarly debate has intensely discussed the practical implications of 'vulnerability'. What are the responsibilities, duties, or claims that follow from the designating of individuals, groups and populations as vulnerable? And who in particular will have those responsibilities or duties? Following the policy documents discussed in the previous chapter, in the context of research ethics three consequences have been identified: justification for inclusion in research, special protection, and sharing benefits. The Belmont Report was the first to point out the need of extra justification for involving vulnerable persons and groups. Since these subjects and populations are easy to recruit for research, the appropriateness of involving them should therefore specifically be justified. In his textbook, Robert Levine devotes several pages to the need to justify the inclusion of vulnerable subjects in biomedical research [100]. It is also the primary concern of the 1991 CIOMS Guidelines: ethical review committees should be 'particularly vigilant' when these populations are included [101]. Given the recent history of abuse and exploitation of vulnerable populations the emerging field of research ethics in the 1970s and 1980s adopted a defensive approach. Researchers should be very cautious about including such populations and they should only do so if they can provide adequate justification (for example, in the case of children or prisoners), or they should not include them in clinical research at all (for example, pregnant women). This approach changed in the 1990s [102]. It was argued that vulnerable subjects will not benefit from scientific progress if they are excluded from research. Especially AIDS activists demanded access to clinical trials. Rather than assuming that vulnerable subjects should preferably not be involved in research unless specific justifications are provided, it was now assumed that they should be involved as long as special protection can be given to avoid abuse and exploitation. The definition of vulnerability that became popular since the 1993 CIOMS Guidelines was of course helpful for such change of perspective [103]. In general, people are able to protect themselves, but if vulnerable persons are those who cannot protect themselves, other people will be responsible for providing this protection. Many scholars therefore associate vulnerability with claims to (special or additional) protection [104]. The third consequence of vulnerability formulates positive responsibilities and duties. In 2002 the CIOMS Guidelines indicate that research with vulnerable persons should be responsive to the needs of these persons. When vulnerable subjects are involved, the research should be 'intended to obtain knowledge that will lead to improved diagnosis, prevention or treatment of diseases or other health problems characteristic of, or unique to, the vulnerable class – either the actual subjects or other similarly situated members of the vulnerable class' [105]. The same guidelines stipulate that vulnerable subjects should have 'reasonable access to any diagnostic, preventive or therapeutic products that will become available as a consequence of the research' [106]. This responsiveness to needs and the probability of sharing benefits for research with vulnerable populations is even stronger emphasized in the latest revision of the Declaration of Helsinki [107]. In the UNESCO Declaration vulnerability is connected to the positive duty of solidarity and international cooperation. The International Bioethics Committee elaborating the principle of respect for human vulnerability and personal integrity argues that this principle is closely related to other bioethical principles, not only personal autonomy and justice but also human dignity, solidarity, social responsibility, and benefit sharing. It specifically reiterates that vulnerability may function as a bridge between individual and societal interests [108].

Ideas about implications of vulnerability have therefore evolved over time. However, most of the debate has taken place within research ethics. Here the focus has been very much on protection. Different modalities of protective regimes for vulnerable people have been developed: stricter consent requirements, diminishing the risks of exposure, limiting participation, and careful review by independent boards or committees [109]. At the same time there was growing dissatisfaction. Like the notion of vulnerable populations the implication of protection was deemed too generic and broad to be effective. It was argued that the protection provided is often inadequate and weak. For instance, it is unclear how 'vulnerability' can provide ethical guidance[110]; the protection provided is often too general to be more than merely a 'special consideration' [111]. Rather than concluding pessimistically, as Leavitt did, that protection of vulnerable research subjects is impossible since present-day researchers, drug companies, universities, and review boards are too compromised to provide independent protection, many scholars advocate a more specific approach [112]. For example, Schroeder and Gefenas propose to relate protection to 'identifiable possible harms'. They do not provide a list of those harms but mention four possible 'markers': unfavorable risk-benefit ratio, breach of confidentiality, invalid consent, and lack of access to benefits of research [113]. Andanda has applied this proposal to poor and little-educated women in the Majengo slum of Nairobi involved in research to develop an HIV vaccine. She showed how research usually disregards the last 'marker': post-trial access or essential healthcare are not provided to the Majengo women [114]. The focus on harms has the advantage that it shifts the attention from generic protection to a more specific approach, mitigating and possibly eliminating the possibility of harm, but it does not sufficiently ameliorate the situation of vulnerable subjects. Other scholars have therefore emphasized the need to delineate more detailed, subtler, and differentiated protections for specific persons and groups, following the recommendation of the NBAC to determine 'appropriate safeguards for each type of vulnerability' [115]. However, all efforts seem to focus on addressing one particular dimension of vulnerability, usually compensating for impaired decision-making. Another, more drastic proposal is to apply 'special scrutiny', for example making sure that certain kinds of research protocols receive more focused intensive review or that review boards are asking targeted questions to determine whether any research subjects are likely to incur harm or wrong [116]. Attention is then not solely directed to characteristics of research subjects; it is also directed to the research protocol and the research setting. The criteria mentioned for special scrutiny are mostly concerned with potential risks (e.g. applying a novel intervention or a controversial research design). But adequate risk assessment and the requirement to minimize risk is already part of the usual process of research review, and meant to provide protection for all research subjects. As a result, the specificity of vulnerability has disappeared. This proposal does not take into account that vulnerability is a relational concept that links exposure, sensitivity, and adaptive capacity [117]. If special scrutiny does no longer address the characteristics of vulnerable subjects, or at least is initiated through the acknowledgement of vulnerable populations, it seems to assume that vulnerability is primarily exposure; it is caused by the risks of the research project or its experimental design, independent from the subjects or populations involved [118]. It also seems to ignore that certain populations share a history and a political reality that make them more vulnerable than others, whether or not they are or will be involved in research [119]. Similar concerns arise from the suggestion of Solomon for a more narrow interpretation of additional protections [120]. In her view, not all but only some vulnerable populations require special protection; in most cases the ordinary, general protective mechanism for all research subjects will be sufficient. Usually, when the decision-making capacity is impaired the informed consent process, if optimized, will sufficiently protect most vulnerable subjects. When the existing regulations are followed additional protection is only needed in exceptional cases. The same is true for economically vulnerable populations. Potential exploitation because the risk-benefit ratio is unfair will not be a reason for extra protection since balancing risks and benefits is already within the mandate of the review committee. Additional regulation is only required when research benefits are not fairly distributed among populations; in that case economically vulnerable populations have the burden of research while the benefits go to other populations [121]. Although Solomon is right in critically analyzing the automatic connection often made between vulnerability and protection, her analysis seems to reduce vulnerability to a rhetorical device that in many cases does not have any implication. Whether or not researchers are involving a vulnerable population, most of the time, in her view, they can continue to apply the usual protective mechanisms. However, Solomon's application of 'special protection' raises questions. On the one hand, it implies a too strict interpretation of what protection entails. Special protection for her seems to be equivalent to new regulation [122]. Vulnerable populations can only claim special protection if new measures and regulatory mechanisms must be developed since the ordinary protection is insufficient. This interpretation is too stringent. Often vulnerability requires more care, precision, and rigor in the application of existing mechanisms. If parents bring their young children to the beach, they take more precautions and will be more alert than if they go with friends. They know that children in these circumstances need special protection but they do not have to initiate protective measures that are new. Similarly, if patients with Alzheimer's disease are recruited for a clinical trial, researchers will be more careful and attentive to the ethical requirements but do not need to develop additional regulations. It is not clear why special protection can not imply better implementation of existing regulations. On the other hand, disconnecting vulnerability and protection makes detailed examination of what constitutes vulnerability more, not less important. In order to determine what kind of protection is necessary, and whether new regulations or better implementation is required, the notion of vulnerability is indispensable [123]. The main concern is that protective mechanisms are in place and will be implemented for vulnerable individuals and populations. This can never be an automatic, self-evident response as Solomon correctly points out.

The concerns with protection in the area of research ethics possibly demonstrate an increasing wariness of the notion of vulnerability. If more and more people are regarded as vulnerable subjects then the moral implications of the notion are becoming unfeasible in the practice of biomedical research. It seems that there are only two options. Either protective mechanisms are redefined or more stringently defined, or the harms and risks of research are better scrutinized. In both options the moral significance of vulnerability is downgraded.

But in fact there are other options. In research ethics, the focus on protection is closely related to the interpretation of vulnerability as impaired decision-making. This is less the case in other areas of biomedicine such as care, policy, and public health. People can be powerless, helpless, defenseless, and thus vulnerable, because they are threatened by illness, lack of resources, harmful social conditions, or general insecurity even if there is no compromised decisional capacity. This broader perspective opens up different normative implications of vulnerability.

In health care ethics, and especially in nursing ethics, it is argued that vulnerability should define a new role for health professionals, i.e. advocacy [124]. In order to care for vulnerable persons professionals should also obtain a wider range of competencies, including cultural competence in order to understand the socio-cultural setting of vulnerability [125].

In health policy the need for public assistance for certain populations is stressed. Since vulnerability is primarily the result of social determinants, it can only be resolved through social means [126]. The ultimate responsibility for the well-being of vulnerable groups is with the government and policy-makers [127]. Policies should therefore be more focused on populations than individuals [128]. Empowering individuals, and thus reducing their vulnerability, is only possible through effectively changing social and economic conditions. This does not imply that all vulnerable populations have the same moral claim. Brock for example distinguishes three moral categories of vulnerable populations, depending on whether the vulnerability is caused by conditions that themselves constitute an injustice, conditions that are morally undeserved but not unjust, or conditions for which individuals or groups themselves are responsible. People in the last category do not have the same moral claim as members of the first category [129].

In public health, vulnerability gives rise to formulating protection duties of the state, and specifying international human rights law for particular categories of vulnerable subjects [130]. Health professionals should engage, as Carolyn Ells explains, in 'advocacy action toward governments on behalf of the populations they serve (or should serve)' [131]. But they also should engage and involve vulnerable people and populations in development of policies and in decision-making that is relevant for their health. Doing so will require collaboration with local initiatives, marginalized communities, and NGOs to counter vulnerabilities [132].

These examples from other areas of medicine show that the implications of vulnerability can go far beyond the standard response of protection. Bioethical discourse is increasingly recognizing that protection is only one option in dealing with vulnerability. But as long as vulnerability is primarily conceptualized as impaired individual decision-making it is difficult to consider vulnerable individuals, groups, and populations as anything other than 'candidates for protection' [133]. The emphasis on protection implies that others will be speaking and acting for them. Vulnerable persons are commonly not included in decisions and policies regarding their health and interests. They are also absent in the bioethical debate. As will be discussed in the last chapter, there are currently several efforts to include vulnerable persons in policy development and implementation, and to empower vulnerable populations through giving them the opportunity to express vulnerability, analyzing their experiences, but also through emphasizing their rights, and addressing inequalities with participatory approaches.

Reconciling different perspectives

In this chapter we have discussed how vulnerability as a concept is widely used in bioethics but that opinions differ on whether it is a helpful and productive notion. These ambivalent attitudes are related to the specific framing of vulnerability in mainstream bioethical discourse. Since the concept has first emerged in the context of biomedical research, vulnerability is traditionally framed as impaired capacity of decision-making. The vulnerable subject is essentially regarded as a 'failed' autonomous subject. Bioethical framing therefore locates vulnerability in the individual and regards it as lack of capacity. This particular framing is normatively driven; it is the result of the primacy of the ethical principle of respect for personal autonomy. The moral wrong of vulnerability is that a subject is not able to determine him or herself. The consequences of this bioethical interpretation are twofold. First, it is only relevant for bioethics to focus on the vulnerable subject and to consider how autonomous decision-making capacity can be protected, restored, or substituted. Second, bioethics cannot be concerned with ameliorating or reducing external threats since exposure is not determinative for vulnerability; it depends on historical and contingent conditions. Priority should be given to building capacities of individual subjects rather than building institutions and conditions that foster individual capacities.

This framing enables mainstream bioethics to work with a rather narrow view of the notion of vulnerability. It also prevents seriously considering different aspects and analyzing why the notion has emerged in the first place. Because of its emphasis on personal autonomy, the bioethical framing of vulnerability has difficulties in accepting the philosophical perspective in which vulnerability is a general feature of human beings. Ontological vulnerability implies that the scope of individual decision-making is limited and that human beings share a common predicament. This philosophical perspective is influential in the area of care. It is associated with the moral notions of responsibility and solidarity. The bioethical framing of vulnerability also has difficulties in accommodating the political perspective, because it directs the attention towards the structural determinants that disadvantage people rather than to the autonomous individual. This political perspective emphasizes that vulnerability is not a matter of being in a particular state of weakness but rather the outcome of specific situations. Human beings are made vulnerable and this is precisely what should instigate action to change the social, cultural, economic, and political contexts that determine and produce vulnerability. This perspective is related to the moral notions of justice and equality. The political perspective is influential in public health and health policy, and it entails a broader concept of special vulnerability that goes beyond the focus on individual weakness. At the same time, this perspective is criticized in mainstream bioethics since it encourages the multiplication of vulnerable groups and populations, making the normative implications of vulnerability impractical since not everyone can qualify for special protection.

The bioethical discourse is therefore caught between these two perspectives: vulnerability as a characteristic of human persons or of specific human situations. The framing of vulnerability as compromised autonomous decision-making does not help to overcome this paralysis. The question is how the two perspectives on vulnerability are related: are they incompatible or can they be reconciled to generate a broader normative view that associates vulnerability with ethical principles of responsibility, solidarity and justice, rather than narrowly connecting it with respect for individual autonomy? This will be the challenge for the next two chapters.

Notes

1 United States Senate *Quality of health care – human experimentation,* 1973, 793.

2 CIOMS. *International Ethical Guidelines for Biomedical Research Involving Human Subjects.* 2002 Preamble; WMA, *Declaration of Helsinki, fifth revision,* specific provision 5.

3 UNESCO, *Universal Declaration on Bioethics and Human Right,* Preamble.

4 Shi and Stevens, *Vulnerable populations in the United States,* 1 ff.

5 Rendtorff and Kemp, *Basic ethical principles in European bioethics and biolaw. Volume I. Autonomy, dignity, integrity and vulnerability,* 46; Rendtorff, 'Basic ethical principles in European bioethics and biolaw: Autonomy, dignity, integrity and vulnerability – Towards a foundation of bioethics and biolaw', 237; UNESCO, *Universal Declaration on Bioethics and Human Right,* Article 8.

6 See Aday, *At risk in America: The health and health care needs of vulnerable populations in the United States.*

7 Hurst, 'Vulnerability in Research and Health Care; Describing the Elephant in the Room?' 196 ff.

8 According to Korsgaard: Ethical standards are normative because: 'They do not merely *describe* a way in which we in fact regulate our conduct. They make *claims* on us; they command, oblige, recommend, or guide. Or at least, when we invoke them, we make claims on one another'. (Korsgaard, *The sources of normativity,* 8).

9 Williams, *Ethics and the limits of philosophy,* 130, 140.

10 Schachter, 'Human dignity as a normative concept', 848 ff.

11 Nickel, 'Vulnerable populations in research: The case of the seriously ill', 247.

12 National Commission for the Protection of Human Subjects of Biomedical and Behavioral Research, 'The Belmont Report', 23197; see also Chapter 3.

13 Nicholson, 'Who is vulnerable in clinical research'? 19.

14 Haugen, 'Inclusive and relevant language: the use of the concepts of autonomy, dignity and vulnerability in different contexts', 210.

15 See Ruof, 'Vulnerability, Vulnerable Populations, and Policy'.

16 Solbakk and Vidal, 'Research ethics, clinical', 782-3.

17 Rogers, Mackenzie, Dodds, 'Why bioethics needs a concept of vulnerability', 12.

18 Goodin, *Protecting the vulnerable. A reanalysis of our social responsibilities.*

19 Callahan, 'The vulnerability of the human condition', 119.

20 Rendtorff, 'Basic ethical principles in European bioethics and biolaw: Autonomy, dignity, integrity and vulnerability – Towards a foundation of bioethics and biolaw', 237.

21 Reich, 'The power of a single idea', 380.

22 Morawa, 'Vulnerability as a concept in international human rights law', 150.

23 Coleman, 'Vulnerability as a Regulatory Category in Human Subject Research', 13.

24 Füssel, 'Vulnerability: A generally applicable conceptual framework for climate change research', 156.

25 Hurst, 'Vulnerability in Research and Health Care; Describing the Elephant in the Room?', 192.

26 Levine, *et al.* 'The limitations of "vulnerability" as a protection for human research participants', 46; Levine, 'The concept of vulnerability in disaster research', 398.

27 Kipnis, 'Vulnerability in research subjects: A bioethical taxonomy', G2.

28 Vladeck, 'How useful is "vulnerable" as a concept?', 1232.

29 Vladeck, 'How useful is "vulnerable" as a concept?', 1232-4.

30 Levine, *et al.* 'The limitations of "vulnerability" as a protection for human research participants', 46.

31 See, for example, Forster, Emanuel and Grady, 'The 2000 revision of the Declaration of Helsinki: A step forward or more confusion?'; Danis and Patrick, 'Health policy, vulnerability, and vulnerable populations'; Levine, *et al.* 'The limitations of "vulnerability" as a protection for human research participants'.

32 Brown, '"Vulnerability": Handle with care', 316.

33 Dunn, Clare, and Holland. *'To empower or to protect? Constructing the "vulnerable adult"* in English law and public policy', 244 ff. Waiton elaborates the construction of victimhood in contemporary societies with the help of the notion of vulnerability (Waiton, *The politics of antisocial behaviour. Amoral panics*, 73 ff).

34 Schonfeld argues that the categorization of pregnant women as vulnerable has led to exclusion from clinical trials and has therefore perpetuated their vulnerability. (Schonfeld, 'The perils of protection: vulnerability and women in clinical research', 203).

35 Coleman, 'Vulnerability as a Regulatory Category in Human Subject Research', 14.

36 See Shivas, 'Contextualizing the vulnerability standard'; Eckenwiler, Ellis, Feinholz, and Schonfeld, 'Hopes for Helsinki: reconsidering "vulnerability"'.

37 Schroeder and Gefenas. 'Vulnerability: Too vague and too broad?' 117; Hurst, 'Vulnerability in Research and Health Care; Describing the Elephant in the Room?' 195.

38 See Kipnis, 'Vulnerability in research subjects: A bioethical taxonomy'; Kipnis, 'Seven vulnerabilities in the pediatric research subject'.

39 See Macklin, 'A global ethics approach to vulnerability'; Rogers, Mackenzie, and Dodds. 'Why bioethics needs a concept of vulnerability'.

40 Luna, 'Elucidating the concept of vulnerability: Layers not labels', 135.

41 DeMarco, 'Vulnerability: A needed moral safeguard', 83.

42 Patrão Neves, 'The new vulnerabilities raised by biomedical research', 186 ff.

43 Declaration of Istanbul on organ trafficking and transplant tourism, 1014; see also: Budiani-Saberi and Delmonico, 'Organ trafficking and transplant tourism: A commentary on the global realities'; Bagheri and Delmonico, 'Guest editorial: organ trafficking and transplant tourism: a call for international collaboration'.

44 See Scheper-Hughes, 'The global traffic in human organs'; Scheper-Hughes, 'Rotten trade: millennial capitalism, human values, and global justice in organs trafficking'; Moazzam, Zaman, and Jafarey. 'Conversations with kidney vendors in Pakistan. An ethnographic study'.

45 Moazzam, Zaman, and Jafarey. 'Conversations with kidney vendors in Pakistan. An ethnographic study', 34–35, 39.

46 Declaration of Istanbul on organ trafficking and transplant tourism, 1014.

47 Moazzam and colleagues call this argument 'misleading' (Moazzam, Zaman, and Jafarey. 'Conversations with kidney vendors in Pakistan. An ethnographic study', 39). Scheper-Hughes is more outspoken: there cannot be an ethical choice in 'an intrinsically unethical context'. She explains: 'The social and economic contexts make the "choice" to sell a kidney in an urban slum of Calcutta or in a Brazilian *favela* or Philippine shantytown anything but a "free" and "autonomous" one' (Scheper-Hughes, 'Rotten trade: millennial capitalism, human values and global justice in organs trafficking', 221).

48 Bielby, *Competence and vulnerability in biomedical research*, 55.

49 Levine, *et al.* 'The limitations of "vulnerability" as a protection for human research participants', 45.

50 Hutchins, Truman, Merlin, and Redd. 'Protecting vulnerable populations from pandemic influenza in the United States: A strategic imperative', S244.

51 Uscher-Pines, Duggan, Garoon, Karron, and Faden. 'Planning for an influenza pandemic. Social justice and disadvantaged groups', 33.

52 Homeland Security Council. *National strategy of pandemic influenza implementation plan: one year summary,* 30.

53 Frohlich and Potvin. 'Transcending the known in public health practice'.

54 Frohlich and Potvin. 'Transcending the known in public health practice', 216.

55 Frohlich and Potvin. 'Transcending the known in public health practice', 218.

56 O'Neill, *Towards Justice and Virtue,* 192.

57 Fineman, 'The vulnerable subject: Anchoring equality in the human condition', 8.

58 See, for example: vulnerability versus susceptibility (Kottow, 'The vulnerable and the susceptible'); intrinsic versus contingent (Silvers, 'Historical vulnerability and special scrutiny: Precautions against discrimination in medical research'); broad versus restrictive (Hurst, 'Vulnerability in Research and Health Care; Describing the Elephant in the Room?'); broad versus narrow (Jecker, 'Protecting the vulnerable'); primary versus secondary vulnerability (Schramm and Braz, 'Bioethics of protection: A proposal for the moral problems of developing countries'?); internal versus external (Schroeder and Gefenas, 'Vulnerability: Too vague and too broad?'); being vulnerable versus making vulnerable (Luna, 'Elucidating the concept of vulnerability: Layers not labels'); alterable versus unalterable (Solbakk, 'Vulnerability: A futile or useful principle in healthcare ethics?'); general versus special vulnerability (IBC, *Report of the IBC on the principle of respect for human vulnerability and personal integrity*); universal versus context-dependent (Rogers, Mackenzie, and Dodds, 'Why bioethics needs a concept of vulnerability'); and universal versus special vulnerability (Rogers, 'Vulnerability and bioethics').

59 Kahn and Bryant, 'The vulnerable in developed and developing countries – A conceptual approach', 57.

60 Schroeder and Gefenas, 'Vulnerability: Too vague and too broad?' 116.

61 Bielby, *Competence and vulnerability in biomedical research,* 54.

62 Dunn, Clare, and Holland. 'To empower or to protect? Constructing the "vulnerable adult" in English law and public policy', 237.

63 I will argue in the next chapter that it is more appropriate to talk about 'anthropological' than 'ontological vulnerability'.

64 See: Solbakk, 'Vulnerability: A futile or useful principle in healthcare ethics?'; Rendtorff, 'Basic ethical principles in European bioethics and biolaw: Autonomy, dignity, integrity and vulnerability – Towards a foundation of bioethics and biolaw'; Kemp, Rendtorff, and Johansen, eds. *Bioethics and Biolaw. Vol. II: Four ethical principles.*

65 Rendtorff, 'Basic ethical principles in European bioethics and biolaw: Autonomy, dignity, integrity and vulnerability – Towards a foundation of bioethics and biolaw', 237.

66 Kottow, 'The vulnerable and the susceptible', 463.

67 Luna, 'Elucidating the concept of vulnerability: Layers not labels', 129.

68 Rendtorff mentions that vulnerability has been misunderstood in modern society because of the 'vulnerability reducing agenda' (Rendtorff, 'Basic ethical principles in European bioethics and biolaw: Autonomy, dignity, integrity and vulnerability – Towards a foundation of bioethics and biolaw', 237). Hoffmaster signals that vulnerability does not mean much for morality because contemporary morality emphasises individual control and rationality, and ignores the body (Hoffmaster, 'What does vulnerability mean?', 42–4).

69 Hurst, 'Vulnerability in Research and Health Care; Describing the Elephant in the Room?' 192.

70 CIOMS. *International Ethical Guidelines for Biomedical Research Involving Human Subjects,* 1993, 10.

71 Schroeder and Gefenas. 'Vulnerability: Too vague and too broad?' 116–117; see also Eckenwiler *et al.* who argue in favor of a broad conception of vulnerability but then indicate that it should refer to 'individual and contextual factors that impede the protection of one's own interests…' (Eckenwiler, Ellis, Feinholz, and Schonfeld, 'Hopes for Helsinki: reconsidering "vulnerability"', 766).

72 See CIOMS. *International Ethical Guidelines for Biomedical Research Involving Human Subjects.*

73 See WMA, *Declaration of Helsinki*, 5th revision, and WMA, *Declaration of Helsinki*, 6th revision. The current new revision does not mention any particular vulnerable groups (WMA Declaration of Helsinki, October 2013).

74 See UNESCO, *Universal Declaration on Bioethics and Human Rights.*

75 See for example: Kipnis, 'Vulnerability in research subjects: A bioethical taxonomy', G-12; Kipnis, 'Seven vulnerabilities in the pediatric research subject', 108 ff; Levine, *et al.* 'The limitations of "vulnerability" as a protection for human research participants', 46; Iltis, 'Introduction: Vulnerability in biomedical research', 8.

76 Resnik, 'Research subjects in developing nations and vulnerability', 64.

77 See for example: Levine, *et al.* 'The limitations of "vulnerability" as a protection for human research participants', 47; Levine, 'The concept of vulnerability in disaster research', 399.

78 See Chapter 3; CIOMS. 'Proposed International Guidelines for Biomedical Research Involving Human Subjects'.

79 Wild, 'How are pregnant women vulnerable research participants?' 85 ff; see also: Rogers and Ballantyne, 'Gender and trust in medicine: Vulnerabilities, abuses, and remedies'; Rawlinson, 'Women and special vulnerability'; Schonfeld, 'The perils of protection: vulnerability and women in clinical research'.

80 ' …rather than focusing primarily on categorizing groups as vulnerable, investigators and IRBs should also recognize and avoid situations that create susceptibility to harm or coercion'. National Bioethics Advisory Commission, *Ethical and policy issues in research involving human participants,* iv. See also 85–92.

81 Among others, Morawa, 'Vulnerability as a concept in international human rights law', 150; Eckenwiler, *et al.* 'Hopes for Helsinki: reconsidering "vulnerability"',766; Luna, 'Elucidating the concept of vulnerability: layers not labels', 129.

82 For a discussion of the analytical approach of Kipnis see the next chapter.

83 See Afolabi, 'Researching the vulnerables: Issues of consent and ethical approval'; Beauchamp, Tom L., Bruce Jennings, Eleanor D. Kinney, and Robert J. Levine. 'Pharmaceutical research involving the homeless'. Leavitt even defines a vulnerable population as 'a population whose ability to exercise autonomous decision-making with respect to research is restricted' (Leavitt, 'Is any medical research population not vulnerable?' 'Bioethics, vulnerability, and protection', 475–77. See also: Stone, 'The invisible vulnerable: The economically and educationally disadvantaged subjects of clinical research'; Jegede, 'Understanding informed consent for participation in international health research'; Annas, 'Globalized clinical trials and informed consent'.

84 Lott, "Module three: Vulnerable/special participant populations," 31; Macklin, "Bioethics, vulnerability, and protection," 475–77. See also: Stone, "The invisible vulnerable: The economically and educationally disadvantaged subjects of clinical research"; Jegede, "Understanding informed consent for participation in international health research"; Annas, "Globalized clinical trials and informed consent."

85 Aday, *At risk in America: The health and health care needs of vulnerable populations in the United States,* 8; Shi and Stevens, *Vulnerable populations in the United States,* 17–18; De Chesnay, 'Vulnerable populations: vulnerable people', 4 ff. See also: Flaskerud and Winslow. 'Conceptualizing vulnerable populations health-related research'.

86 De Chesnay, 'Vulnerable populations: vulnerable people', 3.

87 Aday, *At risk in America: The health and health care needs of vulnerable populations in the United States,* 8.

88 'We have defined vulnerability as a multidimensional construct reflecting convergence of predisposing, enabling, and need attributes of risk at both individual and ecological levels'. (Shi and Stevens, *Vulnerable populations in the United States,* 30).

89 Shi, 'The convergence of vulnerable characteristics and health insurance in the US', 520; Shi and Stevens, *Vulnerable populations in the United States,* 3, 17.

90 Higgins, Hoffman, and Dworkin. 'Rethinking gender, heterosexual men, and women's vulnerability to HIV/AIDS', 436.

91 Chennells, 'Vulnerability and indigenous communities: Are the San of South Africa a vulnerable people?' 150.

92 Lear describes the cultural annihilation of the Crow people, demonstrating that civilizations themselves are vulnerable. See: Lear, *Radical hope. Ethics in the face of cultural devastation.*

93 Nickel, 'Vulnerable populations in research: The case of the seriously ill', 247-8; Gbadegesin and Wendler. 'Protecting communities in health research from exploitation', 250.

94 Shi and Stevens, *Vulnerable populations in the United States,* 17.

95 Resnik, 'Research subjects in developing nations and vulnerability', 64. This point is also stressed by DeMarco ('Vulnerability: A needed moral safeguard', 83).

96 Nicholson, 'Who is vulnerable in clinical research?', 22.

97 Levine, 'The concept of vulnerability in disaster research', 398.

98 Forster, Emanuel and Grady, 'The 2000 revision of the Declaration of Helsinki: a step forward or more confusion?' 1451.

99 Danis and Patrick, 'Health policy, vulnerability, and vulnerable populations', 312.

100 Levine, *Ethics and regulation of clinical research,* 61–66.

101 CIOMS, *International Guidelines for Ethical Review of Epidemiological Studies,* Article 43.

102 Beauchamp, 'The Belmont Report', 152–53.

103 CIOMS, *International Ethical Guidelines for Biomedical Research Involving Human Subjects,* 1993; see previous Chapter 3.

104 Kottow, 'The vulnerable and the susceptible', 463; Macklin, 'Bioethics, vulnerability, and protection', 473 ff; Morawa, 'Vulnerability as a concept in international human rights law', 147 ff; Hurst, 'Vulnerability in Research and Health Care; Describing the Elephant in the Room'? 195 ff; Luna, 'Elucidating the concept of vulnerability: Layers not labels', 133 ff; Solbakk, 'Vulnerability: A futile or useful principle in healthcare ethics?' 234; Solomon, 'Protecting and respecting the vulnerable: existing regulations or further protections?' 18 ff.

105 CIOMS, *International Ethical Guidelines for Biomedical Research Involving Human Subjects,* 2002, Commentary on Guideline 13.

106 CIOMS, *International Ethical Guidelines for Biomedical Research Involving Human Subjects,* 2002, Commentary on Guideline 13.

107 See WMA, *Declaration of Helsinki,* 2008, provision 17; WMA, *Declaration of Helsinki,* 2013, provision 20 (see Chapter 3 note 66).

108 IBC, *Report of the IBC on the principle of respect for human vulnerability and personal integrity,* 1-2.

109 Brody, *The ethics of biomedical research. An international perspective,* 49–50; Kopelman, 'Research policy: II. Risk and vulnerable groups', 2365-66.

110 Coleman, 'Vulnerability as a Regulatory Category in Human Subject Research', 14.

111 Levine *et al.*, 'The limitations of "vulnerability" as a protection for human research participants', 46.

112 Leavitt, 'Is any medical research population not vulnerable?', 87.

113 Schroeder and Gefenas, 'Vulnerability: Too vague and too broad?', 119.

114 Andanda, 'Vulnerability: Sex workers in Nairobi's Majengo slum', 143–44.

115 DeBruin, 'Reflections on "vulnerability"', 7; Luna, 'Elucidating the concept of vulnerability: layers not labels', 134; National Bioethics Advisory Commission, *Ethical and policy issues in research involving human participants,* 91.

116 Levine *et al.*, 'The limitations of "vulnerability" as a protection for human research participants', 48; Hurst, 'Vulnerability in Research and Health Care; Describing the Elephant in the Room'? 197-98. See also Levine *et al.* '"Special scrutiny": A targeted form of research protocol review'.

117 Henderson, Davis, and King, 'Vulnerability to influence: A two-way street', 50; see also Chapter 1.

118 This point is argued by DeMarco ('Vulnerability: A needed moral safeguard', 83).

119 These objections are raised by Campbell ('"Vulnerability" in context: Recognizing the sociopolitical influences', 59) and Silvers ('Historical vulnerability and special scrutiny: Precautions against discrimination in medical research', 57). Eckenwiler makes the point that vulnerable groups are 'social groups for whom self-development, self-determination, and equality are threatened independent of research' (see Eckenwiler, 'Moral reasoning and the review of research involving human subjects', 51).

120 See Solomon, 'Protecting and respecting the vulnerable: existing regulations or further protections?'.

121 Like the example of research subjects in India, discussed in Chapter 1.

122 In her text 'special protections' transform from 'additional regulatory protections' into 'additional regulations' (Solomon, 'Protecting and respecting the vulnerable: existing regulations or further protections?' 18, 20).

123 Solomon also overestimates the proper functioning of existing review mechanisms. Especially in developing countries research review processes are often deficient. But also in the U.S. the review system is affected by weaknesses as the recent study of the Government Accountability Office has shown. See: GAO, *Human subjects research;* Hyder *et al.*, 'Ethical review of health research: A perspective from developing country researchers'; London, 'Ethical oversight of public health research: Can rules and IRBs make a difference in developing countries?'; Petryna, *When experiments travel. Clinical trials and the global search for human subjects.* Jill Fisher for example concludes from her study of the clinical trials industry in the United States that the system of clinical development and federal regulation 'is set up in such a way that it preys upon vulnerable populations'. (see Fisher, A. *Medical research for hire. The political economy of pharmaceutical clinical trials,* 214).

124 Erlen, 'Who speaks for the vulnerable?', 134. See also Baldwin, 'Patient advocacy: a concept analysis'.

125 De Chesnay, Wharton, and Pamp, 'Cultural competence, resilience, and advocacy', 31 ff.

126 Shi and Stevens, *Vulnerable populations in the United States,* 2–3.
127 Flaskerud and Winslow, 'Vulnerable populations and ultimate responsibility', 299.
128 Mechanic and Tanner, 'Vulnerable people, groups, and populations: Societal view', 1221.
129 Brock, 'Health resource allocation for vulnerable populations', 290 ff.
130 Morawa, 'Vulnerability as a concept in international human rights law', 150.
131 Ells, 'Respect for people in situations of vulnerability: A new principle for health-care professionals and health-care organizations', 182.
132 Macfarlane, Racelis, and Muli-Musiime. 'Public health in developing countries', 841 ff.
133 London, 'Ethical oversight of public health research: Can rules and IRBs make a difference in developing countries?', 1080.

5

WE ARE ALL VULNERABLE

Philosophical perspectives on vulnerability

Bioethical analyses often interpret vulnerability within the context of the dominant liberal tradition of individual liberty and respect for personal autonomy. Vulnerability means diminished autonomy, loss of control, lack of self-determination. Efforts to widen the perspective of vulnerability frequently return to the overarching notion of autonomy. The 1991 CIOMS Guidelines suggest that vulnerability is a fundamental principle; but at the same time they subordinate this principle under the principle of respect for persons. The 1993 CIOMS Guidelines go beyond the traditional view of vulnerability as inability to consent but the emphasis on the ability to protect one's interests continues to presuppose the central role of individual agency. The problem with this interpretation of vulnerability is that it has difficulties accounting for the various dimensions of the notion. More importantly, it does not seriously consider the idea that all human beings share vulnerability whether or not they are autonomous persons. Philosophical perspectives emphasize that vulnerability is a general characteristic of being human. Such perspectives, however, are often not considered to be relevant for the practical discourse of bioethics.

This chapter will examine philosophical perspectives on vulnerability. It will argue that it is necessary first to understand vulnerability before the notion can be applied. Prior to making the notion operational in bioethical discourse as a deficit of individual autonomy, we need to know what it means that people are considered as vulnerable. Only then can we discuss how and when the notion can be applied. Analyzing the general characteristics of vulnerability demands not only a broad view, taking into account the various dimensions of the notion, but also a more fundamental conceptual exploration, leaving aside at least for the moment the question of its practical implications. The chapter begins with arguing that conceptual approaches to clarify vulnerability in connection to circumstances, sources, and layers are primarily driven by pragmatic concerns to apply the concept in research and health care. Concentrating on how human beings are *made* vulnerable, these approaches do not take into account that human beings *are* vulnerable. To understand vulnerability as a general feature of being human, it will be helpful to examine philosophical perspectives that are outside of mainstream bioethics. The chapter then proceeds with critically analyzing a broad range of such perspectives. They all include the basic concern that vulnerability is a defining characteristic of what it means to be human. The justifications for this fundamental status of vulnerability are clearly different; they can be naturalistic, arguing that human nature is essentially imperfect or that the human body is corruptible; they can be non-naturalistic, arguing that vulnerability is entailed in human existence and experience, or in the relatedness of human beings as social beings, in the interdependence of humans, or in the unavoidable encounter with the other person. The next section of the chapter concludes from the analysis of philosophical perspectives that the term 'ontological vulnerability', often used in the scholarly literature and also in the previous chapter, is not appropriate to express the idea that vulnerability is a defining characteristic of the human species. Ontology suggests that there is an essential reality prior to ethical discourse. The same ontological presupposition governs mainstream bioethical discourse: the autonomous subject is prior to its situatedness in particular circumstances; the purpose of bioethical discourse is then to safeguard that the subject is using its power to relate to other individuals and to surrounding circumstances. This section argues that the philosophical perspectives on vulnerability demonstrate that ontology and ethics cannot be separated. These perspectives do not defend a vulnerable subject in opposition to an autonomous subject but argue that vulnerability is constitutive for individual autonomy. For these reasons, the section concludes that it is preferable to use the term 'anthropological' rather than 'ontological' vulnerability. Being vulnerable is not a status prior to ethical debate, but is characterized by moral challenges and possibilities that inspire action. It is not a fact that can be described, accepted, or ignored; but it is an anthropological quality that allows human beings to become human, take care of each other, and to develop into autonomous, responsible subjects. The subsequent section of the chapter elaborates three implications of the notion of anthropological vulnerability: universality, passivity, and positivity. First of all, vulnerability is not an individual feature but a generalized condition of the human species. Second, the human condition is characterized by passivity and susceptibility rather than agency. Third, vulnerability is a positive qualification; rather than referring to weakness and lack of power, it signifies openness and potentiality. The next section discusses the normativity of vulnerability. How does vulnerability as a generalized condition of the human species give rise to ethical responsibilities? Is it an ethical principle itself or is the principle that human vulnerability should be respected? This section argues that since anthropological vulnerability does not prioritize ontology we are immediately immersed in the realm of ethics. Vulnerability imposes responsibilities. This is not a matter of choice. However, ethical responses to vulnerability may differ. They can range from harmful to beneficial. Even if the same condition of vulnerability is shared, the possibilities of harm will vary. The inequal allocation of vulnerability is connecting the notion of anthropological vulnerability with political perspectives of vulnerability. This is the focus of the final section of this chapter.

Pragmatic approaches of vulnerability

In recent years several proposals have been made to clarify the notion of vulnerability and its implications. For example, Kenneth Kipnis has proposed an analytic framework distinguishing types of vulnerability and circumstances creating them [1]. However, understanding vulnerability is still modeled on the paradigmatic subject able to make autonomous choices. Particular (economic, institutional, or social) circumstances make individuals vulnerable because they invalidate consent or compromise the ability to consent. Normally, individuals protect themselves but sometimes they cannot. This is often not the result of failures of their own but due to limiting and restraining circumstances. If these limits and restraints can be removed then the autonomous individual will again have its decision-making capacities restored. This approach has the practical advantage that the determination of who is in need of special protection can be more refined. It also does not assume that vulnerability refers to some general characteristic of human beings. The same pragmatic interest is shared by other approaches. In an effort to create more conceptual distance from the focus on autonomous choice, it has been argued that vulnerability should be explained in relation to various sources, for example competence, voluntariness, and fairness [2]. Asking why persons and populations are vulnerable will produce a broader model of vulnerability that takes into account the principle of autonomy as well as justice. However, identifying sources of vulnerability concentrates almost exclusively on special vulnerability. The dynamic approach of layers of vulnerability, proposed by Florencia Luna, has a similar restrictive objective [3]. It argues that vulnerability depends on situations and circumstances. For example, a woman can become vulnerable in a situation of lack of respect for her rights but also in situations of poverty or illiteracy. Vulnerability can be produced in multiple layers at the same time and detailed analysis should determine how different layers can be avoided or minimized. But the point is that vulnerability is not a fixed or permanent condition of an individual person.

Proposing taxonomies of vulnerability and explaining the concept in connection to sources and layers is driven by the practical interest to apply the notion in bioethical areas. The focus of the analysis is on identifying possible actions, primarily protective measures to remediate the vulnerability of subjects. The more specific the indications for protection and the more targeted the protective interventions, if needed, the better. Exposure, as one of the three components of the functional definition of vulnerability, is therefore the most important determinant. The idea that vulnerability is a shared condition of all human beings (underlying for example a common sensitivity) is not very helpful in such pragmatic approaches. Since bioethics is first of all regarded as applied ethics, the philosophical perspective on vulnerability is often either not deemed relevant at all, or interpreted from the standpoint of the political perspective. How the notion can be understood depends on how it can be applied. The assumption is that we can only comprehend what vulnerability is if we can explain what are its implications and possible uses. What it is, especially as an ethical notion, is always determined by what it directs us to do. Without cash value the concept is pointless. A major concern of the above conceptual approaches is to respond to the criticism of vulnerability that if it cannot be specified what actions or obligations are demanded by the ethical concept of vulnerability or even the principle of respect for human vulnerability, the notion and principle are without any teeth or force. This pragmatic interest is often combined with the specific bioethical discourse that emphasizes the principle of respect for autonomy as one of the basic ethical principles (and maybe the most important one), and that therefore primarily underlines the significance of individual decision-making and informed consent. Interpreting vulnerability as a failure or challenge in the application of this basic principle will to a certain extent assign it some usefulness.

In order to understand vulnerability as a relatively new concept in bioethics, it is necessary to take some theoretical distance from mainstream bioethical discourse as well as immediate practical concerns. If this notion cannot easily be incorporated in the current models and theories of bioethics, the reason might be that it needs another framework for bioethical discourse. If it is not immediately clear what this notion can do or deliver, it should not be concluded that it has no precise meaning before a philosophical perspective to clarify its meaning has been developed. What then is exactly meant when philosophical perspectives of vulnerability consider it as a general characteristic of being human?

Understanding vulnerability: A peripheral approach

To answer this question we have to examine a variety of 'peripheral' approaches that are inspired by Continental European philosophy, feminist ethics, philosophy of medicine, and non-Western bioethics. These approaches are 'peripheral' in comparison to mainstream bioethics that is dominated by analytic philosophy. These 'peripheral' approaches might be particularly useful since the notion of vulnerability has emerged precisely in peripheral countries due to processes of globalization [4].

In the philosophical perspective, vulnerability is often understood as one, perhaps even the most basic characteristic of the human condition. This is what is commonly called 'ontological vulnerability'. It is aptly described by Sam Gorovitz in the 1994 CIOMS conference in Mexico: 'The ultimate vulnerability is not that which we each have as fragile humans, but that in which we all share – that is, the vulnerability of our species…' [5]. Fundamental vulnerability is part and parcel of what it means to be human. It is, in the words of Jacob Rendtorff, 'a universal expression of the human condition' [6]. At this basic level, a subject is vulnerable since the human condition necessarily implies vulnerability [7]. Rather than a negative or positive qualification of individual persons, ontological vulnerability is a qualification of the human species. The idea that vulnerability for human beings represents what it is to be human is introduced and elaborated from various points of view. These will be examined in the next section.

Variations on the same theme

Arguments that vulnerability is a defining characteristic of humanity have a long tradition in philosophy and ethics. It is not just the *prima facie* idea that this notion refers to the possibility of damage, harms, and threats. Rather, the argument is that the notion reflects the essential nature of human beings as frail and weak. This basic theme is affirmed in different ways since the arguments can be justified on various grounds.

Vulnerability as natural imperfection

Reflection on the nature of human beings is typical for the tradition of philosophical anthropology. The human being is vulnerable, according to scholars like Scheler and Gehlen, because he is fundamentally 'undetermined'. He is, in the famous words of Arnold Gehlen, a deficient being ('Mängelwesen') [8]. Following the idea of the eighteenth century German philosopher Johann Gottfried Herder, it is stressed that the human species is biologically not well developed in the sense that humans are rather defenseless against threats in comparison to most animals. They are primarily characterized by 'lack of adaptation, lack of specialization, primitive states, and a failure to develop' [9]. They have no natural protection, like other animals, against violent weather, nor are they biologically equipped for defense or attack. They do not have true instincts that can protect them against imminent danger. And they also need a very long time of protection and care from birth until adulthood. The question is how such a weak and vulnerable species is able to survive. As Gehlen remarks: '…under natural conditions, among dangerous predators, man would have long ago died out' [10]. To answer this query Gehlen is using the idea of Max Scheler that this essential vulnerability is compensated because human beings are 'world-open'. The deficient constitution that exposes the human person to threats is at the same time the productive and positive dimension of human nature. Precisely because humans, contrary to other living beings, are not tied or adapted to a specific environment, they can focus their energies on what is not present in time and space. 'World-openness' means that human beings are not enclosed in their environing world but can transcend it and make it the object of acting and thinking, thus modifying and recreating the surrounding environment and nature into their world. Humans can therefore transform nature and can construct a 'second nature' or culture within which they can survive [11].

The constitutional defects of human nature are compensated for through intelligence, language, and abstract reasoning. For Gehlen, deficiencies (such as lack of adaptation, specialization, and instinct) are neutralized through cultural and social institutions such as law, family, ethics, and state. These institutions are involved in 'mediating' human agency; they 'relieve' the human person of the deficiencies that would otherwise hamper his growth and flourishing. Although vulnerability is initially seen as imperfection, it is ultimately positive for human beings as the basis for freedom, plasticity, and capacity for development.

Vulnerability as bodily corruption

The above tradition of philosophical anthropology relates human vulnerability primarily to the biological nature of human beings. It builds on older views that regard the human body as a source of decay and corruption, as the seat of passion and lack of reason. Rabbinic literature, for example, addresses vulnerabilities built into our bodies [12]. For the apostle Paul human beings are earthen vessels; they are made from the dust of the earth and therefore perishable. Humans are vulnerable not because nature has provided them with a deficient biology but because they are creatures of flesh and spirit, body and mind [13]. The duality of the human being implies that the body is regarded as 'a force of negativity' [14]. Since the body is in the realm of matter it is necessarily determined by quantity, extension, and changeability; it is subjected to time and space, and thus finite and mortal. The physicality of the body explains why humans are vulnerable to diseases, ageing, disabilities, death, and injuries. The body therefore should be regarded as a threat; it is associated, as Leder points out, with dysfunction and otherness [15]. As a complicated mechanism the body can have some imperfections but more importantly, because of its vulnerable constitution it can be damaged or broken due to inborn errors, wear and tear, or traumas. Like all machinery it has a limited lifespan.

This separation of the human subject in two components has persisted for a long time, particularly in the context of medicine [16]. The Cartesian dualism positioning the body as a mechanical entity aims at facilitating and promoting scientific medicine that could prolong life and eliminate functional disorders [17]. At the same time it safeguards the true self of human beings. The self is not threatened since it is the body that represents the vulnerable component of humans. The biomedical model has been criticized for considering the patient as a mechanical body rather than an embodied subject. The emergence of bioethics has been associated with criticism of this biomedical model, presenting an alternative view of the human being as an incorporated subject, and emphasizing that the patient is a person. However, the increasing influence of the moral principle of respect for individual autonomy has reiterated the dualistic view of the human body, though in a different manner. Empowering the person seems to be accomplished through conceptualizing the body as individual property. The individual person is regarded as an autonomous subject who is in control of his or her body. Since the body is the private property of this person, it is at his or her disposal. Emphasizing that decision-making authority in healthcare is connected with the autonomous individual is usually associated with property language of the human body: the individual person owning his body is the only one who can decide what will happen to it [18]. In this view, vulnerability is located at the level of the autonomous subject who should normally have mastery over the body that is his or her property; the subject is then vulnerable if his or her autonomy is threatened or compromised. But in daily experience, the body cannot be forgotten. In a more fundamental way, vulnerability is located at the level of the body itself since it is susceptible to decline, decay, and damage even if the owner and master is fully autonomous.

Vulnerability as experience

The view that the human subject is not composed of separate parts but an integrated unity of body and mind characterizes various attempts to overcome the Cartesian model of embodiment. Especially Merleau-Ponty's theory of the human body has been influential [19]. Human experience teaches us that the body is different from physical entities. A distinction should be made between the objective, physical body and the lived body. Focusing on the experience of embodiment shows that there is always an ambiguity between two dimensions, namely the objective and subjective, the biological and the existential. Though medicine is usually interested in the body as object of examination, diagnosis, and intervention, the human person can only experience him or herself as a unity. It is impossible to dissociate myself from my body. Neither do I have a relation of ownership with my body. I have my body and at the same time I am my body. Similar ideas are expressed in the European tradition of anthropological medicine. German physician Herbert Plügge, for example, argues that the body mediates between my world and I. My body is not an instrument by means of which I perceive, act, and exist, but I exist through my body [20]. The essential characteristic of the human being therefore is 'the unity of a mental and bodily subjectivity' [21]. If medicine wants to be a science of the human person it needs to respect the integrity of human reality, thus presupposing an integrated vision of the human person.

Vulnerability in this perspective is not located in the bodily machinery but at the level of personal existence. It does not just refer to possible biological disruptions but to potential existential injuries. Threats do not jeopardize my body, but they happen to me as an embodied person [22]. The lived body is the locus of vulnerability. Having a body makes us vulnerable but the vulnerability is experienced as my vulnerability because I also am my body. Since being a body is my entrance to the world, personal existence is always susceptible to possible threats. When the body is actually harmed, our fundamental vulnerability is not only demonstrated in disordered and failing body machinery but first of all in the finding that our world is transformed [23]. Plügge explains that normally, the mediating role of the body is not noticed [24]. When we are healthy and active, we are not focused on our body but we are 'yonder', at the side of things and events in the world, with the other. But when we are ill or suffering the body will present itself, and we will be aware of our embodiment. The openness of the body, and thus me, towards the world will disappear [25]. Vulnerability is the experience that our world is finite and fragile. The individual is not an autonomous, demarcated entity able to protect itself against the outside world, but it is fundamentally susceptible to threats since it is an embodied subject. Vulnerability therefore is an existential experience of all human beings. It is precisely this idea of vulnerability as experience that has been prominent in the nursing literature, as discussed in Chapter 2. Here it is regarded as a universal experience since the human condition challenges personal integrity. It is therefore a shared experience for both nurses and clients [26].

Vulnerability as relatedness

The argument that the human person is vulnerable due to its biology or its body has evolved into the broader perspective that all human beings experience vulnerability. But this wider view can itself be expanded. Humans have as incarnated beings a dynamic relation with the world. When the human body is experienced as a 'lived body' it necessarily implies openness towards the world. It constitutes the world as experienced. It also is the medium for interacting with the world [27]. Through the body, human beings are always in relationships. The lived body is, as Leder has argued 'a being in relationship to that which is other: other people, other things, an environment' [28]. Being therefore is always being-in-the-world. This is what Plügge has in mind with his statement that the human being normally is 'out there', with the things in the world, not focused on himself or his physical body [29]. The effect of this expanded argument is that the focus of shared vulnerability moves from experience to relatedness. What is characteristic of human beings is that they are always connected and situated. They do not merely experience the world but are necessarily within the world. This predicament inevitably implies vulnerability. It is the consequence of the message of Jean-Paul Sartre: 'to exist and to be situated are one and the same' [30].

That the human person is essentially a social being is also an important motive in the modern revival of the philosophy of medicine. Here it is argued that medicine should leave behind not only the Cartesian dualism of mind and body, and the notion of purely physical body but also the supposition that the individual is independent from his context [31]. The same point is made in communitarian criticism of the Western liberal discourse. We are not 'unencumbered selves' [32].

We cannot conceive ourselves as separate, independent selves, but we are situated in the world, connected to other people, and defined by the community we inhabit. The interconnectedness of human life is finally strongly articulated in non-Western philosophies. For example, in African philosophy the individual is embedded in relationships and defined by the community [33].

Vulnerability in this perspective is not a biological or existential deficit but a positive phenomenon. The human body is the basis for exchange and reciprocity between human beings. It guarantees openness towards the world. Our world necessarily includes others. We cannot come into being, flourish, and survive if our existence is not connected to the existence of others. But exactly this essential relatedness and sociality makes persons vulnerable. Interacting and communicating with our world offers possibilities for change but it may also include negative effects. Vulnerability is located in the encounter with others. Interrelating with other persons is uncertain and ambiguous. We are given over to others. If we are open to the world we may at the same time be negatively affected; there always is the possibility to be hurt. The point is that vulnerability is not within the individual alone; it is in the relation between the self and alterity that vulnerability emerges. Vulnerability therefore is characterized by mutuality. It is a common human condition because of the relationality of embodied life. Human beings are, as Judith Butler puts it, 'from the start and by virtue of being a bodily being, already given over, beyond ourselves, implicated in lives that are not our own' [34].

Vulnerability as dependency

The view that human beings are necessarily embedded in relationships has been modified into another perspective on vulnerability, especially in feminist ethics and care ethics. The emphasis here is on fundamental dependency. Human beings are relational but it is important to acknowledge that relations are not only between equals. Feminist legal scholar Martha Fineman has emphasized that vulnerability is 'universal and constant, inherent in the human condition' [35]. Human existence is characterized by 'a fragile materiality that renders all of us constantly susceptible to destructive external forces and internal disintegration' [36]. Vulnerability therefore defines what it is to be human. Though universal and unavoidable, vulnerability will not be the same for every human being. Fineman points out that it is a complex human feature that can be manifested in multiple forms. It is also at the same time particular: every individual is 'positioned differently' [37]. Our embodiment differs as well in the ways in which we are situated within economic and institutional networks. The result is that human beings are not vulnerable in the same degree. For Fineman vulnerability inevitably arises from embodiment, and through the body our existence is marked by dependency. This is not merely a developmental stage that we leave behind as we grow into an autonomous subject, but it defines us because we are continuously dependent on relations to other people. Human beings need each other. However, dependencies can differ; they do not refer to the same set of relationships and circumstances. Fineman argues that inevitable dependency, although universal, is episodic and sporadic, 'shifting in degree over the lifetime of an individual' in contrast to vulnerability that is constant [38]. The main point is that recognizing the inevitability of dependency will lead to a redefinition of responsibilities [39]. This point shows that the autonomous and independent subject is a fiction used in the liberal tradition. With this fiction society and social policies can perpetuate inequalities and unfair societal arrangements. On the other hand, the focus on the vulnerable subject compels us to rethink social and individual obligations to provide assistance. Vulnerability not only reflects relations between human beings but also between individuals and states [40].

The thesis that human life is precarious since it inevitably depends on anonymous others is central in the recent work of feminist philosopher Judith Butler. We cannot escape the fact that as humans we are necessarily related: we are always exposed to each other. This relatedness is not under our control [41]. Whether Butler's view is 'ontological' and in what sense, is the subject of vigorous debate [42]. It is argued that hers is a 'philosophy of radical finitude' rather than a relational ontology [43]. Human beings are exposed to 'a transcendence that forces us to relate to the other' [44]. Although humans are vulnerable because they are incarnated beings, their 'bodiliness' is not interpreted as vulnerable as a biological qualification but as an experiential category; it is 'the experience of exposure to one another and of the ultimate impossibility of self-control' [45]. Butler herself simply concludes that we are 'given over to the other' [46]. She speaks about a common human vulnerability but for her that does not imply a common notion of the human. This vulnerability emerges with life itself; it precedes the formation of the self. Vulnerability is more like a condition for the development of the individual. As exposure to social conditions that are constitutive for us as singular individuals, it should be evaluated in a positive manner. The problem with vulnerability is that it is not equally distributed because the interdependency is not symmetrical. This means that the shared vulnerability that is typical for humans is not merely mutual. Even if we are all vulnerable, one person might be more vulnerable to me than I am to her. The body is not merely open to other people but submitted to power differences. Because of such inequalities there is the possibility of violence all the time. The awareness of vulnerability arises in a context of violence and power in which people fear to be abused and exploited.

For both Fineman and Butler, dependency is not a negative condition. Rather than regarding it as a limitation, it should be considered as a resource. Dependency is, in the words of Eva Kittay 'an aspect of what it is to be the sorts of beings we are...' [47]. All human bodies can be subjected to injury, pain, and violence. If being human means being vulnerable, then the body necessarily implies vulnerability, but at the same time it can be the site of humanity. This is also the positive outcome of the argument of Alasdair MacIntyre that human vulnerability shows how dependence, rationality, and animality are connected. Reasserting and acknowledging our animal condition, dependencies, and our vulnerabilities will help to (re)discover the distinctive virtues of dependent rational animals that we need for human flourishing and for rational agency [48]. For Butler, the emphasis on vulnerability serves as a platform to develop an ethics of non-violence, a new humanism [49]. What will be first of all necessary is recognition of dependency on others. In ethical and political discourse the idea of the sovereign individual should be replaced by the view of the vulnerable subject. For Fineman, vulnerability points to the need to restructure social institutions and to challenge social inequalities. The significance of vulnerability is that it requires us to examine how power, social goods, and wealth are distributed, and to go beyond a mere focus on characteristics of individuals and groups [50].

Vulnerability as being hostage

The most radical view on vulnerability is communicated by the French philosopher Emmanuel Levinas. European bioethicists commonly refer to Levinas as one of the first to emphasize the ethical significance of vulnerability in the 1960s [51]. His work is often used to argue that vulnerability is ontologically prior to other ethical principles and that it is a strong incentive to develop an ethics of care [52]. Rendtorff for example argues that although the principle of autonomy is most frequently mentioned in bioethical debates, it is in fact vulnerability that is more fundamental as a principle [53]. We cannot understand the human condition nor morality itself without the concept of vulnerability. This is not simply a reflection of European perspectives (presenting merely a different descriptive context), but the outcome of a philosophical analysis of the moral concepts of autonomy and vulnerability.

For Levinas, human beings are fundamentally connected to each other. But this should not be understood in a theoretical or abstract way, assuming that subjectivity or the individual self comes first, and that it then decides to relate and interact with other selves. Human beings are not self-conscious, rational, autonomous entities who can relate to each other and the world. The relatedness of humans is for Levinas not a matter of knowing, thinking, reflecting, or deciding. Although philosophy has continuously tried to create a synthesis, a totality in which one can understand oneself and the world rationally, it is in his view not possible to objectify the outside world and individualize the subject. The experience of relationship is not the result of synthetic thought but given in the face-to-face of humans. The ego and the other are facing each other: 'The true union or true togetherness is not a togetherness of synthesis, but a togetherness of face to face' [54].

In Levinas' thought, relatedness is ethical. The sociality of humans has a moral significance before we are able to know, articulate, and explain it. Through the concrete encounter with the face of another we are immediately in an ethical condition, or rather, as Levinas calls it, a 'non-condition'. Ethics is preliminary, antecedent to ontology. It is 'first philosophy' [55].

The face is the expression of the transcendence of the other; it disturbs the self. Before one can describe or analyze it (specifying for example the color of the eyes or the hair) it commands the self, imposing an unconditional ethical imperative [56]. The face is always unique and individual but it is also abstract and naked [57]. The face shows that humans are embodied. Here the notion of vulnerability comes into play. The face is the epiphany of the other; it shows the otherness of the other. This 'alterity' cannot be reduced to the countenance or appearance of the other person. But the vulnerability of the face invites to reduce the other to his countenance, his social role, or his illness, making the other into an object that I can try to control and dominate for my own interests and projects. This is why Levinas argues that the encounter with the other is founded in violence. Meeting another person can result in aggression and domination. But the human face is naked, weak, defenseless, and thus the expression of human vulnerability. The nakedness of the face is destitution, desolation, poverty. The face provokes a temptation to kill and at the same time the interdiction to kill [58]. Its vulnerability is the reason why the human face appeals and commands, why it obliges me and makes me responsible. This is why the encounter with the other is demanding. The revelation of the human face imposes on me the obligation not to commit violence, not to kill the vulnerable other in front of me, not to reduce him to his countenance. Most of all, it imposes the responsibility to take care of the other person, to recognize and respect him in his otherness. Vulnerability upsets 'the very egoism of the Ego', it allows me to free myself from my self-centered preoccupations [59]. In the relation and encounter with the other I learn that vulnerability opens up humanity. It makes me responsive to the suffering of the other person (for example, the widow, the orphan, the poor, and the stranger). As Levinas puts it: 'No one can stay in himself: the humanity of man, subjectivity, is a responsibility for others, and extreme vulnerability' [60]. The ability to recognize the moral priority of the other person is exactly what makes up our humanity. Vulnerability evokes the power of the powerless.

The ethical demand revealed by the vulnerability of the other person constitutes the self. In the encounter with the other, I myself am exposed and vulnerable. Recognizing the other person and responding to his needs defines me as a subject. The self is created by the relationship with other persons. However, the proximity to the other is outside ontological categories [61]. The ethical relation to the other is therefore prior to ontology, i.e. before I have being as an individual and before the things surrounding us as our world. Levinas explains: 'I am defined as a subjectivity, as a singular person, as an "I", precisely because I am exposed to the other. It is my inescapable and incontrovertible answerability to the other that makes me an individual. So that I become a responsible or ethical "I" to the extent that I agree to depose or dethrone myself – to abdicate my position of centrality – in favor of the vulnerable other' [62]. Levinas formulates here a powerful implication of vulnerability: it immediately brings us into an ethical relation. Ethics here is 'disruption of our being-in-the-world which opens us to the other' [63]. Being-for-the-other is more important than being for-itself.

For Levinas, the moral priority of the other person is not a choice. The ethical relation with the vulnerable other is not symmetrical since I am for-the-other before I am myself. The face of the other constitutes the responsible self. This means that I am responsible whether I want it or not; it is not based on agreement, free engagement, or consent. This responsibility for the other cannot be cancelled; this would be, according to Levinas an 'impossibility more impossible than jumping out of one's skin' [64]. The other appears or reveals himself as face, and the face simply enters my world prior to consciousness and before the autonomy of subjective freedom emerges [65]. The proximity of the face ('visage' in French) is a 'visitation'. It is addressing me and it makes me unique. I am able to respond to the needs of this vulnerable, defenseless stranger, and this makes me responsible. I cannot be indifferent to the other [66]. This is why Levinas explains vulnerability as being hostage. As a subject I am the 'irreplaceable *hostage* of others' [67].

Anthropological rather than ontological vulnerability

The various philosophical perspectives show that vulnerability can be explained in multiple ways. Human beings are vulnerable because they are biologically not sufficiently equipped to defend themselves: as autonomous persons they have a body that is necessarily finite and perishable; as embodied persons they experience that personal existence is fragile; as social beings they are always related to the other and thus susceptible to harm; as fundamentally dependent beings they are subjected to inequalities and power differences; due to the proximity of the other they are ethically charged by vulnerability as expressed in the face of the other person. However, all these explanations share the same basic argument: vulnerability is a characteristic of the human species that is manifested in every individual. It is a primary, universal trait that signifies what it means to be human, although the foundations for this trait in the human species can differ. Vulnerability demarcates the boundaries with other living organisms but also with what should be considered as inhuman. Organisms that are invulnerable, like the steel dolls mentioned by Anthony, cannot be human beings [68]. It is this primary vulnerability that is impossible to ignore [69]. This is what makes vulnerability a fundamental notion in ethics. In Levinas' view it is not something to contemplate; in the frailty of others we discover ethics and we are obliged to respond. For Fineman, recognizing dependency on others is opening up new forms of sociality and political community. For Butler, vulnerability is a precondition of humanization, but it should be perceived and recognized in order to have an ethical impact. In this sense, vulnerability can become a platform for humanism. It immediately brings us into the field of ethics and confronts us with the challenge to what it is to be human.

The ethical significance of vulnerability posits at the same time a challenge to the conception of vulnerability. If it is conceived as a basic human condition how can it be argued that vulnerability is a true ethical principle; how can we move from a descriptive approach to a normative one? Talking about 'ontological vulnerability' immediately suggests that this is how human beings essentially are, prior to ethics or social and political situatedness. Grounding ethical obligations in the ontological fact of vulnerability will then be disqualified as 'naturalistic fallacy'. Being vulnerable means describing how human beings are. One can agree or disagree with the factual description but the universal trait of vulnerability does not directly produce normative conclusions. This problem is particularly evident with the naturalistic arguments that associate vulnerability with the imperfections of human nature or the human body. It is less evident, however, with the non-naturalistic arguments. Vulnerability explained as experience, dependency, or relatedness is not based on an ontologically prior condition. On the contrary, it produces the predicament that is characteristic for the human species, and it allows the emergence of the autonomous individual that is necessarily a vulnerable subject. For scholars as Butler, Fineman, and Levinas the relationship between ontology and ethics is more nuanced. Emphasizing the vulnerable subject, they introduce a new understanding of the subject, not as a sovereign and independent power that has being, that exists before it relates to other individuals. Rather they argue that the subject in fact comes into existence through the encounter with the other. It becomes a subject through the confrontation with the vulnerable other person. Vulnerability therefore constitutes subjectivity. This philosophical perspective does not simply replace the autonomous subject by the vulnerable subject as another ontological framework for ethics. On the contrary, it assumes that it is impossible to disentangle ontology and ethics.

Judith Butler uses the expression 'common human vulnerability' and explains that this is 'a condition of being laid bare from the start and with which we cannot argue' [70]. She also underlines that it does not refer to a common notion of the human. Shared vulnerability is not common humanity. Similar tension exists when she argues that corporeality means vulnerability to powers that the subject cannot control; as embodied subjects we *are* vulnerable; therefore we need a 'new bodily ontology' in order to redirect social and political policies [71]. But at the same time, 'ontology' for Butler is not a description of fundamental structures of being. We can only talk of 'being' within the context of social and political interpretation. Being a body means being vulnerable but this cannot be intelligible outside a social context. Humans come into being because they are given over to others. For Butler, vulnerability is a generalized human condition but also a precondition of humanization; it has ethical consequences when it is recognized. In this sense it is not merely the fact that being vulnerable is a common attribute of being human that generates ethical obligations. Rather, it is the recognition of shared vulnerability, the apprehension that we are from the start given over to each other, and thus exposed to violence, that makes us human. Humanity is not a state of being but instead of becoming. The social relationality of humans produces ethical responsibility [72]. This view reframes vulnerability: it is not an intrinsic quality of a subject but 'an inalienable condition of becoming' [73]. The terminology of ontological vulnerability is therefore no longer adequate. The concept of vulnerability refers to a generalized human condition that calls into question the autonomous subject and the ontology of liberalism without itself proposing an alternative ontology.

The inevitability of dependency is for Martha Fineman the defining feature of the human condition. This explains why human beings are vulnerable and in need of care. This universal trait of the human condition generates the responsibility to provide support. Again, the normative implication seems to be based on the ontological fact. Human beings are vulnerable because they are fundamentally dependent; thus care should be provided. For Fineman it is the state that has the moral responsibility to support dependency and provide care. While this particular claim is disputed, it is not uncommon to derive moral obligations from the shared condition of dependency [74]. The argument connecting vulnerability and responsibility is that the shared condition of vulnerability directly generates moral obligations for other people if they have the capacity to help the vulnerable subject. These obligations are not the result of voluntary choice or commitment but arise from vulnerability [75]. Robert Goodin uses the notion of vulnerability to question the common idea that moral obligations can only be self-assumed. Members of a family, for example, are dependent on each other. This membership is neither a voluntary choice nor a contract but the resulting vulnerability is nonetheless producing responsibilities and reciprocal duties between spouses, children, and parents [76]. However, for Goodin not all dependency is problematic. This basic human condition is only morally problematic if it is asymmetrical and if basic needs are at stake. When dependency involves unequal power relations, there is a risk for abuse and exploitation. Thus it seems that it is not the fact of dependency that generates ethical implications but the inequality of dependency. Prior to the ontological condition, there is an ethical assessment of the others' condition that leads to responsibilities. A similar reversal of the relation between ontology and ethics seems to underlie the theory of Fineman. For her, vulnerability is a universal and at the same time a particular condition. While everybody is vulnerable, some people are more vulnerable due to social inequities and limited access to resources that render them more or less dependent. Why then is there a moral obligation to provide care to the vulnerable subject? Fineman seems to assume that everybody is entitled to some level of human flourishing, and that it is necessary to ensure equality of opportunity and equitable access to institutional assets so that everyone will be able to realize individual capabilities as fully as possible [77]. She uses the notion of vulnerability to argue that it is not sufficient to provide equal protection to vulnerable people because the social conditions are not equal. The degree of dependency implicated in vulnerability can vary. All human beings need each other but some have greater needs than others. This is the result of existing social and institutional arrangements that can privilege some people and disadvantage others. Again this would imply that it is not the condition of dependency as such that gives rise to ethical responsibilities and obligations, but that there is first an ethical judgment that makes certain 'ontologies' acceptable or not. It is therefore not ontological vulnerability as such that is the basis for an ethics of vulnerability.

Likewise, in the philosophy of Emmanuel Levinas, there is not a priority of ontology over ethics. The ethical demand that arises in the encounter with the other, and that constitutes the humanity of the human, is what Levinas calls 'a rupture of being' [78]. The relatedness of humans indicates that there is something better than being. In the encounter with the human face the fundamental ethical experience begins. Ontology is not a neutral condition that is outside of ethics [79]. I am constituted as a subject through the responsibility imposed on me in meeting the face of the other. I do not exist in myself before this ethical relationship. I am subjected to the other. The proximity of the other, whether or not I know the person, or whether I am close or distant to him, makes me a unique being-for-the-other [80].

Two conclusions follow from this analysis of philosophical perspectives. The first is that they all develop the argument that vulnerability is a general characteristic of the human condition. Not only is it important for ethical discourse to recognize that human beings are vulnerable in a fundamental sense because they are human. Vulnerability is not an attribute of individuals but a basic feature of the human species. But the argument makes a more fundamental claim: it is vulnerability that constitutes humanity and its ethical discourse. This is not, at least not in all perspectives, *ipso facto* an ontological claim. For philosophers such as Levinas and Butler, ontology is intertwined with ethics. Ontology is definitely not prior to ethics. Speaking about 'ontological vulnerability' will therefore convey the wrong message. Vulnerability is not the structure of human existence before I am addressed by its demands, but it actually brings me into being as a human being-for-the-other. The second conclusion therefore is that it will be more appropriate to talk of 'anthropological' rather than 'ontological' vulnerability [81]. This terminology expresses that as human beings we are characterized by a general fragility or precariousness because we have a finite, mortal body and because we are unavoidably socially related and dependent on others. This is not the result of a pre-given and static condition. It rather signifies the continuous interrelations of human beings open to and responsible for each other. Only beings that have no body and are completely alone would be invulnerable [82]. The qualification 'anthropological' signifies that vulnerability is not merely associated to the biology of the body or to human corporeality but also in a broader sense to embodied existence, the fragility of human life, and to basic dependency. This human predicament entails being vulnerable. The goods envisioned by human beings (health, well-being, or personal autonomy) are only possibilities; they can always be threatened, eliminated, or modified beyond our control. They are never guaranteed.

Why is anthropological vulnerability important for bioethics?

The argument that being human means being characterized by anthropological vulnerability has three implications for bioethical discourse: universality, passivity, and positivity.

Universality

Vulnerability as a shared human condition means that it is a characteristic of the human species not of the individual. It is a generalized condition, as Butler pointed out, not an individual attribute. The human body is the site of this universal feature that marks us as a species. Anthropological vulnerability therefore emphasizes that all humans are equally vulnerable. This is why violating the body can be regarded as a crime against humanity [83]. This universality does not imply that people are homogeneous. The human body is at the same time a manifestation of our individuality and subjectivity. All humans are embodied but individual bodies are not similar. Vulnerability can also have different locations: in the body, in personal existence, in relationships, and in dependencies. The notion of anthropological vulnerability does not deny that individuals differ. But it stresses that everybody is in the same anthropological position. We are all equally susceptible to harm. Nobody can avoid vulnerability. The possibility to be harmed is a constant threat for every human being and we do not know who will be hurt and when. The implication is that the possibility of stereotyping or discriminating is eliminated if anthropological vulnerability is recognized. It is at odds with the idea that only some categories of people are vulnerable because they are dependent on others (like children or the very old). Dependency is not a contingent state. All human beings are dependent from the start and throughout their lives. Vulnerability as a condition or characteristic is primarily a relational concept. It demonstrates that fundamentally we are connected with others and with the surrounding world. Anthropological vulnerability focuses our attention on what unites us as human beings.

Awareness that every human being is vulnerable is important for professional relationships in healthcare. The confrontation with disease, disability, and death may activate the experience that all human beings are vulnerable, not only patients but also healthcare providers. Medical practitioners may experience their own vulnerability. Acknowledging and sharing this may be beneficial for the patient [84]. Recognizing one's own vulnerability in the vulnerability of another person may enhance our authenticity but will also enhance trust in the professional relationship. It is argued, for example, that authentic nursing care requires the recognition of mutual vulnerability [85].

The emphasis on common humanity implied in the notion of anthropological vulnerability also has implications for health policy. If everybody shares the same fundamental vulnerability it is not, at least initially, possible to make distinctions and identify categories of vulnerable people. Anthropological vulnerability is incompatible with the categorization of vulnerable groups. Every person should be considered in his or her vulnerable condition, and not as a member of a specific category. If distinctions are made, they may actually increase vulnerability through allocating it to specific populations and groups, and through undermining mutual responsibilities. Titmuss' famous study of blood donation illustrates this. The basic situation is that every person can equally be in need of blood. His vulnerability is the same as everybody's. Richer people have no more advantage than the poor in regard to the supply of blood. Everybody is dependent on the altruism of blood donors. Because vulnerability is shared universally and everybody is in the same position of possible need, the ethics of altruism will be encouraged and nourished. When a market is introduced and blood is regarded as a commercial commodity, then the vulnerability of some is addressed but for many others it will be increased [86].

Passivity

A second implication of anthropological vulnerability is a critical examination of the role of human agency. Human existence is fundamentally passive since human frailty implies that humans are exposed to forces beyond their control. Vulnerability indicates that the language of human agency is limited. Human beings do not have full control of their existence. This implication is expressed in various ways. In philosophical anthropology and phenomenology human beings are not regarded as autonomous subjects in charge of a vulnerable body but as incarnated vulnerable persons. They have a dual nature as acting subject as well as sentient and perceiving subject. German physician Viktor von Weizsäcker has argued that the practical context of medicine shows that life is not an event but an experience, passivity rather than activity. The human person is a pathic subject that is suffering and enduring existence [87]. For Judith Butler, vulnerability means in the first place that we have to undergo the exposure to others, with all the risks involved. It signifies the collapse of the idea of self-mastery. We are dependent on social conditions that we cannot control but that constitute us as moral beings. Martha Fineman argues that the vulnerable subject comes first; only when society attends to its needs we can expect that the autonomous subject will emerge. Autonomy needs to be cultivated; it is not an inherent human characteristic. Vulnerability therefore is not the deficit of autonomy but the condition that make the development of the autonomous subject possible. This is most radically expressed in the philosophy of Emmanuel Levinas. For him, vulnerability is more than weakness and fragility; it is not the breakdown of autonomy since the autonomous subject only comes into being through vulnerability. In his words, the exposure to damaging 'identifies me as the unique one' [88]. Face to face with another human being who is soliciting my response, I become conscious of myself. It is the vulnerability of the face (Levinas also speaks of the 'nudity' of the other) that generates ethical responsibility. This is not an activity of an autonomous subject. On the contrary, it happens despite myself; I am taken hostage by the other person. As exposedness to the other vulnerability implies that the obligations to the other person are more important than self-interest and self-concern. This ethical responsibility for the other is not based on a rational, individual decision but it already exists before we are rational individuals. Personal autonomy starts from the vulnerability of the other. This responsibility is a pre-original challenge, it commands me prior to my freedom. This is what Levinas calls the passivity of the vulnerable [89]. Vulnerability precedes autonomy.

It is evident that these philosophical views differ substantially from the discourse of contemporary bioethics. Kipnis and Luna have pointed out that research ethics discussions assume the existence of a paradigmatic research subject who has all the characteristics of a well-informed, rational, competent, self-determining agent [90]. The same assumption seems to guide many discussions in other areas of bioethics. Precisely this assumption is questioned by the notion of anthropological vulnerability. Such questioning does not imply that there is an antithesis between autonomy and vulnerability so that the latter exists only when the first is absent or impaired. Rather, the latter is the condition that makes the first possible. In the encounter with vulnerability (our own and that of others) autonomous decision-making and responsibility is constituted.

The notion of anthropological vulnerability implies that most significant in our experience are susceptibility and passivity rather than agency and activity. At first sight, vulnerability appears as inability, i.e. lack of power and capabilities. But it is more. According to Jacques Derrida, elaborating on the ideas of Levinas, vulnerability is equivalent to passion and suffering. It can no longer be expressed in the language of power since it is a 'non-power': 'Being able to suffer is no longer a power; it is a possibility without power, a possibility of the impossible' [91]. Vulnerability is beyond the discourse of power and capacity, otherwise it would still be regarded as a characteristic or condition that we can cultivate or eliminate. It refers instead to 'the unchosen and the unforeseen', the existential susceptibility to what is beyond our abilities [92]. The only answers to this fundamental predicament are respect, care, and compassion, or in the words of Derrida, the sharing of the possibility of non-power. Action and intervention should not be the primary responses; respect and care should be. The notion of anthropological vulnerability is relevant for bioethics, not only because it demonstrates the inadequacy of the language of human agency, and therefore of protection, but also because it seriously questions the dominating role of personal autonomy in bioethical discourse. It affirms the conviction that 'heteronomy is the secret of humanity' [93].

Positivity

In previous chapters it was mentioned that vulnerability often has a negative connotation. Being vulnerable implies weakness, failure, lack of action and power. It is an unfortunate condition that needs to be overcome. But the idea of anthropological vulnerability enables a more positive interpretation of vulnerability. Because it is a characteristic which is universally shared by all humans, it is no longer reasonable to oppose the autonomous subject to the vulnerable one. Negative associations of vulnerability often arise because the moral self is assumed to have the capacity to decide and to act. Individual autonomy represents the normal and morally desirable situation, while the vulnerable subject is marked by incapacity and powerlessness. The notion of anthropological vulnerability emphasizes that this opposition is false. In fact, vulnerability as a fundamental human condition is outside the realm of choice and will; it is prior to the emergence of the autonomous subject. It is not exceptional but rather the normal situation from which human beings could develop into autonomous subjects.

As passivity, anthropological vulnerability is furthermore openness to alterity, susceptibility to the unchosen; it means that human beings are not steel dolls but can be affected. Vulnerability is never fixed and static; it is fundamentally a potentiality, not merely of being harmed or damaged but also of being open to change and transformation. We discussed earlier that one aspect of this positive dimension has been mentioned by Gehlen and Scheler when they relate vulnerability to 'world-openness'. Because human beings are vulnerable they need relationships and social institutions to mediate human agency and to compensate for weakness and fragility. Vulnerability as passivity therefore means that humans can be affected in negative and positive ways in their interactions with the world. In a more neutral way it refers to openness to influences, the possibility to change and the ability to transform. Passivity does not mean absence of activity; it is the condition of future activity [94]. The capacity to be affected is what instigates human development and what makes human life challenging.

Associating vulnerability with inability is therefore too limited. Vulnerability is primarily potentiality. This positive interpretation has significant implications. First, recognition of vulnerability will enable new forms of cooperation and community. Solbakk argues that vulnerability can bridge the gap between moral strangers since they are all in the same predicament [95]. This is also what Levinas means when he points out that humans can be each other's brother because they are vulnerable, i.e. open to the suffering of others [96]. Fineman focuses on the role of societal institutions. They have been created in response to vulnerability and they should take their role more seriously. Particularly nowadays, many persons are vulnerable and they are affected by circumstances beyond their control, 'caught in systems of disadvantage that are almost impossible to transcend' [97]. As governments and societies no longer are responsive to vulnerability (for example, due to processes of globalization, as will be discussed in Chapter 7), other ways to address vulnerability must be developed. Second, the language of protection is insufficient. It presupposes the dichotomy between autonomous and vulnerable subjects, and is therefore reflecting the existing power structures that continuously threaten our shared vulnerability. In the perspective of anthropological vulnerability it is illusionary language, assuming that the person confronted with vulnerable subjects is superior in the sense that he is invulnerable, fully in control, independent, and with a perfect body. Third, since it is a universal feature of humans, vulnerability cannot be eliminated; but it should be compensated, diminished, and transformed [98]. Goodin has pointed out that it is not desirable to remove or reduce all vulnerabilities [99]. Personal relations, for example, can only exist because we are vulnerable. Love would be impossible if we do not make ourselves vulnerable to another person. What makes vulnerability problematic is the possibility of abuse and exploitation. Fourth, the problem with modern medicine and modern culture in general is that they continuously try to reduce and close the space of vulnerability [100]. The moral universe that is presupposed is rather simple: there are strong, capable, and independent agents who continuously encounter weak, dependent, and helpless others. Instead of recognizing shared vulnerability, the sovereign subject of the dominant bioethical discourse denies its own vulnerability and locates it in other persons. Instead of recognizing human interdependency, the bioethical discourse emphasizes decision-making of independent rational subjects; there is no weakness or fragility within the sovereign subject. Instead of acknowledging that humanity is produced through heteronomy, the ideal is an autonomous subject that is as invulnerable as possible. While acknowledging anthropological vulnerability facilitates a positive discourse, the denial of vulnerability is ethically dangerous since it cannot address the continuous possibility of violence, domination, and oppression [101]. But this issue raises the question of the connection between vulnerability and normativity.

Vulnerability as ethical principle

What are the normative implications of the notion of anthropological vulnerability? As discussed earlier in this chapter, it is often assumed that vulnerability as a common and inevitable feature of the human condition has normative force: it demands a response, produces responsibility, and implies obligations. For others, this is problematic. Vulnerability may be a description of the human condition, but it does not as such imply any normative requirements. Samia Hurst, for example, argues that the claim to protection cannot be based on vulnerability itself. Identifying a subject as vulnerable shows that there is a need for protection but does not itself provide an argument that protection should be provided [102]. Michael Kottow even has a stronger opinion: describing vulnerability as a universal feature of being human does not have an ethical dimension. Deriving an ethical principle from vulnerability as a descriptive category is committing a naturalistic fallacy [103].

That the ethical implications of anthropological vulnerability are not straightforward is reflected in bioethical terminology. Human vulnerability itself is not considered as a principle—the ethical principle is protection or respect for vulnerability. Goodin is careful to speak about the principle of *protecting* the vulnerable. It grounds the duty to prevent harm from occurring to people [104]. The Unesco Universal Declaration on Bioethics and Human Rights presents the principle of *respect* for human vulnerability and personal integrity. This formulation expresses that perhaps human vulnerability cannot always and should not always be protected since it is not necessarily a negative or bad condition. Respect therefore indicates that this condition has to be acknowledged and recognized as a fundamental and shared predicament. Respect however does not mean acceptance, resignation, or denial. Kemp and Rendtorff are the only ones who directly speak of the principle of vulnerability. They call this principle 'the strongest expression in our time of an ethics of care and concern' [105]. However, they speak alternatively of the principle and the notion of vulnerability. They refer to vulnerability as a principle but also as the 'object of a moral principle requiring care for the vulnerable' [106]. The principle expresses a concern for human fragility and implies an appeal for care [107]. They argue that recognition of vulnerability is important and refer to respecting vulnerability [108]. It seems therefore likely that the principle is used as shorthand for respect for vulnerability. This emphasis articulates that ethics is concerned with recognition and respect rather than comprehension and understanding of vulnerability [109].

It is clear that the criticism of the normative force of vulnerability presupposes a different connection between ontology and ethics than most interpretations of the notion of anthropological vulnerability discussed in this chapter. The critical view argues that the autonomous subject recognizes vulnerability and feels the moral appeal to respond. As moral subject he or she then assumes responsibilities and takes action to protect the vulnerable subject. The notion of anthropological vulnerability, however, implies a radically changed perspective. It dismisses the presupposition that there is an autonomous subject confronted with vulnerable others who then experiences ethical responsibilities. Rather, as dependent, inter-related, and social beings we are immersed in vulnerability from the beginning. Relating, connecting, and interacting with other people means being exposed and thus vulnerable. This brings us immediately into the realm of ethics. We are irre-sistibly confronted with responsibilities and obligations rather than self-assuming them. It is in this perspective impossible to position ourselves outside of ethics. Initially, ethics affirms more what we are than what we do. Vulnerability means that the other person is not primarily regarded as a threat, as a possible danger for our own existence. The face of the other makes us experience our own as well as the other's vulnerability. The appearance of the other, the stranger as a vulnerable subject interrupts my own life since it imposes responsibility upon me. The reason is that vulnerability is ambiguous; it may lead to care and compassion but also to abuse and violence, it may produce transformation or devastation, healing or suffering [110]. Human beings are capable of harming others but also of offering care to them. The encounter with the other person therefore poses the question how do I respond to him; this will make me aware of possible violence, espe-cially since the relationship is asymmetrical. In the radical perspective of Levinas, there is not first a sovereign subject that subsequently decides to relate to others and the world. The unique, autonomous subject only comes into being through the ethical claims of the vulnerable, defenseless other. Not ontology but ethics is first philosophy. Vulnerability forces upon us a reorientation from a focus on ourselves towards the needs of the other [111]. Also for Butler shared human pre-cariousness has normative consequences, although she argues that anthropological vulnerability is not ethically prescriptive in itself since it is basically marked by ambiguity. But it cannot be dismissed; we must attend to it [112]. How then do moral obligations emerge from vulnerability? How can an anthropological condition commit us ethically? The answer of the discussed philosophical perspectives is that anthropological vulnerability as a ubiquitous and shared feature of the human species already implies moral responsibilities. The ethical predicament of vulnerability brings us into being as humans.

Linking philosophical and political perspectives

Therefore, vulnerability as a shared human condition is not simply a factual characterization that has normative implications. Anthropological vulnerability is not the opposite of the view that the autonomous subject is ontologically prior and that ethical responses to vulnerability are the result of rational arguments and voluntary decisions of a moral entity that is in control of his existence. It does not argue that vulnerability is ontologically prior and that we then have to identify and justify what ethical responsibilities emerge from this fundamental condition. Such a separation between ontology and ethics cannot be made. Anthropological vulnerability signifies that human beings are already within the realm of ethics and that this brings them into being. Vulnerability is the expression of human relationality and dependency. Being vulnerable challenges us to become human.

Anthropological vulnerability affirms that human existence always begins in a normative context. At the same time, ethical responses to vulnerability may differ. Though vulnerability makes us aware that we are all in the same predicament and that we should liberate ourselves from our self-interested point of view, responses may be negative or positive, ranging from violence to care. This is why vulnerability is upsetting: it confronts us with the powerlessness of being human (including the defenselessness of our own being) while simultaneously it offers the opportunity to exploit the weakness of others.

The ambiguity of the normative context provides the opportunity to connect the philosophical perspective of anthropological vulnerability with the political one. Because of this connection, the criticism can be addressed that this notion of vulnerability does not enable helpful responses. If every human being is vulnerable, protecting subjects in research is unfeasible. In a pragmatically orientated bioethics, the idea that vulnerability is a general characteristic of the human species may be theoretically interesting but practically useless. The notion of anthropological vulnerability, however, is more complicated than the criticism assumes. It certainly is associated with a general sense of moral responsibility; it explains why ethics is primordial, why we are engaged in ethical concerns, and why autonomous, moral subjects emerge. But it is also associated with special responsibilities. Exposure to harm, even if every human being is equally susceptible, is not equally distributed. While anthropological vulnerability matters, it should also be recognized that vulnerability has a 'differential allocation' [113]. When the conditions for human flourishing are hampered for some people more than for others, there is a special obligation to assist. Vulnerability may differ because the possibilities to satisfy human needs are unequal. Though all human beings are in the same vulnerable predicament, the possibility of harm is not the same. The concept of needs is used as an explanation. A vulnerable person will be harmed if his or her needs are not satisfied. But not all human needs are similar. When vital needs are neglected or restricted, vulnerability will be damaged more. The focus on different needs links the universal and contextual dimensions of vulnerability. Since all human beings have vital needs, they are all equally susceptible to serious harm; this is the philosophical perspective of anthropological vulnerability. But some vital needs depend on the situation of the subject making it impossible to satisfy them without assistance [114].

The emphasis on recognition, equality, and need demonstrates that anthropological vulnerability generates not only individual but also social responsibility. Although his philosophy seems to highlight first of all individual responsibility for the other person, Levinas argues that moral obligations are not restricted to the specific other who I am encountering. In the face of the other I see the virtual presence of all men and women [115]. Moral obligations therefore have a more general character; this is reflected in social arrangements that are (and should be) organized to demonstrate mutual respect between individuals and equality of rights. The emphasis on social responsibilities as consequence of anthropological vulnerability is perhaps strongest in the work of Fineman. When dependency is universal and unavoidable, it can only generate in her view a collective responsibility. It would be a distortion to only talk about responsibility of individuals and families without articulating the broader societal responsibilities. Vulnerability should be mitigated through creating a more equal society [116]. This emphasis demonstrates that the notion of anthropological vulnerability is not antithetical to the political perspective of vulnerability. As will be discussed in the next chapter, it is even fundamental for the development and application of this specific perspective. The association of anthropological vulnerability with social responsibility and the requirement of equality furthermore demonstrate the crucial role of this notion of vulnerability in global bioethics. Particularly in a global perspective it is necessary to identify the mechanisms of distribution and allocation of vulnerability. This effort should go beyond the focus of contemporary mainstream bioethics on individual autonomy.

Conclusion

The argument put forward in this chapter is that understanding vulnerability as a generalized and shared characteristic of human beings opens up wider and richer perspectives for bioethical debate than the usual framing in terms of failed or restricted individual autonomy. The notion of anthropological vulnerability has been developed in various philosophical approaches, relating vulnerability to imperfections in the nature of the human species or the human body, or to the existential experience, relatedness, and dependency of persons as social beings. All approaches, though justifications and explanations differ, have in common that vulnerability is considered as a primordial predicament of human beings that is universally and equally shared, that precedes and enables human agency, and that emphasizes potentialities and human transformation. Anthropological vulnerability therefore presents a moral perspective that reaches beyond the discourse of autonomous subjects. It introduces a language that is not merely concerned with weakness and powerlessness, and with questions of protection and damage control. Focusing on the relational context of human existence it furthermore articulates the moral significance of respect, care, solidarity, and responsibility. Vulnerability therefore entails a broad moral vision that cannot be reduced to deficient autonomy. However, arguing that vulnerability is a general characteristic of the human species does not exclude that, especially at the global level, vulnerability can be unequally distributed. On the contrary, anthropological vulnerability is providing the foundation for political perspectives that emphasize that vulnerability is often produced by specific circumstances that require remediate action. The next chapter aims to analyze these political perspectives on vulnerability.

Notes

1 See Kipnis, 'Vulnerability in research subjects: A bioethical taxonomy'. See: Kipnis, 'Seven vulnerabilities in the pediatric research subject'; Kipnis, 'Vulnerability in research subjects. An analytic approach'; Kipnis, 'The limitations of "limitations".' See also, National Bioethics Advisory Commission, *Ethical and policy issues in research involving human participants. Volume I: Report and recommendations of the National Bioethics Advisory Commission*, 85–92. Kipnis is often praised for having introduced a contextual view of vulnerability. See, for example, DeBruin, 'Looking beyond the limitations of "vulnerability": Reforming safeguards in research', 77; Shivas, 'Contextualizing the vulnerability standard', 84–85; Iltis *et al.* 'Federal interpretation and enforcement of protections for vulnerable participants in human research', 39.

2 The expression 'sources of vulnerability' is used by Nickel as well as Ganguli-Mitra and Biller-Andorno (see: Nickel, 'Vulnerable populations in research: The case of the seriously ill', 249; Ganguli-Mitra and Biller-Andorno. 'Vulnerability in healthcare and research ethics', 245). See also: Coleman, 'Vulnerability as a regulatory category in human subject research', 15–16.

3 Luna, 'Elucidating the concept of vulnerability: Layers not labels'. See, Macklin, 'A global ethics approach to vulnerability', 69 ff and Luna, *Bioethics and vulnerability. A Latin American view*.

4 The notion of 'peripheral bioethics' is introduced particularly in the publications of the Brazilian bioethicist Volnei Garrafa (see Diniz, Guilhem, and Garrafa, 'Bioethics in Brazil'; Garrafa and Porto. 'Intervention bioethics: A proposal for peripheral countries in a context of power and injustice'). Haugen has furthermore pointed out that vulnerability is more recognized in the Continental European tradition than in the Anglo-American tradition (Haugen, 'Inclusive and relevant language: the use of the concepts of autonomy, dignity and vulnerability in different contexts', 209).

5 Gorovitz, 'Reflections on the vulnerable', 64. Also Onora O'Neill in her distinction between persistent and variable, selective vulnerabilities calls the first category 'species vulnerabilities' (O'Neill, *Towards justice and virtue*, 191). Lydia Feito is using the expression 'anthropological vulnerability' (Feito, 'Vulnerabilidad', 8–10).

6 Rendtorff, 'Basic ethical principles in European bioethics and biolaw: Autonomy, dignity, integrity and vulnerability – Towards a foundation of bioethics and biolaw', 237; see also Rendtorff, 'The second international conference about bioethics and biolaw: European principles in bioethics and biolaw. A report from the conference', 164–165.

7 The International Bioethics Committee of UNESCO formulates this as: 'The human condition implies vulnerability. Every human being is exposed to the permanent risk of suffering "wounds" to their physical and mental integrity. Vulnerability is an inescapable dimension of the life of individuals and the shaping of human relationships' (IBC, *Report of the IBC on the principle of respect for human vulnerability and personal integrity*, 2).

8 Gehlen, *Man. His nature and place in the world*, 13.

9 Gehlen, *Man. His nature and place in the world*, 26.

10 Gehlen, *Man. His nature and place in the world*, 26.

11 Gehlen, *Man. His nature and place in the world*, 29.

12 See Wyn Schofer, *Confronting vulnerability. The body and the divine in rabbinic ethics*.

13 See Turner, *The body and society. Explorations in social theory*, 64 ff; Culp, *Vulnerability and glory. A theological account*, 13 ff.

14 'Within the Western philosophical tradition the body has often been regarded as a force of negativity, an obstacle to the soul's attempt to secure knowledge, virtue, or eternal life' (Leder, *The absent body*, 127).

15 Leder, *The absent body*, 126 ff.

16 Marcum, *An introductory philosophy of medicine. Humanizing modern medicine*, 49–61.

17 Leder, *The absent body*, 140. See also Leder, *The body in medical thought and practice*.

18 Ten Have and Welie, 'Medicine, ownership, and the human body', 2. See also Gracia, 'Ownership of the human body: Some historical remarks'; Campbell, 'Body, self, and the property paradigm'.

19 Merleau-Ponty, *Phenomenology of perception*. See also: Leder, *The body in medical thought and practice;* Toombs, *Handbook of phenomenology and medicine;* Carel, 'Phenomenology and its application in medicine'.

20 Plügge, 'Man and his body', 295.

21 Buytendijk, *Prolegomena to an anthropological physiology*, 27. See also Dekkers, 'F.J.J. Buytendijk's concept of an anthropological physiology'.

22 Toombs, *Handbook of phenomenology and medicine*, 7.

23 Leder, *The body in medical thought and practice*, 5.

24 Plügge, *Wohlbefinden und Missbefinden*, 65. This statement is a reminder of the famous expression of René Leriche, surgeon in nineteenth century France: 'health is life lived in the silence of the organs'. See, Van Wolputte and Meurs, 'Lichaam, zorg, kwetsuur en verzorging. Een antropologie van de menselijke kwetsbaarheid', 177; Svenaeus, 'Illness as unhomelike being-in-the-world: Heidegger and the phenomenology of medicine'.

25 Plügge, *Wohlbefinden und Missbefinden*, 85.

26 Spiers, 'New perspectives on vulnerability using emic and etic approaches', 719. See also Sellman, 'Towards an understanding of nursing as a response to human vulnerability'. For an application of these ideas in medicine, see Malterud and Hollnagel, 'The doctor who cried: A qualitative study about the doctor's vulnerability'.

27 The lived body according to Toombs is 'the medium through which, and by means of which, I apprehend the world and interact with it' (Toombs, *Handbook of phenomenology and medicine*, 5).

28 Leder, *The body in medical thought and practice*, 25.

29 Der Mensch is gewöhnlich *draußen*, bei der Sachen' (Plügge, *Der Mensch und sein Leib*, 25).

30 Sartre, 'The body', 223. The view that the human being is essentially an embodied, relational being is also emphasized in personalist approaches in bioethics. See, for example, Schotsmans, 'Ownership of the body: A personalist perspective'.

31 For example, Gaille, *Philosophie de la Médecine*.

32 Sandel, *Public philosophy. Essays on morality in politics*, 153.

33 See, for example, Ikuenobe, *Philosophical perspectives on communalism and morality in African traditions*.

34 Butler, *Precarious life. The powers of mourning and violence*, 28. See also Cadwallader, '(Un)expected suffering: The corporeal specificity of vulnerability'.

35 Fineman, 'The vulnerable subject: Anchoring equality in the human condition', 1.

36 Fineman, 'The vulnerable subject: Anchoring equality in the human condition', 12.

37 Fineman, 'The vulnerable subject and the responsive state', 269.

38 Fineman, 'The vulnerable subject and the responsive state', 265; Fineman, 'The vulnerable subject: Anchoring equality in the human condition', 9 ff; Fineman, *The autonomy myth. A theory of dependency*, 35 ff.

39 See Eichner, 'Dependency and the liberal polity: On Martha Fineman's the autonomy myth'; Fineman, *The autonomy myth. A theory of dependency*.

40 Fineman, 'The vulnerable subject and the responsive state', 255.

41 Butler, *Precarious life. The powers of mourning and violence*, 22–23.

42 Murphy argues that Butler proceeds with an ethical ontology, a philosophy based on the 'ontological fact of vulnerability' (Murphy, 'Corporeal vulnerability and the new humanism', 589). Vlieghe argues that her approach is not ontological (Vlieghe, 'Judith Butler and the public dimension of the body: Education, critique and corporeal vulnerability', 163). For Butler herself, ontology is 'an effect of power' (Butler, 'Reply from Judith Butler to Mills and Jenkins', 184).

43 Vlieghe, 'Judith Butler and the public dimension of the body: Education, critique and corporeal vulnerability', 158.

44 Vlieghe, 'Judith Butler and the public dimension of the body: Education, critique and corporeal vulnerability', 158.

45 Vlieghe, 'Judith Butler and the public dimension of the body: Education, critique and corporeal vulnerability', 163.

46 Butler, *Precarious life. The powers of mourning and violence*, 31.

47 Kittay, 'The ethics of care, dependence, and disability', 57.

48 MacIntyre, *Dependent rational animals. Why human beings need the virtues*, 5, 63 ff.

49 See, McRobbie, 'Vulnerability, violence and (cosmopolitan) ethics: Butler's *Precarious life*'; Mills, 'Normative violence, vulnerability, and responsibility'; Jenkins, 'Towards a nonviolent ethics: Response to Catherine Mills'.

50 Fineman, 'The vulnerable subject and the responsive state', 253.

51 Kemp, 'Four ethical principles in biolaw', 21–22; Patrão Neves, 'The new vulnerabilities raised by biomedical research', 185; Solbakk, 'Vulnerability: A futile or useful principle in healthcare ethics'? 234.

52 'The principle of vulnerability … is perhaps the strongest expression in our time of an ethics of care and concern' (Rendtorff and Kemp, *Basic ethical principles in European*

bioethics and biolaw. Volume I. Autonomy, dignity, integrity and vulnerability, 51). See also Botbol-Baum, 'The necessary articulation of autonomy and vulnerability'; Moulin, 'La vulnérabilité entre sciences et solidarité'.

53 Rendtorff, 'European perpectives', 297, 304–305; see also, Rendtorff and Kemp, 'Vulnérabilité (Principe de)' and Rendtorff and Kemp, 'Vulnérabilité (Personne)'.

54 Levinas, *Ethics and Infinity*, 77.

55 Levinas, *Ethics and Infinity*, 77. 'Passivity of the vulnerable, condition (or incondition) by which beings shows itself creature' (Levinas, *Humanism of the Other,* 64). In another text, Levinas speaks about this noncondition: '…proximity is responsibility for the Other, substitution for the Other, expiation for the Other; condition – noncondition – of serving as hostage…' (Hand, *The Levinas reader,* 246).

56 'The other man commands by his face, which is not confined in the form of its appearance; naked, stripped of its form, denuded of its very presence…' (Levinas, *Humanism of the Other,* 7).

57 'It is denuded of its own image' (Levinas, *Humanism of the Other,* 32).

58 'The face is exposed, menaced, as if inviting us to an act of violence. At the same time, the face is what forbids us to kill'. (Levinas, *Ethics and Infinity. Conversations with Philippe Nemo,* 86). See also: Burggraeve, 'Violence and the vulnerable face of the Other: The vision of Emmanuel Levinas on moral evil and our responsibility'.

59 Levinas, *Humanism of the Other,* 33. See also Peperzak, *To the other. An introduction to the philosophy of Emmanuel Levinas,* 164.

60 Levinas, *Humanism of the Other,* 67.

61 Levinas, *Otherwise than being or beyond essence,* 15.

62 Levinas, 'Ethics of the Infinite', 78. See also Goodman who argues that in this perspective the needs of others are necessarily more important than my own (Goodman, *The demanded self. Levinasian ethics and identity in psychology*).

63 Levinas, 'Ethics of the Infinite', 74.

64 Levinas, *Humanism of the Other,* 7.

65 Levinas explains that 'the epiphany of the face is visitation' (Levinas, *Humanism of the Other,* 31). The face enters our world 'before the creature collects himself…to make himself essence' (Levinas, *Humanism of the Other,* 67). The face is obliging me because it 'presses the neighbor up against me'. (Levinas, *Otherwise than being or beyond essence,* 91).

66 Levinas points this out as follows: 'In the face of the other man I am inescapably and consequently the unique and chosen one'.(Hand, *The Levinas reader,* 84).

67 Levinas, *Humanism of the Other,* 51. Peperzak explains that as substitute and hostage of the other I am vulnerable myself; I do not have any control of the situation. 'I am hostage even before I may know it' (Peperzak, *To the other. An introduction to the philosophy of Emmanuel Levinas,* 26).

68 See Chapter 2, note 30.

69 It is this primary vulnerability, as Butler observes, that 'one cannot will away without ceasing to be human'. (Butler, *Precarious life. The powers of mourning and violence,* xiv).

70 Butler, *Precarious life. The powers of mourning and violence,* 30–31.

71 Butler, Judith. *Frames of war. When is life grievable?* 2.

72 As Butler puts it: '…the human comes into being, again and again…' (Butler, *Precarious life. The powers of mourning and violence,* 49).

73 Shildrick, 'Becoming vulnerable: Contagious encounters and the ethics of risk', 226; Shildrick, *Embodying the monster. Encounters with the vulnerable self,* 85.

74   The claim is disputed by Eichner, 'Dependency and the liberal polity: On Martha Fineman's The Autonomy Myth', 1310 ff.

75   See Rogers, Mackenzie, and Dodds. 'Why bioethics needs a concept of vulnerability'. The same point is made in several contributions in Mackenzie, Rogers, and Dodds, *Vulnerability. New essays in ethics and feminist philosophy*.

76   Goodin, 'Vulnerabilities and responsibilities: An ethical defense of the welfare state', 76–77, 83. See also: Goodin, *Protecting the vulnerable. A reanalysis of our social responsibilities*; and Goodin, 'Relative needs'.

77   Fineman, 'The vulnerable subject and the responsive state', 274.

78   Levinas, *Ethics and Infinity. Conversations with Philippe Nemo*, 87.

79   In Levinas' words: 'Ethics…does not supplement a preceding existential base' (Levinas, *Ethics and Infinity. Conversations with Philippe Nemo*, 87).

80   '…the first fact of existence is neither being in-itself (*en soi*) nor being for-itself (*pour soi*) but being *for the other* (*pour l'autre*); in other words that human existence is a creature' (according to Levinas in Hand, *The Levinas reader*, 149).

81   The adjective 'anthropological' is used here as referring to the Continental European tradition of 'philosophical anthropology', rather than to social or cultural science, as is more usual in Anglo-Saxon scientific traditions. The term 'anthropological vulnerability' is used by Michael Kottow but without explanation; he seems to refer to anthropological descriptions of the human condition (Kottow, 'Vulnerability: What kind of principle is it?' 283). As an alternative for ontological vulnerability the term is introduced by Nathalie Maillard. She describes it as 'the intrinsic fragility of the body, of the mental and emotional integrity, as well as the important human capacities' (Maillard, *La vulnérabilité. Une nouvelle catégorie morale?* 198; my translation).

82   It is interesting that nowadays in theology it is argued that God and the Church are also vulnerable. See Gulp, *Vulnerability and glory. A theological account*; Koopman, 'Vulnerable Church in a vulnerable world? Towards an ecclesiology of vulnerability'; Placher, *Narratives of a vulnerable God, Christ, theology, and scripture*.

83   See Bergoffen, 'February 22, 2001: Toward a politics of the vulnerable body'.

84   Malterud and Hollnagel, 'The doctor who cried: A qualitative study about the doctor's vulnerability', 350 ff; Malterud, Fredriksen and Gjerde, 'When doctors experience their vulnerability as beneficial for the patients', 85 ff.

85   Daniel, 'Vulnerability as a key to authenticity', 191.

86   See Titmuss, *The gift relationship: from human blood to social policy*.

87   Ten Have, 'Bodies of knowledge, philosophical anthropology, and philosophy of medicine', 27.

88   Levinas, *Otherwise than being or beyond essence,* 49. See also Burggraeve, 'Violence and the vulnerable face of the Other: The vision of Emmanuel Levinas on moral evil and our responsibility'.

89   Levinas, *Humanism of the Other*, 64.

90   See Kipnis, 'Vulnerability in research subjects: A bioethical taxonomy'; Kipnis, 'The limitations of "Limitations;" 'Luna, 'Elucidating the concept of vulnerability: Layers not labels'.

91   Derrida, 'The animal that there I am (more to follow)', 396.

92   Harrison, 'Corporeal remains: vulnerability, proximity, and living on after the end of the world', 427.

93   Bergoffen, 'February 22, 2001: Toward a politics of the vulnerable body', 133.

94 Bluhm, 'Vulnerability, health, and illness', 155 ff; Gunaratnam, 'From competence to vulnerability: Care, ethics, and elders from racialized minorities', 36; Gilson, 'Vulnerability, ignorance, and oppression', 310.

95 Solbakk, 'Vulnerability: A futile or useful principle in healthcare ethics'? 230.

96 Levinas clearly has linked vulnerability and solidarity: 'Responsibility for the other, this way of answering without a prior commitment, is human fraternity itself, and it is prior to freedom' (Hand, *The Levinas reader*, 106). This is implied in vulnerability as the condition of being hostage: 'The unconditionality of being hostage is not the limit case of solidarity, but the condition for all solidarity' (Hand, *The Levinas reader*, 107).

97 Fineman, 'The vulnerable subject and the responsive state', 257.

98 Fineman, 'The vulnerable subject and the responsive state', 269.

99 Goodin, *Protecting the vulnerable. A reanalysis of our social responsibilities,* 193.

100 Callahan, 'The vulnerability of the human condition', 119 ff; McRobbie, 'Vulnerability, violence and (cosmopolitan) ethics; Butler's *Precarious Life',* 79.

101 Gilson, 'Vulnerability, ignorance, and oppression', 324.

102 Hurst, 'Vulnerability in research and health care; describing the elephant in the room'? 201.

103 Kottow, 'Vulnerability: What kind of principle is it?' 284.

104 Goodin, *Protecting the vulnerable. A reanalysis of our social responsibilities,* 109, 111. His formulation also correctly points out that what needs protection is not vulnerability as such but vulnerable subjects.

105 Rendtorff and Kemp, *Basic ethical principles in European bioethics and biolaw. Volume I. Autonomy, dignity, integrity and vulnerability,* 51.

106 Rendtorff and Kemp, *Basic ethical principles in European bioethics and biolaw. Volume I. Autonomy, dignity, integrity and vulnerability,* 398.

107 Kemp, 'Four ethical principles in biolaw', 21; Rendtorff, 'The second international conference about bioethics and biolaw: European principles in bioethics and biolaw. A report from the conference', 158.

108 Rendtorff and Kemp, *Basic ethical principles in European bioethics and biolaw. Volume I. Autonomy, dignity, integrity and vulnerability,* 46; Rendtorff, 'The second international conference about bioethics and biolaw: European principles in bioethics and biolaw. A report from the conference', 164. Rendtorff calls respect for vulnerability 'the foundation of ethics in our time' (Rendtorff, 'Basic ethical principles in European bioethics and biolaw: Autonomy, dignity, integrity and vulnerability – Towards a foundation of bioethics and biolaw', 237).

109 Nortvedt, 'Subjectivity and vulnerability: reflections on the foundation of ethical sensibility', 227.

110 'The face is exposed, menaced, as if inviting us to an act of violence. At the same time, the face is what forbids us to kill'. (Levinas, *Ethics and Infinity. Conversations with Philippe Nemo,* 86). See also Burggraeve, 'Violence and the vulnerable face of the Other: The vision of Emmanuel Levinas on moral evil and our responsibility', 37 ff.

111 Goodman explains: 'since we are exposed we can no longer be orientated towards ourselves' (Goodman, *The demanded self. Levinasian ethics and identity in psychology,* 108–109).

112 Butler makes a difference between implying and entailing: 'The postulation of a generalized precariousness … implies, although does not directly entail, certain normative consequences' (Butler, *Frames of war. When is life grievable?* 33). See also Butler, *Precarious life. The powers of mourning and violence,* 29.

113 Butler, *Frames of war. When is life grievable?* 3, 22; Murphy, 'Corporeal vulnerability and the new humanism', 581.

114 Rogers, Mackenzie and Dodds. 'Why bioethics needs a concept of vulnerability'. See also Goodin, 'Relative needs'; Braybrooke, *Meeting needs*; Brock, *Necessary goods. Our responsibilities to meet others' needs.*

115 As explained by Peperzak, *To the other. An introduction to the philosophy of Emmanuel Levinas*, 31. See also Diedrich, Burggraeve and Gastmans, 'Towards a Levinasian care ethic: A dialogue between the thoughts of Joan Tronto and Emmanuel Levinas', 45 ff.

116 Fineman, 'The vulnerable subject and the responsive state', 269.

6

SOME OF US ARE MORE VULNERABLE

Political perspectives on vulnerability

Anthropological vulnerability expresses the commonality of human beings. Basically, everyone is vulnerable and nobody is an exception. Since this characteristic is necessarily shared among all members of the human species, some individuals cannot be regarded as more vulnerable than others. It also means that vulnerability is not a negative qualification. Rather than referring to humans as victims or helpless subjects, vulnerability demonstrates the interdependence and connectedness of human beings and the moral responsibility humans take for each other. Precisely because it is not an individual characteristic, vulnerability is the basis of human solidarity. This explains why vulnerability is difficult to understand in mainstream bioethical discourse. It cannot be considered as lack or failure of individual autonomy, but it is on the contrary the condition that makes the autonomous subject possible. Without vulnerability, this subject could not come into existence and could not continue to exist.

These positive, general qualifications do not preclude, as I have shown earlier, that the philosophical discourse of vulnerability is challenged because in daily life, and especially in health care settings, vulnerability is not an abstract notion but a concrete reality that is very different for individual persons. In everyday existence, some human beings are clearly more vulnerable than others, for example the girls in India in the HPV vaccine trial mentioned in Chapter 1. Entire populations, like the inhabitants of the Pacific island of Tuvalu, are vulnerable. But also, vulnerability is especially experienced in certain stages of life, for example if one grows older. Obviously, vulnerability is not the same condition for everyone. It seems that the idea of anthropological vulnerability falsely suggests that we are all in the same predicament, and that it disregards the different situations in which some individuals are more vulnerable than others. From a political perspective this idea of shared vulnerability is therefore ideological; it justifies the differences in human conditions by arguing that all individuals basically are similar, and it eliminates the urgency to change conditions producing vulnerability. Vulnerability, on the contrary, implies differences in power, and is therefore related to questions of justice and human rights. The notion of anthropological vulnerability hides these differences. It also denies that identifying subjects as 'vulnerable' implies that action is necessary to prevent or remediate exploitation, mistreatment, abuse, and stigma. These criticisms are grounded on what has been called the political perspective on vulnerability. This perspective will be discussed in the current chapter. Vulnerability is not a given and universal condition of the human species but it is made, produced, or generated within specific conditions within which individual human beings happen to exist. This chapter argues that there is in fact no opposition between the two perspectives. Rather, the political perspective can be more effective by building on the ideas of the philosophical perspective. Interconnecting both perspectives will create the possibility to transcend the mainstream bioethical approach to vulnerability.

Vulnerability exacerbated

David Thomasma has argued that all human beings are vulnerable because they are subject to illness and decay but that there is an additional vulnerability of the sick [1]. Anthropological vulnerability therefore can be combined with special vulnerability. The first is the precondition for the second. Though everyone can be damaged because of being a vulnerable human being, somebody has already been damaged by a specific condition. The point is that this particular person can be further harmed or damaged, which is what makes him or her specially or additionally vulnerable. The already given anthropological vulnerability can thus be exacerbated. This has been the reason that in the scholarly literature distinctions have been made between general and special vulnerability. But both types of vulnerability are closely connected. The additional, second-order vulnerability is only possible because of the primary vulnerability of human beings; they can be hurt in specific circumstances because they are vulnerable as a human characteristic. Precisely because anthropological vulnerability is not a negative feature or a biological deficit but determines the positive phenomenon of 'world-openness' there is always the possibility of additional vulnerability. The two types of vulnerability emphasize in fact different components of the notion. Anthropological vulnerability focuses primarily on sensitivity. Humans are embodied, dependent, relational, and responsible beings, and therefore necessarily sensitive to possible harms and damage. Special vulnerability emphasizes the components of exposure and adaptive capacity. Human beings will be additionally vulnerable when they are in threatening and risky conditions and when they do not have adequate capabilities to respond and cope. The next section explores this special vulnerability.

Special vulnerability

We all share, as argued in the previous chapter, a common human vulnerability, but it is always articulated in different ways because it is situated within fields and operations of power. Anthropological vulnerability then 'becomes highly exacerbated under certain social and political conditions' [2]. It gives rise to special vulnerability that has a differential allocation; some populations are more subjected to possible harm and violence than others. The normative implication is that this notion of special vulnerability implies that we need to uncover the mechanisms of this unequal distribution and to find ways for a more inclusive and egalitarian redistribution. Recognizing that vulnerability is allocated and articulated differently should therefore lead to concrete social policies [3]. The interesting aspect of this point of view is not only that it presents a clearly political perspective on vulnerability, but also that it associates this perspective with the philosophical one. It is the notion of anthropological vulnerability itself that underlies the political view that the vulnerable subject is produced through social processes and that vulnerability can only be reduced by attending to the social and economic conditions that sustain it [4].

That the philosophical and political perspectives on vulnerability are not contradictory but interconnected is argued by Eva Kittay. Dependency is an inevitable part of the human condition; it is not an exceptional circumstance. But persons are not 'symmetrically situated' [5]. Dependencies may vary according to social and economic conditions. Dependency relations furthermore imply inequality of power. The dependent needs care, the dependency worker provides care, but both are vulnerable to domination and exploitation. The inequalities of situations, capacities, and relationships can never be adequately addressed, according to Kittay, if equality is not redefined on the basis of inescapable human interdependence [6].

The idea of special vulnerability particularly highlights the inequality in human relations. This inequality manifests itself in the medical context when the human person has become a patient. Illness necessitates the person to seek assistance and treatment. But engaging in a relationship with healthcare professionals can itself produce special vulnerability. The medical relationship is imbalanced because there is inequality of power. The patient who seeks healing and care can actually be helped or hurt by the power of medicine. Vulnerability can be exacerbated or amplified within (professional) relationships. In this relational context that is asymmetrical the consideration of special vulnerability is therefore an ethical requirement. This is the reason why a principle of respect for vulnerability of patients has been proposed [7]. David Thomasma has formulated it in more general terms as follows: 'In human relations generally, if there are inequities of power, knowledge, or material means, the obligation is upon the stronger to respect and protect the vulnerability of the other, and not to exploit the less advantaged' [8]. The connection between special vulnerability and medical power is elaborated in the phenomenological work of Richard Zaner [9]. To be a patient for Zaner means to be vulnerable. It is not merely the inequality of the condition (being ill) but also the structural asymmetry of power and resources in favor of the physician that creates vulnerability. Recognition of the special vulnerability of patients and the imbalance of power in the professional relationship is according to scholars as Pellegrino, Thomasma, and Zaner the main reason why medicine has been regarded as a moral enterprise from its beginning [10].

However, special vulnerability is not limited to a medical context. It can manifest itself in disadvantaged conditions such as poverty, homelessness, or gender discrimination. Because it is produced in relations and situations that are characterized by asymmetry and inequality, the notion of special vulnerability has a much wider significance beyond the medical relationship itself. As the next chapter will demonstrate, it therefore provides a critical lens to examine current processes of globalization that are associated with growing power differences across the world. But first we have to examine how vulnerability introduces an ethical perspective that goes beyond the focus on autonomous individuals.

Social ethics

The interconnection between anthropological and special vulnerability is two-dimensional, not in the superficial way that there are two levels, i.e. general and special, but in the more complex manner of dimensions that necessarily implicate each other. Anthropological vulnerability refers to a generalized condition that produces the rhetoric of universalism and that brings forth the general discourse of bioethics as it is concerned with human life, welfare, and health. This is the predicament of passivity, the phenomenon of 'givenness', when human beings experience that they are susceptible to harm and violence. But at the same time, particularly in health care, special vulnerability is manifested in various ways, demonstrating that what is potential harm to all is already actualized in some. This exacerbated vulnerability of some human beings shows the phenomenon of 'taking'. Because human conditions are localized, particularistic, and relational, not only in health care but in everyday human life, these conditions are asymmetrical and unequal. The exposure to the other means that life, health, and well-being can be taken from us. The possibility to take and to be taken is the potential damage or violence that threatens our vulnerability when we are sick, ageing, or living in disadvantaged circumstances [11].

Because special vulnerability is produced in relations and situations, it can never be properly understood as long as it is regarded merely as an individual characteristic, i.e. as a failure of autonomy or compromised informed consent. If vulnerability is emerging within a relational context, then the vulnerable subject is simply one element to consider. Like the notion of anthropological vulnerability that considers vulnerability as a species qualification, special vulnerability refers to humans as social and relational beings. This type of vulnerability is primarily an indicator of social inequality or differences in the allocation of power.

The consequence is that the discourse of vulnerability, whether anthropological or special, is fundamentally linked to social and political concerns. However, this view assumes that society is more than simply an aggregate of equal and autonomous individuals. When subjects are vulnerable due to interconnectedness and dependency these determining conditions need to be taken into account in our conception of society and justice. It also implies that political action is required beyond individual agency.

The political perspective associated with the emphasis on special vulnerability is generally concerned with two issues: (1) how vulnerability is exacerbated, and (2) what can be done to mitigate exacerbated vulnerability. The first issue focuses on circumstances that make people vulnerable and that may lead to paternalism, exploitation, abuse, mistreatment, discrimination, or violence. The second issue leads to formulation of ethical and political responses, connecting vulnerability with human rights, theories of justice, and global ethics of care. But both issues presuppose that vulnerability should be regarded as a notion of social rather than individual ethics. And this is precisely contested in mainstream bioethics that usually has a predominantly or exclusively individualistic take on vulnerability.

Beyond individual agency and autonomy

As discussed in earlier chapters, the idea that groups, populations, and communities can be vulnerable has met major criticism and even rejection. The prevailing argument is that the notion of vulnerability should only be applied to the individual person and his/her capacities. This emphasis on the individual person is perhaps one of the most salient characteristics of the current discourse in bioethics; it is not comfortable with a perspective on vulnerability that puts more emphasis on the social nature of humans. Nonetheless, it is precisely this perspective that is introduced with the notion of anthropological vulnerability. Persistent vulnerability, according to Onora O'Neill, is typical for the human species [12]. Human beings have a long infancy, they need the support of other people to develop essential capacities, and they depend on long-term interactions with others. It is a myth to assume that personal autonomy can exist in isolation. A similar relationship between self-determination and being determined is elaborated by other philosophers arguing that human autonomy is always dependent on social and cultural conditions. For example, Edgar Morin points out that we only become an autonomous self if we learn a language and a culture [13]. Autonomy therefore is nourished by dependency. The development and exercise of competent and effective agency of individuals is intrinsically connected to other people. Personal autonomy is acquired and sustained in interaction. Human beings are quintessentially social beings. Precisely this connectedness with other people (and thus our vulnerability to them) makes us into human individuals. The upshot of this argument is that vulnerability cannot be framed in terms of deficient autonomy. This point is further elaborated by Paul Ricoeur. Human beings are always situated within the dialectics of self-determining and being determined. The personal autonomy characteristic for human beings is the autonomy of a fragile, vulnerable being. Furthermore, autonomy would never exist without the prior sociality of vulnerability. The moral vocabulary of abilities and capacities depends on the moral experience of heteronomy, inability, and incapacity. Our powers and capacities are correlated to modes of incapacity, to fragility. At the same time, this argument does not undermine the importance of autonomy. As Ricoeur correctly points out, it primarily implies that the significance of autonomy is relative, dependent on conditions of sociality. Vulnerability would be pathological if human beings were not called on to become autonomous. Autonomy is never given, never certain; it is always a goal to attain and therefore always threatened [14]. It furthermore implies that vulnerability is always provoking action, instigating efforts to overcome or remediate it. Without the perspective of resisting it, the notion of vulnerability will result in misery and fatalism. Of course, resistance should not be the only or principle response [15]. The struggle against vulnerability can easily result in the fiction that one can create invulnerabilities, with the result that medicine becomes unsustainable, chronic illness and disability meaningless, and death intolerable. Vulnerability can never be eliminated since otherwise humanity itself would be erased.

Human interaction and cooperation

Going beyond the perspective of individual autonomy has implications for the kind of action that is required in regard to vulnerability. Two different views on human beings and social cooperation are at stake. One view assumes that humans are primarily self-interested individuals. The basic vulnerability of the human condition can only be ameliorated or reduced through rational cooperation in the interest of everyone. This view takes for granted that social interactions are secondary, i.e. assuming ethical responsibility for each other can only be the result of free decisions of autonomous subjects [16]. In this view, it can be accepted that there are two types of vulnerability (anthropological and special) but both are reduced to individual decision-making and action. The vulnerability inherent in the human condition, as assumed in the Hobbesian state of nature, necessitates individuals to cooperate for the sake of self-preservation. Vulnerability therefore can be overcome as the result of rational decisions of individual actors. The point, however, is that vulnerability is primarily an individual concern; it is my own vulnerability that is threatened by others and that can only be overcome through voluntary cooperation.

The other view assumes that individual beings are human because they are social beings. Human existence is not primarily characterized by competition and conflict, but by interconnectedness. For philosophers like Levinas and Butler, our being in the world is always a being-together. The ethical relationship of responsibility between human beings is prior to self-interest and individual choice.

The fundamental connection between oneself and others comes first. Vulnerability in this view can never be an individual affair. It is what is first encountered in human existence. It makes us human and constitutes individual subjectivity. We cannot eliminate it without abolishing our humanity.

The social context of special vulnerability

The same differences in point of view are noticeable in connection to the political perspective on vulnerability. This perspective specifically focuses on special vulnerabilities produced by particular circumstances. Although the notion of vulnerability refers to determinants beyond individual agency, the conventional view, expressed especially in official documents, continues to associate vulnerability with the primacy of individual autonomy. The idea that there are vulnerabilities affecting entire populations is reframed in terms of individual incapacities. This is the paradox of the 2002 CIOMS Guidelines. On the one hand, they present an expanding number of vulnerable groups, on the other hand they use a definition of vulnerability in reference to the individual incapacity to protect one's own interests even if such lack of capacity is produced by more generic circumstances as insufficient power, intelligence, education, and resources. This is also the stance of the National Bioethics Advisory Commission in the United States. Criticizing current regulations for its focus on vulnerable groups, the NBAC proposes a more refined approach, based on Kipnis' analytical framework, differentiating between individuals in certain conditions. Such an approach is appealing precisely because it 'better expresses respect for persons by allowing people to be treated as individuals rather than as members of a group' [17]. Another example of the ultimate primacy of an individual perspective is the definition of vulnerability advocated by Schroeder and Gefenas. Although integrating external and internal dimensions and maintaining the language of groups, their definition continues to associate vulnerability with deficient individual ability [18].

A slightly different view takes the social context of vulnerability more seriously. This view argues that one should specifically look at 'spaces of vulnerability' or 'situations of vulnerability' [19]. It is furthermore stressed, for example in Luna's layers approach, that vulnerability should not be considered as a characteristic of groups or a label on a population [20]. It is more appropriate to focus attention on the specific features of the context that have produced and continue to produce vulnerability. But in arguing that one should abandon the language of populations and groups, this situational view ultimately reiterates the significance of the individual. The identification of vulnerability-producing features of the context is relevant as long as they impact on individual decision-making. This means that the political perspective on vulnerability in the end does nothing more (nor less) than accentuate the importance of personal autonomy and individual empowerment. This is also the disappointing conclusion from Kottow's effort to separate vulnerability from susceptibility [21]. If modern society is creating special vulnerabilities (or what he proposes to call 'susceptibilities') for its citizens and in fact makes them increasingly vulnerable, what specific *social* actions can be taken when it is argued that vulnerability as such is an attribute of mankind (according to Kottow, an anthropological condition without normative force) while susceptibility is a specific condition of individuals?

A more radical view argues that vulnerability is 'antithetical to the ethos of individualism' [22]. This is also the view taken in this book. Vulnerability refers to commonality and solidarity. When human beings are primarily social beings, groups, and communities cannot be reduced to aggregations of individuals. This view articulates that the language of vulnerable populations has a specific purpose [23]. It directs attention to another, political level of analysis: if certain social conditions do not sufficiently flourish, any talk about individual autonomy and personal control is ephemeral. Words can be deceptive. Talking about 'members' of vulnerable populations suggests that 'membership' is the result of choice and that people can decide to cancel it. But, when human beings are quintessentially social, we are always already within groups, communities, and populations whether we like it or not; this embeddedness has determined who we are. The point here is that being included in particular populations is associated with being vulnerable. People are living in conditions of poverty, deprivation, hunger, oppression, and corruption, not as a result of individual decisions; they are marginalized even if they are fully autonomous persons – for example because of illiteracy [24]. These persons do not necessarily have problems with consent or assessment of risks. However, if they participate in research they cannot be sure that the possible benefits of such research will be extended to the population they are part of. Research is carried out in a global context of injustice and this makes some populations vulnerable. The economic and social conditions are asymmetric. Inequalities are accentuated by processes of globalization. For this reason, more vulnerable people are living in less developed countries, as Michael Kottow rightly points out [25]. The vulnerability of populations does not mean that all individuals in the population are equally affected. It also does not follow that belonging to a population implies that one is *qualitate qua* vulnerable. Injustice or corruption in a country does not imply that every citizen is unjust or corrupt. There will always be just persons within an unjust environment. But the argument here is that justice can only be restored at a more systemic level, through an institutional approach. The same goes for vulnerability. If special vulnerability is produced at a systemic level it will be impossible to overcome by empowering individuals. A more categorical approach will be needed – for example strengthening the role of communities to provide opportunities and resources to preserve health or changing the conditions that give rise to disparities in health and healthcare [26]. This is presumably what makes vulnerability in the context of internationalization of research troubling, since it refers to 'broader concerns about inequalities of power and resources', as postulated by Carol Levine and colleagues in their critical analysis of the notion [27]. Addressing vulnerability is no longer an individual enterprise that can be governed with the usual principles of bioethics. The political perspective, if taken seriously, should locate and address special vulnerabilities beyond the realm of individual agency and decision-making. The challenge is to identify and scrutinize the relations and situations that produce special vulnerability. This will be the aim of the next section.

The production of vulnerability

Special vulnerabilities are the result of exacerbation of anthropological vulnerability. On the one hand, the latter, as underlined for example by Arnold Gehlen, explains the human effort to create culture and society [28]. The special vulnerabilities, on the other hand, are consequences of power differences and inequalities within the cultural and social environments that have been created. They are therefore of our own making, although at the same time determining us while we are continuously shaping and reshaping them by remodeling and restructuring these environments. The main question in the political perspective is how special vulnerabilities are produced. Only if the producing mechanisms are identified and analyzed, it can be clarified what actions will be required.

Using the notion of vulnerability, many authors have emphasized the disadvantages of African Americans as vulnerable population in the United States [29]. Social, economic, and cultural determinants have systematic negative impacts on their health and health care. Ken Fox describes the case of 17-year-old Hotep Johnson, an African American male from a Boston ghetto. Race, class, poverty, and urban violence all produce vulnerability. His individual agency is not sufficient to overcome the harms produced by these determinants affecting his health and healthcare. Because vulnerabilities are not evenly distributed across populations, Hotep like many other African Americans is a vulnerable subject embodying harmful social forces. His case illustrates, as Fox concludes, that disparities between groups and inequities in harm to health are 'beyond the agency and power of any individual sufferer' [30].

In a political perspective, vulnerability is an indicator of social inequality [31]. This is described in the detailed study of medical anthropologist Nancy Scheper-Hughes of the poor and illiterate people of the hillside shantytown Alto do Cruzeiro in northeastern Brazil [32]. They are exposed to everyday violence. The continuous threats of sickness, hunger, and death are determined by a context that is marked by a colonial history of slavery and sugar plantation economy, class and racial differences, marginal life in a *favela* without water, electricity, sewerage system and paved roads, an economic system of 'unequal exchanges and dependencies', a bureaucratic and indifferent state, and criminal and police brutality [33]. The population therefore lives in a chronic state of vulnerability with 'routinization of human suffering' [34]. The majority however regard the circumstances as unalterable. Because they have no control over the conditions in which they are living, they project vulnerability upon themselves. They consider their vulnerability as the result of individual weakness rather than systemic exploitation [35].

Detailed cases and elaborated studies like those of Fox and Scheper-Hughes help to identify and explain the mechanisms that systematically produce vulnerability. It is clear that the focus is not on vulnerable subjects but rather on the social, economic, and political processes to which they are subjected. Understanding these processes will require another language than that of individual autonomy and decision-making. Particularly the vocabulary of violence, inequality, and power ought to imagine and explain how special vulnerabilities are produced.

Structural violence

Paul Farmer, another medical anthropologist who has extensively described the connections between poverty and disease, attributes the plight of his patients in Central Haiti to structural violence. People suffering from AIDS and tuberculosis in these rural areas are not only poor and hungry, but are also subjected to political violence and discriminated against on the basis of gender, race, and ethnicity. Their suffering is structured by historical and economic processes and forces that constrain agency [36]. In order to improve their health, it is important to analyze the mechanisms and conditions that produce vulnerability.

'Structural violence' occurs when there is no acting subject committing violence. Individuals may be harmed or exploited but the violence is indirect; it is 'built into the structure' [37]. This type of violence is not visible in concrete actions; it is not personal since it is not the result of individual acts of violence; it is often not intended; and most importantly, it is not readily evident, and thus difficult to perceive. Though always individual persons are hurt, violence cannot merely be understood as individual experience. It is the manifestation of a harmful and injuring social context [38]. If people die from a treatable disease such as tuberculosis because medication is not accessible or affordable, they are harmed from structural violence, even if no specific person is committing any violence. Another example is starvation. If people starve to death while this is avoidable because there is no lack of food at a global level, violence is committed. This is the consequence of the global organization of economic relations; in other words: structural violence [39]. The basic mechanism is unequal distribution of power and resources.

Inequality

Structural violence is exposed in social structures and conditions that are not egalitarian. Farmer provides many examples of the pathology produced by inequality [40]. In Chiapas a substantial part of the population is excluded from the benefits of the development of the country, Mexico. In the Russian Federation, and after *perestroijka*, money is no longer available for young prisoners to treat tuberculosis. And in Haiti, people who need surgery must travel to the capital city and bring money for services and equipment. Farmer is convinced that inequality is the fundamental ethical problem of our time [41].

Over the last few decades, extensive scholarship has time and again demonstrated the correlations between inequality, poverty, and deterioration of health. Many of these studies show that health conditions are determined by social circumstances and class structures [42]. Within countries morbidity and mortality rates among socio-economic groups are significantly different. Especially extreme poverty leads to avoidable death and disease. Increasing interest for global health has also focused on social inequalities in health between countries. It is generally assumed that the role of medical care in generating and addressing these inequalities is rather limited [43].

The focus on social inequality as an underlying mechanism for structural violence producing special vulnerability emphasizes primarily that resources are not evenly distributed. Populations or groups of people are made vulnerable because they are without access to education, housing, and healthcare. They are poor and hungry because wealth, income, and jobs are not evenly distributed. There are several determinants, often interacting, that contribute to social inequalities, such as race or ethnicity, gender, religious belief, and social class [44]. The implication is that the benefits of scientific and technological advances are not available for everyone. While the notion of anthropological vulnerability can be used to explain why all human beings are susceptible to threats such as infectious diseases, the notion of special vulnerability emphasizes that vulnerability is only exacerbated for those populations that are in disadvantaged situations created by inequalities in social, economic, and political contexts.

Power

Although Johan Galtung agrees that the underlying mechanism of structural violence is inequality, he has also pointed out that equality will not be enough to remove the violence. As long as the process of decision-making about distribution is not addressed, more equal distribution of resources will not have any real effect on structural violence [45]. The notion of power is therefore more fundamental in explaining how vulnerability is produced, particularly the difference in power to decide over the distribution of resources. But it is important to note that the notion of power, like that of violence, is ambivalent. Different types of inequality result from different types of power. On the one hand is the inequality of capacities; in this case the power of persons to take charge of their own interest is limited due to impaired rationality and decision-making capacity. On the other hand is the inequality of situations; here the power to utilize material and immaterial resources is different (for example, between the rich and poor, and between strong and weak states) [46].

In connection to the production of special vulnerability, it is not the power of individuals that matters. Certain people have power as a result of an official position, a particular personality or a profession such as medicine. What is relevant are the sources of unofficial power that may produce structural violence.

Social, economic, and political circumstances can be such sources of power. They determine that men have power over women, landowners over shantytown people, wealthy persons over the poor, and researchers over research subjects. This power is not formally assigned to persons but it is embedded in the informal arrangement of societies and cultures. But as all power, it affects the actions of other people, it provides leverage in 'influencing the other to act or refrain from acting in a certain way' [47]. The moral question should be how this power is exercised.

That differences in power rather than in equality of distribution explain how special vulnerabilities are produced, is from another point of view also argued by Iris Young. Her well-known distinctions between five faces of oppression assumes that there are structural phenomena that make groups vulnerable [48]. People suffer disadvantages not because of powerful individuals or because of particular acts but because norms of distinction, privilege, hierarchy, status, and authority are built into everyday practices. These power relations determine that groups of people are treated unfairly and unequally. The phenomenon of exploitation is discussed in connection to special vulnerability and is a particular focus of the next section. It is regarded as the major threat for vulnerable people and populations.

Exploitation

Identifying persons, groups, or populations as vulnerable implies that they are at risk of being harmed. They can be subjected to many different forms of wrongful treatment: neglect, abuse, discrimination, exclusion, and oppression. One particular concern, frequently associated with vulnerability, is exploitation. Taking advantage of someone's vulnerability to extract benefits from him is intuitively regarded as morally wrong. Instead of using vulnerable subjects for a particular selfish purpose, one should help them.

The language of exploitation is used in many bioethical debates, for example on commercial surrogate motherhood, selling of organs, non-therapeutic medical experiments on prisoners, cancer patients hoping for cures, and clinical research in developing countries [49]. The question of exploitation is raised when healthcare is regarded as a commodity that is only available for people who can afford to pay [50]. Entire communities can be exploited through international trade agreements that regulate prices of pharmaceutical products making them unaffordable to many people [51]. As discussed in earlier chapters, exploitative practices, especially in medical research, have instigated the development of ethical guidelines at the national and international level. The occurrence of actual cases has also nourished the concern that particular individuals, groups, and populations are more at risk of being exploited than others. One might argue therefore that the increasing awareness of the possibility of exploitation is associated with the growing interest in the notion of vulnerability.

The extensive scholarly ethics literature on exploitation is rather recent; it started almost simultaneously in business ethics (with sweatshop labor as a paradigm case) and bioethics [52]. Historically, the concept was used to analyze economic interactions, especially following Marxist and neo-Marxist theories, but more recently it was applied to non-economic interactions for example within the professional contexts of research and healthcare as well as personal and intimate relationships. This shift is significant for the political perspective of vulnerability since it resulted in focusing the issue of exploitation at the individual rather than social level. But first it will be important to examine how the relationship between exploitation and vulnerability is generally conceived.

Wide agreement exists that it is morally problematic to use vulnerable people for purposes that are not beneficial to them. At the same time, there is not one single theoretical account that explains all types of exploitation and different exploitative practices. There is not agreement on the concept itself nor its implications [53]. Four explanations are commonly offered in the numerous publications for the wrongfulness of exploitation: injustice, lack of freedom, disrespect, and vulnerability.

Injustice

Frequently cited is Alan Wertheimer's description of exploitation as a transaction in which one party takes unfair advantage of another [54]. The focus is on the distributive outcome of a transaction. Whether the outcome is fair is measured with the hypothetical competitive market price. According to this market approach, an impoverished vendor of an organ is not exploited when her vulnerability is neutralized by offering a fair price for her body parts [55]. In this theory, harm or invalid consent are not relevant considerations. Many exploitative practices are consensual and mutually advantageous agreements. The people in Alto do Cruzeiro, for example, would starve without work offered by plantation owners, even if the wages are low. Participating in clinical trials is for many persons in developing countries the only way to receive potential treatment and some health care, even if only for a limited time. The relevant question in this perspective is whether such transactions are fair. The problem with this view is that the focus on transactional fairness is thin [56]. Regarding exploitation as taking advantage of injustice implies that harm is actually done, even when there is mutual advantage. One party is gaining at another's expense so that a loss is inflicted on this other party because the exploiter fails to benefit the exploited party. This loss cannot be compensated by paying a higher price. Furthermore, unfair exploitation cannot be confined to exchange interactions. Justice requires that people have equal access to healthcare and that labor will at least lead to a decent standard of living.

Lack of freedom

A second explanation why exploitation is wrong emphasizes lack of freedom. It is not the distribution of the product of a transaction that matters but the conditions of the transfer. Wrongful advantage is taken of others in social arrangements that are coercive, dominating, and alienating. Exploitation is explained in reference to the process rather than the outcome of a transaction [57]. It usually occurs in a context in which there is not a free exchange between equal parties or in circumstances that are primarily controlled by one party [58]. Redistribution of benefits will not help as long as freedom is denied.

Disrespect

Disrespect is a third factor that makes exploitation wrong. According to this theory, exploitation is wrong, even if no injustice or lack of freedom is involved, because it is an affront to human dignity, a failure of respect for persons. Advantage is taken of other people or their situations, and they are merely treated as means for the purposes and interests of a stronger party. What is necessary for their flourishing or for fulfilling their basic needs is ignored. Such treatment is humiliating and degrading [59].

Vulnerability

The fourth explanation relates exploitation explicitly to vulnerability. Crucial for exploitation is that someone's vulnerability is used in order to derive some benefit for oneself. Because of his vulnerability the exploitee is not able to influence the conditions of the interaction. His weakness or desperate situation enables the exploiter to extract benefits [60]. The most known theory is Goodin's [61]. In his view, exploitation is violation of the duty to protect the vulnerable. Vulnerability creates special responsibilities because the vital interests or needs of the vulnerable subject are not met. Goodin assumes that in most cases vulnerability is created by social arrangements. Exploitation therefore must be determined in relation to special vulnerability. It depends on the context and circumstances whether the interests of some people are affected by the actions of others, and whether someone's vital interests or basic needs are at stake [62].

Exploitation and power differences

The four theoretical accounts offer different explanations of exploitation but they all more or less assume the relevancy of vulnerability [63]. Exploitation may only occur in a setting where one person has an advantage over another. There always is an initial need, insufficiency, or dependency to begin with that provides an opportunity to the exploiting party. Such unequal balance can best be interpreted in terms of power. Differences in power explain exploitation because people are dependent due to unequal standing in society or social exclusion [64]. Exploitation can briefly be defined as 'self-interested exercise of power' [65]. Thus, the prime concern is to eliminate dominance so that people become less vulnerable.

Exploitation refers to situations in which people often have little choice; they are elements in a system where they have no control over the process and outcome of interactions. Exploitation as lack of freedom and degradation thus also implies powerlessness [66]. Even unfairness accounts of exploitation assume that vulnerability exists, whether or not it can be neutralized. Mutually advantageous exploitation, as in the case of Alto do Cruzeiro or clinical research with the poor, is still a choice between two threats, i.e. being utilized as a tool for production or knowledge and being crippled by starvation or untreated illness [67].

The emphasis on power in explanations of exploitation provides a different perspective on vulnerability. Exploitation is in fact an ambiguous term. Initially used in political discourse, liberal theories interpreted it in terms of distribution and inequality, while (neo)Marxist interpretations emphasized differences in power. This ambiguity allows that the term may be used to divert attention from the structural to the personal level [68]. One might argue that this is exactly what is happening when the term is transferred from the political into the bioethical discourse. The focus is on particular transactions between two parties, not on the relationship between them and the context in which they are interacting. When the notion of exploitation is used in bioethics it is effectively decontextualized. Whether the relationship or the context is exploitative is not regarded as relevant for determining whether an interaction is exploitative. Wertheimer commonly is taken as the major exponent of this view. His theory of exploitation, focusing on distribution rather than power, makes a distinction between taking unfair advantage and taking advantage of unfair circumstances. The moral assessment of the particular transaction should be distinguished from the moral assessment of the background conditions of that transaction. The transaction can be fair while the conditions may be unjust [69]. The problem with this view is that it acknowledges the asymmetry of the exchange relationship but disregards the source of the inequality [70]. The emphasis on distribution as the main challenge is pinpointing the discussion on the question whether the benefits and burdens of the exchange are fairly distributed. The emphasis on power would have elicited the question why the exchange is unequal in the first place, and would then have directed attention towards the context of transactions.

Structural exploitation

The debate whether and how exploitation is determined by social, economic, and political background conditions is directly relevant for the political perspective on vulnerability. The emphasis on power in connection to exploitation enables this perspective to articulate the role of 'structural exploitation'. According to the political perspective, vulnerability is manifested in individuals but not a characteristic of individuals since it is produced by the circumstances in which they live. The background conditions determine the vulnerability and therefore the possibility of exploitation. In order to clarify the debate Jeremy Snyder has proposed a distinction between micro and macro-levels of exploitation [71]. The first level exclusively analyzes the exchange between worker and employer or between research sponsor and research subject without including concerns about the structural setting (sweat-shop labor or clinical research in poverty-ridden populations). The second level is concerned with systemic background disadvantages. This distinction has allowed more focus on 'structural exploitation', i.e. on exploitative systems and practices in which discrete transactions take place [72]. Of course, the major question is how the two levels are related. There are many examples where background conditions do generate exploitative transactions [73]. The tuberculosis patients of Farmer in Haiti suffer under exploitative conditions. If there is pervasive inequality, marked by a history of racism, colonialism, or marginalization that is systematically disadvantaging certain populations, it seems to make all transactions with individuals from these populations exploitative. But in practice there might be exceptions. Not all transactions might be affected. Wertheimer is correct that structural exploitation does not necessarily make every particular transaction exploitative.

But if political, social, and economic structures themselves can be exploitative, this is a relevant consideration in the moral assessment of transactional exploitation. It will be impossible to maintain a moral firewall between the background conditions and individual transactions for at least two reasons.

Exploitative arrangements

First, focusing on individual transactions masks the existence and continuation of systematic injustice. Concentrating on vulnerable individuals ignores that powerful parties benefit from situations that produce special vulnerability but also often perpetuate them to their own advantage. The people in Alto do Cruzeiro will starve without work; plantation owners offer them work but with a payment hardly enough to survive. No alternative employment is available. People will be better off by accepting the offer but still they are at the same time exploited. The transaction may be played according to the rules of the game, but the game itself is unfair. Scheper-Hughes gives an example of how a focus on individual transactions may keep out of sight that exploitation is built into the structure of the situation [74]. The squatters of the Alto do Cruzeiro make a difference between good and bad bosses. The abusive and exploitative boss is not threatening the system of exploitation; he simply is an aberration. The good patron is worse. He supports and assists the exploited worker but he also upholds the system. Because he is so benevolent and humane it becomes even more difficult for the workers to escape from the power relations and to transform the exploitative system. This example illustrates that analyzing power (and thus vulnerability) is more important to understand structural exploitation than concentrating on unequal distribution of resources among individuals. Power differences have created and are reinforcing such distribution. A more equal distribution of resources will not change these differences in power [75].

Similar arguments can be used for clinical research in developing countries. Research subjects may benefit and voluntarily agree to participate. In the micro-level account, these transactions therefore cannot be assessed as exploitative, even if the subjects are not offered the normal standard of care, or the results of the research are not made reasonably available when the trial is over. Why should research sponsors address structural exploitation if their transactions with particular research subjects are not exploitative?[76] This argument ignores that research sponsors are often responsible for the background conditions in which research takes place. Thomas Pogge, for example, has shown how multinational pharmaceutical companies themselves have contributed to an economic order in which medication is so expensive that it is unaffordable to the poor. They can therefore not use the argument that the normal standard of care is not available in many countries. The unavailability of medication is not a fact of nature but the result of their own policies and lobbying efforts [77]. Off-shoring clinical research to developing countries is taking advantage of the special vulnerability produced by desperate conditions to which companies themselves have contributed. At the same time, the possible benefits of such research are often not shared between participants and beneficiaries who are usually patients in developed countries. Vulnerable populations in developing countries participate in research efforts to improve global healthcare, but in many cases they themselves will not benefit from their participation [78]. The conclusion of this first consideration is that a focus on transactional exploitation alone restricts the ethical analysis. If exploitation can happen at micro as well as macro levels, a comprehensive ethical assessment should examine exploitative arrangements at both levels and not ignore one level by focusing exclusively on individual transactions.

Social connectedness and political responsibility

The second consideration concerns the issue of responsibility. Scholars who attempt to separate background conditions from discrete transactions argue that even if it is granted that structural exploitation occurs, it does not follow that there is any ethical obligation, let alone any practical opportunity to change this. If there is systemic disadvantage, in the words of Wertheimer, '...it is unreasonable to expect the better-off party to repair those background conditions by adjusting the terms of a particular transaction' [79]. The unfortunate conditions are not caused by individuals. There are several ways to respond to this issue. One is that the focus on transactional exploitation still assumes that it is based on individual interactions. In practical reality, the more powerful party is not one individual, but he usually represents a network of actors while vulnerable subjects are often isolated and not embedded in a supportive network. In clinical research for example the person who is conducting research is part of a team of researchers, is supported by institutions, and is funded by sponsors. In healthcare (at least in the U.S.), the majority of individual physicians are associates of a practice or hospital, and are contracted by health care organizations [80]. The more powerful party can therefore mobilize many more resources than an individual actor.

It is also argued that the interaction is often initiated by the more powerful party, especially in the case of research. In healthcare it is usually the other way around; it is the vulnerable party who initiates the interaction. But in both cases the context is different from the commercial market setting. Instead of an exchange between two parties, there is a fundamental difference in position, and in power. As discussed earlier, this asymmetry in relationships gives rise to obligations that are stronger for one party than for the other.

Finally, the emphasis on the relational setting can be made more intense by broadening the context. Structural exploitation is often the result of global injustice. One cannot expect that individuals will be able to redress such systemic injustice. A collective response will be required. Lack of macro-fairness demands engagement and action at macro level [81]. In this context reference is made to the idea of social connectedness [82]. If one is interacting with vulnerable subjects and populations in research or care, one has not only an individualistic responsibility to promote the good of the individual subjects, but also a political responsibility to address the disadvantaged context. This latter responsibility is a shared one and it seeks collective responses to institutional injustice. Special vulnerability produced by social, economic, and political conditions can provide opportunities for structural exploitation; individual interactions within such settings will confer political responsibility to these individual actors (and the networks, institutions, agencies, and companies they are representing) in order to address background conditions so that people are less vulnerable to exploitation.

A global perspective

The aim of this chapter has been to analyze and clarify the political perspective on vulnerability as relevant for present-day bioethics. Special vulnerability is often contrasted with anthropological vulnerability. This chapter has argued that both notions are connected. The political perspective on vulnerability combines the basic idea of anthropological vulnerability that human beings are necessarily dependent, relational, and connected, with a particular view of the social context. On the basis of the first idea, vulnerability cannot be understood as only an individual characteristic. Because humans are primarily social beings, individual autonomy will emerge and flourish within a nurturing social environment. But this context can also impede and restrict personal flourishing, producing special vulnerabilities. It is therefore useful to examine the social, political, and economic conditions that exacerbate human vulnerability in order to know how best to address these special vulnerabilities. The discourse of vulnerability, and therefore also of exploitation, is inevitably linked to social, political, and economic concerns. Individual ethics is not sufficient to understand and address vulnerability, but social ethics will be required at the same time.

The bioethical debate on exploitation illustrates the challenges that will be encountered. Bioethics usually assumes the unquestionable primacy of the individual point of view, emphasizing personal autonomy. Interpreting exploitation as a transaction between A and B reflects the preference for an individualist perspective. This is the kernel of the critique of Wertheimer's interpretation of exploitation: he is positioning each of the actors as rational and independent subjects that are distinct from the social and cultural context [83]. The relation between individual and society is therefore already formatted upon a specific but unarticulated political choice [84]. Societal interest or the common good are reduced to preferences of individuals or discrete groups [85]. One of the implications of this liberal individualist approach is that it makes the production of vulnerability less visible. Concentrating on individual transactions gives the impression that they are free and beneficial while the constraints, dependencies, and vulnerabilities in the background are ignored [86]. It also promotes the view that background conditions are only relevant if they are manifested in the vulnerability of individuals. These conditions only produce vulnerability if they create internal psychological disability. Exploitation is then an individual problem rather than a social or economic one [87].

Another implication of the focus on individual autonomy is that it privileges policies in a specific direction [88]. Harm to be avoided is determined by negative liberty, i.e. not interfering with the choices of autonomous persons. For example, if poor women in India voluntarily contract for surrogate motherhood because they need money, it is a mutually advantageous agreement. It is not exploitative as long as the compensation is adequate. As such, these choices should not be prevented through national or international regulations. The idea that policies can aim at preventing commodification or lack of respect vanishes into irrelevancy.

What can be done to address vulnerability? How to cope with anthropological vulnerability and how to mitigate special vulnerabilities? Most efforts are aimed at protecting vulnerable subjects and populations. As discussed in Chapter 3, following guidelines and applying constraints promulgated by a range of national, regional, and international organizations can make sure that in individual interactions in research and care the vulnerability of subjects, groups, and populations will be respected and taken into account as significant moral concerns. But from a political perspective this will not be sufficient. If vulnerability is exacerbated, and is often produced and sustained through social practices and circumstances, important efforts to mitigate, reduce, and eliminate vulnerability should be aimed at these conditions. These efforts will contribute to prevent the production of vulnerability. But many of the social, political, and economic circumstances in which people live are damaging and empowering at the same time [89]. This is clear when the global context is taken into account. Vulnerability in this context manifests another dimension that is often not addressed since the typical focus is on individual vulnerable persons. This will be the subject of the next chapter. What in the processes of globalization is articulating, producing, or exacerbating human vulnerability? And what are the possibilities for protecting vulnerable subjects and populations at this global level, and more importantly, to curb the production of special vulnerability?

Notes

1   Thomasma, 'The vulnerability of the sick', 5.
2   Butler, *Precarious life. The powers of mourning and violence*, 29.
3   Butler, *Frames of war. When is life grievable?* 13; Butler refers here to a normative framework of 'a more radical and effective form of egalitarianism'. (Butler, xxii).
4   Butler makes a distinction between 'precariousness', the existential condition of shared vulnerability, and 'precarity', the differential vulnerability resulting from social and political forces, or in her words, '…the politically induced condition that would deny equal exposure through the radically unequal distribution of wealth and the differential ways of exposing certain populations, racially and nationally conceptualized, to greater violence' (Butler, *Frames of war. When is life grievable?* 28).
5   Kittay, *Love's labor. Essays on women, equality, and dependency*, 15. The difference is nicely summarized by Kittay: 'Among the vulnerable, not all cases are equal'. (Kittay, Ibidem, 174).
6   Kittay, *Love's labor. Essays on women, equality, and dependency*, 50.
7   The presupposition of this argument is that medicine can be defined as a specific kind of relationship: 'the relationship of healing in which one person in need of healing seeks out another who professes to heal, or to assist in healing'. (Pellegrino and Thomasma, *A philosophical basis of medical practice. Toward a philosophy and ethics of the healing professions,* 5; see also 65, 173).
8   Thomasma, 'The vulnerability of the sick', 8.
9   Zaner, *Ethics and the clinical encounter*, 55, 84 ff. See also Zaner, 'Power and hope in the clinical encounter: A meditation on vulnerability' and Zaner, 'Physicians and patients in relation: Clinical interpretation and dialogues of trust'.
10 Pellegrino, *Humanism and the physician*, 117 ff; Pellegrino and Thomasma, *The virtues in medical practice*, 42 ff, 149.
11 The dialectics of taking and givenness is elaborated by Staudigl, who seeks to explain why violence was not sufficiently analyzed in phenomenology (Staudigl, 'The vulnerable body: Towards a phenomenological theory of violence', 268–269).
12 See Chapter 4, note 56; O'Neill, *Towards Justice and Virtue. A constructive account of practical reasoning,* 192.
13 Morin, *Introduction à la pensée complexe*, 89.
14 Ricoeur, *Reflections on the Just,* 72 ff.
15 See Callahan, 'The vulnerability of the human condition'.
16 Hughes, 'The primacy of ethics: Hobbes and Levinas', 86 ff.
17 National Bioethics Advisory Commission, *Ethical and policy issues in research involving human participants. Volume I: Report and recommendations of the National Bioethics Advisory Commission*, 91.
18 See Schroeder and Gefenas, 'Vulnerability: Too vague and too broad?'
19 Delor and Hubert, 'Revisiting the concept of "vulnerability."', 1562; Eckenwiler, Ellis, Feinholz, and Schonfeld, 'Hopes for Helsinki: reconsidering "vulnerability"', 765; Watts and Bohle, 'The space of vulnerability: The causal structure of hunger and famine', 44 ff.

20 Luna, 'Elucidating the concept of vulnerability: Layers not labels'. See also DeBruin, 'Reflections on "vulnerability"'.

21 See Kottow, 'Vulnerability: What kind of principle is it?'

22 Hoffmaster, 'What does vulnerability mean'?", 42.

23 See Flanigan, 'Vulnerability and the bioethics movement'; Fox, 'Hotep's Story: Exploring the Wounds of Health Vulnerability in the US; Stone, 'Race and healthcare disparities: Overcoming vulnerability'.

24 Luna, *Bioethics and vulnerability. A Latin American view*, 31 ff.

25 Kottow, 'The vulnerable and the susceptible', 471.

26 These suggestions are made by Flaskerud and Winslow, 'Vulnerable populations and ultimate responsibility' and Blacksher and Stone, 'Introduction to "vulnerability" issues'.

27 Levine *et al.,* 'The limitations of "vulnerability" as a protection for human research participants,' 45.

28 See chapter 5; Gehlen, *Man. His nature and place in the world.*

29 See, for example, Aday, *At risk in America: The health and health care needs of vulnerable populations in the United States;* Shi and Stevens, *Vulnerable populations in the United States;* Stone, 'Race and healthcare disparities: Overcoming vulnerability'.

30 Fox, 'Hotep's Story: Exploring the Wounds of Health Vulnerability in the US,' 489.

31 Nichiata, Bertolozzi, Takahashi and Fracolli, 'The use of the "vulnerability" concept in the nursing area'.

32 Scheper-Hughes, *Death without weeping. The violence of everyday life in Brazil.*

33 '...slaves on the original sugar plantation of Pernambuco were better fed than today's rural wage laborers because the slave masters had a greater concern for the well-being of their tools of production'. (Scheper-Hughes, *Death without weeping,* 153).

34 Scheper-Hughes, *Death without weeping,* 16.

35 Scheper-Hughes, *Death without weeping,* 173. The author describes how the vulnerability of everyday life is individualized and transformed into medical problems: 'Through the idiom of *nervos,* the terror and violence of hunger are socialized and domesticated, their social origins concealed'. (Ibidem, 214).

36 Farmer, *Pathologies of power,* 40.

37 Galtung, 'Violence, peace, and peace research', 171. Galtung is commonly credited for having introduced the notion of 'structural violence' in the scholarly discourse.

38 In the words of Farmer, 'large-scale social forces crystallize into the sharp, hard surfaces of individual suffering'. (Farmer, 'On suffering and structural violence: A view from below.' 263).

39 Galtung, 'Violence, peace, and peace research', 171.

40 Farmer explains that the central thesis of his study is that 'a rising tide of inequality breeds violence'.(Farmer, *Pathologies of power,* xxvii). With many examples he illustrates what he calls the 'pathogenic role of inequity'.(Farmer, *Pathologies of power,* 20).

41 Farmer, *Pathologies of power,* 209.

42 See, for example, Beauchamp, *The health of the republic;* Mann, Gruskin, Grodin, and Annas (eds.). *Health and human rights.* Marmot, 'Social differentials in health within and between populations'. Marmot and Wilkinson (eds.). *Social determinants of health;* Wilkinson, *Unhealthy societies. The afflictions of inequality.*

43 Marmot, 'Social differentials in health within and between populations', 209. The Brazilian moral theologian Marcio Fabri dos Anjos has summarized this scholarship arguing that contemporary global health is characterized by 'unequal conditions in national health programs, biomedical research and technology, and opportunities for

social progress…' He draws the ethically relevant conclusion that more than anything else it is increasing social and economic inequality that is damaging health, quality of life, and well-being (Dos Anjos, 'Medical ethics in the developing world: A liberation theology perspective', 633).

44 Farmer, *Pathologies of power*, 219.

45 '…it is the way in which decisions about distribution are arrived at and implemented that is basic'. (Galtung, 'Violence, peace, and peace research', 188).

46 Maillard, *La vulnérabilité. Une nouvelle catégorie morale?* 191.

47 Wilson, 'In one another's power', 304.

48 Young distinguishes five structural phenomena of oppression: exploitation, marginalization, powerlessness, cultural imperialism, and violence. Young, *Justice and the politics of difference*, 40 ff.; Young, 'Five faces of oppression', 56. See also Dong and Temple, 'Oppression: A concept analysis and implications for nurses and nursing'.

49 See, for example, Ashcroft, 'Money, consent, and exploitation in research', Ballantyne, 'HIV international clinical research: exploitation and risk', Ehrenreich, 'Conceptualism by any other name', Ganguli-Mitra, 'Off-shoring clinical research: Exploitation and the reciprocity constraint', Gbadegesin and Wendler, 'Protecting communities in health research from exploitation', Hawkins and Emanuel (eds). *Exploitation and developing countries*, Hill, 'Exploitation', Jecker, 'Exploiting subjects in placebo-controlled trials', Kuntz, 'A litmus test for exploitation: Taylor's *Stakes and Kidneys*', Martin, 'Hope and exploitation', Phillips, 'Exploitation in payments to research subjects', Resnik, 'Exploitation and the ethics of clinical trials', Resnik, 'Exploitation in biomedical research', Wilkinson, 'The exploitation argument against commercial surrogacy', Wood, 'Exploitation'.

50 Mayer, 'A Walzerian theory of exploitation', 346-347.

51 Snyder, 'Exploitation and sweatshop labor: Perspectives and issues', 205.

52 On the sweatshop case, see Arnold, 'Exploitation and the sweatshop quandary'; Mayer, 'Sweatshops, exploitation, and moral responsibility'; Snyder, 'Exploitation and sweatshop labor: Perspectives and issues'.

53 For some authors, exploitation is a neutral, purely descriptive notion, referring to using people or taking advantage but not always wrong. See Mayer, 'What's wrong with exploitation?'; Schwartz, 'What's wrong with exploitation?'; Valdman, 'Exploitation and injustice'; Wood, 'Exploitation'. For others, it is a normative notion that already has a negative qualification so that exploitation always implies wrongful use. (Logar, 'Exploitation as wrongful use: Beyond taking advantage of vulnerabilities'; Wertheimer, *Exploitation*; Zwolinski, 'Structural exploitation'.) These meaning differences are related to specific languages. In the Dutch language for example, there is a difference between utilizing and exploiting. Exploitation is only used for a commercial context; for the application for persons a different term is used that is necessarily negative ('uitbuiten', 'uitmelken', or 'uitzuigen').

54 Wertheimer, *Exploitation*, 10.

55 Kuntz, 'A litmus test for exploitation: Taylor's *Stakes and Kidneys*', 558.

56 See, Mayer, 'A Walzerian theory of exploitation'; Mayer, 'What's wrong with exploitation?'; Valdman, 'A theory of wrongful exploitation'; Zwolinski, 'Structural exploitation'.

57 Schwartz, 'What's wrong with exploitation?' 160, 161, 172 ff.

58 Holmstrom, 'Exploitation', (1977) 357–358.

59 'I argue that exploitation is taking advantage of another's vulnerability in a way that fails to respect the personhood of that being' (Sample, *Exploitation: What it is and why*

it's wrong, 164). See also Snyder, 'Needs exploitation'; Snyder, 'Exploitation and sweat-shop labor: Perspectives and issues'.

60 Wood, 'Exploitation', 142 ff. Valdman explains exploitation in terms of leveraging someone's desperation (Valdman, 'Exploitation and injustice', 570).

61 See Goodin, *Protecting the vulnerable. A reanalysis of our social responsibilities*; Goodin, 'Exploiting a situation and exploiting a person'.

62 It can be argued that defining exploitation as taking advantage of vulnerabilities is too broad and general (Logar, 'Exploitation as wrongful use: Beyond taking advantage of vulnerabilities', 333 ff.; see also, Sample, *Exploitation: What it is and why it's wrong*). In such cases, all human interactions would be exploitative. That there is always the danger of exploitation can be due to anthropological vulnerability, but determining when and how exploitation occurs depend on special vulnerability. Carse and Little for example argue that one should distinguish 'actual' vulnerabilities from the more general notion of vulnerability (Carse and Little, 'Exploitation and the enterprise of medical research', 211). But in order to specify what conditions are exploitative, one has to specify what features of special vulnerability are relevant. Sample develops the theory that exploitation implies taking advantage of basic needs. This might work well to explain exploitation in case of illness, hunger, and poverty but cannot explain exploitation in personal and intimate relationships (Logar, 'Exploitation as wrongful use: Beyond taking advantage of vulnerabilities', 340, 343). It furthermore requires an elaboration of relevant basic human needs.

63 The exception is Alan Wertheimer.

64 '...the neediness of the other makes the advantaged party dominant' (Mayer, 'A Walzerian theory of exploitation', 348).

65 Wilson, 'In one another's power', 307. The moral wrong of exploitation 'lies precisely in the fact that one takes advantage of vulnerability'. (Ganguli Mitra and Biller-Andorno, 'Vulnerability and exploitation in a globalized world', 96).

66 That exploitation is the result of power over vulnerable subjects and groups is expressed in different ways. Holmstrom makes a strong statement: 'Force, domination, unequal power and control are involved in exploitation both as preconditions and as consequences'. (Holmstrom, 'Exploitation', (1977) 364; see also Holmstrom, 'Exploitation', (2013). Goodin regards exploitation as 'abuse of power', and McLaughlin as a form of social power (Goodin, 'Exploiting a situation and exploiting a person', 184; McLaughlin, 'The ethics of exploitation', 9). Also Vrousalis interprets exploitation as domination for self-enrichment; the exploiter takes advantage of his power over the exploitee; he is instrumentalizing another's vulnerability (Vrousalis, 'Exploitation, vulnerability, and social domination').

67 Buchanan explains that in a Marxist perspective exploitation is 'the harmful utilization of a person as a mere means to one's advantage'. The point is not the unjustice but the fact that exploitation is a 'form of servitude'. (See, Buchanan, 'Exploitation, alienation, and injustice', 135, 136). The same point is made by Carse and Little: 'Exploitation, at its heart, is not just about unfair shares or garden-variety degradation; it is about wrongfully exacting benefit from another's actual vulnerability' (Carse and Little, 'Exploitation and the enterprise of medical research', 211).

68 Galtung, 'Violence, peace, and peace research', 171, 188.

69 Wertheimer, *Exploitation*, 298; Wertheimer, 'Exploitation in clinical research', 76, 84.

70 As pointed out by Ganguli Mitra and Biller-Andorno, 'Vulnerability and exploitation in a globalized world', 98.

71 Snyder makes a distinction between micro fairness and macro fairness, but there is no reason to limit this distinction only to fairness-based accounts of exploitation (Snyder, 'Exploitation and sweatshop labor: Perspectives and issues', 189 ff.). Commercial surrogacy for example can be degrading for individual women but it can also be regarded as disrespectful because it implies commodification of women as a group (see Ehrenreich, 'Conceptualism by any other name').

72 Zwolinski distinguishes transactional and structural exploitation (Zwolinski, 'Structural exploitation', 158 ff.). The emphasis on structural exploitation in fact resuscitates the initial political discourse of exploitation (considering exploitation as a feature of an economic system).

73 Sample, *Exploitation: What it is and why it's wrong*, 97–98. See also Ballantyne, 'How to do research fairly in an unjust world'. According to Elster, exploitation can take place 'through the impersonal mechanism of the market' (Elster, 'Exploring exploitation', 4).

74 Scheper-Hughes, *Death without weeping. The violence of everyday life in Brazil*, 126.

75 See, Holmstrom, 'Exploitation', (2013).

76 For this kind of reasoning, see Wertheimer, 'Exploitation in clinical research', 95 ff.

77 'U.S. pharmaceutical companies are causally involved both in the high level of drug prices and in the persistence of severe poverty in the so-called developing countries' (Pogge, 'Testing our drugs on the poor abroad', 114).

78 This argument has been developed by Agomoni Ganguli Mitra. She criticizes the micro-fairness approach; she also points out that there is little evidence that clinical research is providing any benefits to participants in developing countries and that it is questionable whether off-shored clinical research contributes at all to strengthening healthcare infrastructure. She concludes that 'we are asking individuals to put their health and well-being at risk for a system that does not attend to their needs as patients' (Ganguli-Mitra, 'Off-shoring clinical research: Exploitation and the reciprocity constraint', 5).

79 Wertheimer, *Exploitation*, 234.

80 In 2008 in the U.S. the vast majority of physicians have managed care contracts (87.6%) and only one-third is working in solo or two-physician practices (see: Boukus, Cassil, O'Malley, 'A snapshot of U.S. physicians: Key findings from the 2008 health tracking physician survey', http://www.hschange.com/CONTENT/1078/) (Accessed 10 October 2013).

81 Snyder, 'Multiple forms of exploitation in international research: The need for multiple standards of fairness', 41; Snyder, 'Exploitation and sweatshop labor: Perspectives and issues', 192 ff.

82 Young, *Responsibility for justice*, 104 ff. See also Young, 'Responsibility and global labor justice'; Young, 'Responsibility and global justice: A social connection model'.

83 The presupposition is that each actor is an 'essentialized subject – a self-creating person separate from social context, who responds to culture rather than being constituted by it'. (Ehrenreich, 'Conceptualism by any other name', 1304).

84 Robert Goodin has criticized Wertheimer's book on exploitation as rooted in 'bourgeois liberal microeconomics' (Goodin, 'Exploitation', 733). He also criticizes his use of 'nonstandard incidents' (Goodin, 'Exploitation', 734).

85 Ehrenreich, 'Conceptualism by any other name', 1300.

86 Holmstrom refers to the Marxist interpretation of this emphasis on individual interaction: 'According to Marx, the exchange between capitalist and worker appears free only if we take a narrow look just at the exchange itself as occurring between an

individual worker and capitalist. Looking at the social background of the exchange, and looking at the relationship, not between an individual worker and an individual capitalist, but between the working class and the capitalist class, we get a very different picture. What looks free and independent is in reality totally dependent and unfree' (Holmstrom, 'Exploitation' (1977), 367).

87 Hill, for example, regards vulnerability as an internal psychological condition. He argues that 'poverty and political oppression only constitute a vulnerability if they create some internal psychological disability' and he concludes: 'Exploitation is not a social or economic construct'. (Hill, 'Exploitation', 699).

88 Ehrenreich, 'Conceptualism by any other name', 1296.

89 As Nancy Jecker has pointed out (Jecker, 'Protecting the vulnerable', 60–61).

7

VULNERABILITY IS EVERYWHERE

Globalization and vulnerability

Vulnerability is applied as a concept in a variety of disciplines. This use is very recent; approximately 90 percent of scholarly publications with this keyword have appeared since 1990. Bioethicists have discovered the concept even more recently; the majority of publications in this area are dated since 2000 [1]. The emergence of the notion in scholarly discourse is often regarded as a reflection of a growing sense of vulnerability at the global level. In the 1990s the risk of nuclear war diminished. Security was redefined because there were new non-military threats. International organizations linked security to human development. They argued that as long as basic human needs were not met, many people would face vulnerability. States as well as the international community were blamed for increasing vulnerability due to persisting poverty and hunger. Simultaneously, individual citizens experienced a growing awareness of mutual vulnerability. They were confronted with relatively new threats such as environmental degradation, climate change, food shortages, and bioterrorism. This heightened sense of vulnerability is particularly manifested in the area of medicine and health care. As discussed earlier, the discourse of vulnerability has expanded in the context of global phenomena such as the pandemics of AIDS/HIV and avian flu, as well as disaster relief. Arguably, the landscape of medical research has significantly changed since it was outsourced and off-shored [2]. Women in developing countries are especially regarded as vulnerable populations in relation to clinical research, but also healthcare in general [3]. Not only is research in developing countries most of the time not focused on prevalent health needs of those countries, but the benefits of these research projects are generally not shared with developing countries, so that the social and economic contexts producing vulnerability are not improved [4]. New developments such as the marketing of medical tourism in countries like India and

Thailand, and the privatization of healthcare in many developing countries, are associated with a reduction of public health expenditures, widening existing inequities, and weakening access and quality of care for the majority of populations [5].

The common denominator of all these changes is the globalized context in which they arise. Processes of globalization have created asymmetries of power of which vulnerability is one of the major expressions.

Vulnerability as a phenomenon of globalization

What exactly are the interconnections of vulnerability and globalization? During the 1990s the term 'globalization' was increasingly used in social sciences and public policy discourse. In the US Library of Congress the number of entries for this term increased twenty-fold from 1994 to 1999 [6]. 'Globalization' is a general concept with different interpretations and evaluations. It covers multidimensional processes. Basically, it is characterized by transformations of human interacting. The concept refers to the 'increasing connectedness and interdependence of relationships across the world' [7]. Manfred Steger distinguishes four main dimensions of globalization [8]. The first is the emergence of a new global economic order with transnational corporations and international economic institutions regarding the world as one market for trade and finance. The political dimension is characterized by the rise of a world without borders and the expansion of global organizations and institutions impacting on the power of the nation-state. Globalization also has a cultural dimension because it facilitates awareness and exchange of cultural diversity but at the same time promotes a global culture founded on dominant Western values. The ideological dimension is the most interesting one from an ethical point of view. Processes of globalization have been promoted, stimulated, reinforced, and sometimes imposed since they are driven by certain norms and values. Normative ideas have guided economic and political policy-making.

Globalism

Assessments of globalization differ. Globalization is applauded for having created multiple opportunities for interactions and social change, but it is at the same time blamed for a range of negative consequences such as environmental degradation, persistent poverty, violence, growing inequalities, human trafficking, exploitation, marginalization, and discrimination. Many authors argue that these harms are not the result of processes of globalization as such but of a particular kind of globalization driven by the specific ideology of neoliberalism [9]. Over the last few decades, regulatory frameworks and governance mechanisms have emphasized the removal of constraints on free market competition, encouraged privatization, deregulation, reduction of public expenditure, tax reform, and protection of property rights [10]. Steger has introduced the term 'globalism' to distinguish processes of globalization from the ideology that drives these processes in particular directions based on specific norms, values, and meanings [11]. Although there are different globalisms, the discourse that came to dominate international policies since the 1980s has been neoliberalism. This worldview makes several ideological claims that will help to explain how vulnerability is produced and interpreted.

Neoliberal claims

Globalization basically is liberalization. When global markets are free, this will foster individual liberty and human well-being. The role of the state should therefore be limited; it should create the institutional framework to secure free trade and private property rights. The power of governments should be used to deregulate and remove constraints and social policies that curtail the flow of commodities. The state should withdraw from social provision and protection because that impedes the proper functioning of markets [12]. Social welfare provisions and public institutions should be privatized. Every domain of human life should be open for market transactions so that individual citizens are free to choose what they want. Genetic tests and preventive as well as therapeutic interventions are consumables. Healthcare is a business that will flourish in a climate of competitiveness and efficiency. Medical research can thrive if it operates in a global market. It is able to provide individuals with a range of choices concerning drugs and interventions, broader than ever before. Another claim of neoliberalism is that globalization is inevitable. The dissemination of free market principles is like a natural force. Furthermore, nobody is in charge. Processes of globalization do not have main agents or leaders who are accountable. Finally, the claim is that everyone benefits, at least in the long run. Inequalities and negative impacts on vulnerable populations will only be transitional and temporary. Economic liberalization will in the end free individuals from the frailties of human existence [13]. Market mechanisms are the best way to reduce and even eliminate vulnerability.

Global production of vulnerability

The market nowadays is regarded as the main source of vulnerability and insecurity [14]. Neoliberal policies exacerbate vulnerability since in the first place the dominant market-driven logic has multiplied insecurities: less and more precarious employment, deterioration of working conditions, financial instability, growth of poverty, and environmental degradation. In terms of the functional definition of vulnerability, one can argue that globalization has increased exposure to harms; individuals, groups, and populations are confronted with more and broader threats. From the point of view of neoliberal globalization, however, freeing the human being from coercion and servitude will liberate new energies and capacities. Exposure means the possibility of competition, the confrontation with challenges. It is only threatening if people do not take their own fate in their own hands. Individual security, as David Harvey puts it, is 'a matter of individual choice' [15].

But it is not simply growing exposure that is creating vulnerability but also the breakdown of protective mechanisms. Liberalization in neoliberal policies means that safety networks and solidarity arrangements that existed to protect vulnerable subjects have been minimized or eliminated. Rules and regulations protecting society as well as the environment are weakened in order to promote global market expansion [16]. The second way in which neoliberal policies produce vulnerability therefore is, again using the functional definition of vulnerability, that adaptive capacity is affected. Because social institutions are less powerful and protective, the social conditions for coping mechanisms are eroded. It is also more difficult to know and learn how to cope when anonymous forces are driving globalization, and when it is impossible to identify responsible agents and recognizable processes. The result is a general expansion of precariousness without the experience that something can be done to ameliorate it [17].

According to these analyses, vulnerability is the result of the damaging impact of the global logic of neoliberalism. As a consequence of this type of globalization, threats to human well-being have increased and coping mechanisms have eroded [18]. These effects are visible everywhere but especially in developing countries. It is not amazing that international and intergovernmental organizations use the language of vulnerability to describe the impact of globalization. For example, the United Nations Department of Economic and Social Affairs describes vulnerability as 'a state of high exposure to certain risks and uncertainties, in combination with a reduced ability to protect or defend oneself against these risks and uncertainties and cope with their negative consequences' [19]. The United Nations Development Program (UNDP) concluded in 1999 that growing vulnerability is the result of globalization: 'People everywhere are more vulnerable' [20]. While acknowledging the vast progress in human development over the past decades, subsequent UNDP reports continue to use vulnerability as a core notion to pay attention to the weakened position of the most disadvantaged people and to advocate more equitable policies. Due to increasing risks and lower resilience, especially people in low-income countries have diminishing abilities to cope with threats and challenges. The 'space of vulnerability' has widened [21]. Present-day society, as famously characterized by German sociologist Ulrich Beck, is a 'risk society' [22]. In this analysis, vulnerability is a generalized phenomenon produced by social, economic, and political changes associated with globalization. It is therefore not an individual concern but is socially produced since society is itself affected by global processes.

There are, however, two more impacts of globalism that are relevant for the ethical analysis of vulnerability, in particular for the context of health care. One is related to the claims of neoliberalism, the other concerns the image of human beings that is dominant in this type of globalism.

Existential insecurity

Lack of safety and security has become a crucial dimension of present-day existence, according to sociologist Zygmunt Bauman [23]. It is not simply, as much data indicates, that daily life has become more uncertain, jobs more unstable, social conditions degrading, and the environment more risky; but there is a more systemic vulnerability that characterizes contemporary human existence, at least for the majority of people [24]. This sense of vulnerability is deepened since globalization is associated with increasing social inequalities. Contrary to ideological claims, neoliberal policies tend to be beneficial to very few people while disadvantaging many. Neoliberal policies, concludes Jan Aart Scholte, 'encourage unfair outcomes' [25]. Research shows that this type of globalism is associated for example with growing health inequalities as well as the failure and near-collapse of healthcare systems in many countries [26]. Privatization of healthcare and reduced public expenditure on health have encouraged medical tourism in developing countries, attracting wealthy patients from other countries, but it has at the same time decreased quality of care and access to healthcare services for the majority of the local population [27]. Inequalities not only grow within countries but also between countries [28]. 'Peripheries' are created that are affected by globalization but that do not participate in the benefits [29]. Existing stratifications based on class, gender, race, and age are reinforced. Globalization processes are therefore connected with growing injustices and progressive exclusion [30]. Robert Cox has shown how a new social hierarchy has emerged worldwide with the integrated at the top (those who are essential to the maintenance of the economic system), the precarious in the middle (those who are not essential to the system and thus disposable), and the excluded at the bottom (the permanently unemployed) [31]. Precariousness, marginalization, and exclusion rather than exploitation are characteristics of this new social order of globalization. In the neoliberal perspective however, this order will necessarily have different effects; there is no need to pay special attention to vulnerability; vulnerable populations can best be disregarded.

The claim that globalization is inevitable and that nobody is in control contributes to the experience of weakness and powerlessness, thus vulnerability. Nobody can escape from being included in processes of globalization but in practice these processes only favor the strong and powerful. When public expenditures for health and education are reduced, subsidies on food decreased, and entitlements for care, social security and pensions lowered, vulnerable persons and groups are disproportionately affected; usually also the number of vulnerable subjects will rise. The market is not an interaction between equal and competing actors but, as Von Wright has pointed out, it is 'the pull and push of manipulated demands, artificially created needs, and desire for quick profit' [32]. These anonymous forces impact everyone. They intensify feelings of abandonment and alienation because it seems that the state and the political system no longer care about the population or the people but are primarily concerned with the economy and the budget. The nature of social regulation and intervention has changed, 'prioritizing the well-being of market actors over the well-being of citizens' [33]. Policies have shifted away from maximization of public welfare to promotion of business, innovation, and profitability. What used to be delivered as public service is now only offered for profit. Society has become subservient to the needs of the economic system.

Neoliberal ideology furthermore decreases the significance of responsibility. Power is extra-territorial but individual life is always localized and situated. If globalization is inevitable, the asymmetrical power relations between the Global North and the Global South are nobody's responsibility. If nobody is in charge of globalization, the elite of the integrated do not have any public accountability. Power is disconnected from responsibilities and obligations. The integrated or powerful do not have any duty to contribute to amelioration of the daily life of the vulnerable. While special vulnerabilities are created, and use is made of the vulnerabilities of people, it is impossible, in the neoliberal perspective, to attribute responsibility for the consequences.

Social disintegration

A second fundamental relation between globalization and vulnerability concerns the image of human beings. According to the market thinking of neoliberal globalization, the human person is primarily *homo economicus*; a rational individual motivated by minimizing costs and maximizing gains for himself. He primarily relates to others through market exchanges. The emphasis is on the individual chooser while the community is absent. The market logic therefore separates economic activities from social relationships. Competition rather than cooperation is the preferred mode of social interaction [34]. Citizenship, the public sphere, and social networks will erode since there are only individuals and commodities that can be traded [35]. The French economist Daniel Cohen argues that the dominance of economic thinking promotes a kind of Darwinian logic according to which the weak are losing and the stronger prosper [36]. Shared communal life is disintegrating. Even the capacity for co-existence is eroding [37].

Considering the human person as *homo economicus* implies that vulnerability is primarily an individual affair. The most efficient antidote to becoming or being vulnerable is the possession of tools for personal decision-making and individual empowerment. This view is often reflected in how vulnerable subjects are perceived. The poor, for example, are perceived as lazy or weak [38]. They are regarded as individuals who do not take the chance to improve their condition rather than as unfortunate fellow human beings. Poverty is the consequence of bad choices. In France, intensive debates have followed the suggestion that homelessness is the effect of a psychiatric condition that develops early in life, so that some individuals later will manifest a 'syndrome of de-socialization' [39]. Neoliberal policies in Portugal, as Pinto shows, have constructed disability as an individualistic concern, primarily the consequence of a biological condition, rather than resulting from the interaction with the environment. Policies emphasize that disabled individuals have to change, take their lives into their own hands and find solutions; they have to show personal responsibility and to make choices. The state's concern is reduced to employability; if they work they are included in society. But even if they succeed in entering the labor market, they are usually underpaid, marginalized, and impoverished [40].

Within a neoliberal perspective, vulnerability therefore originates in specific deficiencies or pathologies within the individual subject. Poverty, homelessness, or disability are transformed into private problems. Even when vulnerability is clearly the result of circumstances beyond individual control, the solution is to articulate personal responsibility and individual agency. Beck gives the example of climate change [41]. As illustrated in the case of Tuvalu (in Chapter 1) global warming is disastrous for certain populations. But the usual approach is to present climate change as a challenge that every individual must address through making choices for a green lifestyle. Vulnerability produced at social, cultural, and economic levels is therefore reduced to a problem of a vulnerable subject. The difficulty with this reduction is that it only leaves room for individual responses and solutions. But even more problematic is that it isolates the vulnerable subject. He or she is no longer connected with people in the same situation. The neoliberal 'denial of social structure' thus contributes to social exclusion [42]. Joseph Stiglitz has underlined that the individualizing approach of neoliberal policies extracts a high price: 'the erosion of our sense of identity in which fair play, equality of opportunity, and a sense of community are so important' [43].

Neutralizing vulnerability

This reduction of vulnerability to individual deficiency is remarkable. The discourse of vulnerability has first of all developed within a systemic context; it has emerged in association with increasing processes of globalization and as a result of neoliberal policies. It is tempting to view the attention for vulnerability as a critique of these processes and policies. The growing interest in vulnerable groups and the identification of expanding lists of vulnerable populations has made clear that neoliberal globalization produced a growing number of 'casualties', i.e. people who do not comply with the dominant image of the rational consumer and autonomous decision-maker. The damage is done to entire categories of people whose existence has become precarious and marginalized as a result of global processes.

But it also seems that this systemic critique could be 'deflected' in a certain way to demonstrate that neoliberal globalism is concerned about vulnerable subjects. The discourse of vulnerability can be appropriated in order to 'humanize' globalism. What started as a critique at the level of social ethics, arguing that groups and populations are made vulnerable due to economic and social circumstances has in that case been transformed into a critique at the level of individual ethics. As discussed earlier, the notion of 'vulnerable populations' initially was often used in bioethics but was also severely criticized. The intention to focus on the conditions producing vulnerability arguably led to ever longer lists of vulnerable populations, and therefore tended to make the notion too vague and ethically self-defeating. Since bioethics was primarily concerned with the ethical principle of respect for autonomy, this focus on populations and social conditions was not conducive to ethical analysis. On the other hand, the notion of vulnerability itself was discarded as too abstract so that a more restrictive interpretation of the vulnerable subject could be salvaged. In this interpretation, as discussed in previous chapters, some subjects do not act like other rational individuals since they lack capacities to protect themselves due to compromised or inadequate personal autonomy. Reinterpreting vulnerability in this manner implies no longer a critique of neoliberal globalism but can serve to provide neoliberal policies with a more human face. The idea is that this type of globalism can be made more acceptable if it pays specific attention to vulnerable subjects, helps them to make informed choices, provides more transparent options, and protects them against abuse. Reducing vulnerability and incorporating the concept into neoliberal globalism may explain the dominant discourse of vulnerability in contemporary bioethics. Being ill, receiving treatment and care, and participating in research are first of all individual affairs. If human beings are primarily autonomous persons, then informed consent, and individual decision-making, are necessary to involve them in care and research. Persons who do not comply with this ideal of autonomy need special attention as 'vulnerable subjects'. Instead of being an expression of the critique of neoliberal globalism, the notion of vulnerability has now been transformed into a useful tool to promote the human face of globalization. But, only at the expense of being strictly interpreted as an individual deficit.

The dimensions of globalism today

After several decades of intense globalization, the disadvantages of the neoliberal ideology have clearly come to light. Nowadays, the disparity between the rhetoric of neoliberalism and its actual consequences is evident [44]. Financial crises and economic recession have increased social inequalities and accelerated defunding of public services, making vulnerability an even more relevant issue. At the same time, there are challenges and alternatives to neoliberal approaches promising a broader and more inclusive ethical vision. Before elaborating what this implies for vulnerability, it is important to articulate two dimensions of neoliberalism as it has developed. September 11, 2001, and the War on Terror brought neoliberalism together with neoconservatism in what Steger calls, 'imperial globalism' [45]. While the emphasis in the 1990s was on economic issues and market ideology, globalism became more militaristic and authoritarian, and more openly focused on the interests of the remaining superpower. Domination is no longer only secured by economic means but with a combination of different forms of power. In the 2010s, more subtle approaches were used. Rather than full-scale military interventions, microsurgical operations were performed with sophisticated technologies of surveillance and annihilation at a distance [46]. The major characteristic today is the merger of market and security discourse, eliminating the distinction between private and public. The market in fact demands a particular order in which private and public needs, interests, and preferences are aligned. It is obvious that individual interests and rights are increasingly subordinated to national security or public safety. In principle all personal information is accessible and exchangeable to security services while public auditing and political supervision is almost absent. Major companies are not only complicit with the security establishment but have similar monitoring programs of their own to determine and target individual preferences and to specifically market their products. This merger of market and security has falsified the neoliberal claims that globalization is liberalization and that nobody is in control.

The second dimension of today's globalism leads to similar conclusions. The liberalization of trade, deregulation of society, and restriction of the centralized power of government promoted by neoliberalism relies not on free market dynamics but rather heavy political control [47]. Instead of the invisible hand of market forces, it was often the visible hand of governments and international organizations that had driven processes of globalization [48]. Neoliberalism implies increasing regulation rather than *laissez-faire*. This is obvious in the enormous growth of bureaucracy. Free markets are created through intervention and interference, often through concerted political action. Similar concerted action has disseminated the norms, rules, procedures, and formalities of business administration to all dimensions of present-day society. Education, research, healthcare, environmental protection, safety, and defense are all submitted to the overarching framework of market logic, not because market ideology is naturally expanding but because it is deliberately promoted to dismantle social structures of welfare and protection [49]. This process also contributes to making the distinction between private and public less relevant. It is the systemic entirety of rules, procedures, and routines in which the figure of the individual businessperson, politician, administrator, or scientist plays a less significant role [50]. Top-down and hierarchical power and authority are increasingly replaced by horizontal networks and practices, governed by experts and elites. Inducements, incentives, stimuli, and best practices substitute explicit orders and directives. Neoliberal bureaucracy has created, as Béatrice Hibou explains, a social form of power, a technostructure of mostly invisible mechanisms of inspection, discipline, control, policing, and surveillance [51]. These tendencies are clearly visible in the new area of the life sciences. Direct-to-consumer genetic testing, human enhancement technologies, and 'personalized medicine' all promise a healthier and longer life, if the individual will make rational use of the opportunities offered at the market and will make responsible decisions [52]. The assumption of course is that we have a choice over our health, and that rational behavior will follow more information, specifically regarding our genome. The promise is that development of disease can be predicted and therapies tailored for individuals, but it requires a genetic surveillance of body and self [53]. The promises of these emerging technologies can only be fulfilled, however, if two assumptions are met. First, responsibility for health is primarily an individual issue. The citizen is regarded as a consumer who has to choose on the basis of his or her perceived self-interest among the many possibilities that are offered by modern genetics. Life is a 'business plan' that requires constant accumulation of genomic data and careful management by its owner [54]. The best approach to health is self-governance; we are autonomous managers of ourselves; health, disease, and death are the result of investment decisions in our biocapital. But it also means that risks are displaced to the individual. Second, the moral debate needs to focus on practical applications and potential uses of new data, devices, and discoveries. Consumers will then be shown that the potential to enhance and manage health, and if necessary disease, is constantly growing so that they can effectively be stimulated, nudged, and incentivized to utilize it. Present-day genomics is like 'speculative capitalism'; as a venture science it is based on hopes and expectations rather than concrete products, just as in the financial world futures and derivatives have replaced sales and profits. This is the reason why the emergence of biotechnology has been connected to the rise of neoliberalism. Biotechnical innovation and the life sciences represent a new face of neoliberal economy: bioeconomy [55].

The consequence of these two assumptions is that it has become increasingly difficult to address the social context of decision-making. The promissory visions of biotechnology encourage individuals to cooperate with scientific endeavors, speculating that the future will always be better. But the promise can only become reality if massive data are collected in biobanks and repositories, and if personal data can be freely exchanged. In other words, the promised benefits can only be accomplished if the boundary between private and public interests is conflated. In the ideal consumer society health benefits and commercial benefits are synonymous; the distinction evaporates between biological life, marked by bodily susceptibilities and dispositions identifiable with genetic technology, and social and political life, focused on collective efforts of societies and communities to cope with disease and disability [56]. Similar effects result from the emphasis on personal responsibility for health. If health is the product of (responsible) choice and if human enhancement technologies can even further improve individual health, why should one bother about social and economic determinants for health [57]? Social problems can be ameliorated through technological innovations, and it is the rational consumer who decides about their use. Empowering individuals is therefore more important than assuming that governments or collectives are responsible for health. But governance by individualization will imply that many individuals will not benefit from new knowledge, not only because they are not responsible but also because 'personalizing' medicine is necessarily associated with progressive exclusion; 'tailoring' means that specific therapies are often not available or too risky [58].

Both dimensions of contemporary globalism remind us that the current emphasis on individual decision-making must be located within a context of permanent surveillance. In this context everything is monitored, registered, and archived; individual behavior patterns are schematized, and anomalies can be detected and become the target of anticipative prevention and pre-emptive intervention [59]. Human vulnerability is not the unhappy, accidental side effect of market forces that needs to be taken into consideration and that requires proper management. It is not true that vulnerability is either the result of the impossibility to make good choices or of the fact that poor choices have been or are being made. It is on the contrary the deliberate product of policies to individualize social ills [60].

Globalism demonstrates how vulnerability is constructed within the context of market relations. It also shows how the main determinant in this construction is the moral value of self-determination. Vulnerability seems to be the unavoidable result provoked by the dominating image of the self-determined and self-interested rational chooser. But, oddly, the notion of vulnerability instead of being the flipside, functions as the death trap of this dominating image. This is evident in the field of healthcare. Vulnerability cannot be demarcated from the universe of rational clients and autonomous decision-makers since they, like everybody else, are also vulnerable. In her final year as Director-General of WHO, Gro Brundtland affirmed that we are all in the same predicament: 'No impregnable walls exist between a world that is healthy, well-fed, and well-off and another that is sick, malnourished, and impoverished' [61]. Failing to view vulnerability within a social, economic, or political setting will reproduce misunderstandings and inefficiencies in the role of health professionals who continue to address individual capacities even in circumstances where it is clear that vulnerability is created by dominant power structures. In Malawi, for example, where over half of the population lives below the poverty line, HIV/AIDS policies ignore structural inequalities and focus on individual behavior and responsibility. The result is that the policies reinforce gender inequality, as well as victim blaming and stigmatization, especially of women [62]. In Mozambique, nurses assist vulnerable individuals without paying attention to the social and economic inequalities and injustices that produce vulnerability [63]. Caring for the vulnerable individuals then results in their adaptation to unjust conditions without efforts to transform the conditions themselves. Hence, framing vulnerability as a consequence of impaired or limited individual autonomy, although in accordance with the neoliberal ideology of globalization, contradicts the fundamental structures and conditions that produce and exacerbate vulnerability in the first place. What does this conclusion mean for addressing vulnerability in bioethics?

A different bioethics

The analysis thus far leads to a paradox. Over the past two decades vulnerability has become an important notion in bioethical debates. The focus on vulnerability is associated with globalization, in particular neoliberal market policies that have exposed more people worldwide to more threats and hazards, and has decreased their capacities to cope. These policies are based on the assumption that human beings are self-interested, rational individuals. In addressing vulnerability, contemporary bioethics demonstrates the concern that everyone's health needs should be addressed and that all persons should benefit from the progress of science and technology. The concern for vulnerability reflects the effort to refocus on the well-being of human beings that is often jeopardized by damaging and unjust structures and policies; it is the expression of humanitarian concerns.

However, bioethical discourse often uses the same basic assumptions as neoliberal globalism, arguing that vulnerability should be addressed and reduced through protecting and empowering individual decision-makers. It is understandable that bioethics is concerned with the fall-out of globalizing processes for individual persons. But using an individual focus abstracted from the social dimension of human existence, and neglecting the impact of market mechanisms on social life will not allow bioethical policies and guidelines to redress the production of vulnerability. What is a symptom of the negative impact of a one-dimensional view of human beings is remedied with policies based on the same type of view. As long as the problematic conditions creating and reinforcing human vulnerability are not properly analyzed and criticized, bioethics will only provide limited palliation.

Two faces of vulnerability

Bioethics as theoretical discourse should take a broader approach to vulnerability and explore the implications of the philosophical and political perspectives elaborated in previous chapters. This exploration will provide not only a rich and multidimensional view of vulnerability as an ethical notion but also highlight a range of possible responses in practices of care and research. Both perspectives outline primarily the positive rather than negative implications of the notion of vulnerability. In most of bioethics literature, vulnerability is regarded as a 'deficit', a weakness, dependency, or imperfection that should be eliminated or at least reduced. If respect for persons as autonomous agents is a basic ethical principle, then the vulnerability of the human condition does not make sense. It does not lead to constructive action and intervention. It implies passivity that is hard to reconcile with moral agency. Therefore, it is difficult to give a positive meaning to vulnerability. As discussed earlier, this negative interpretation of vulnerability motivated several authors to dismiss the notion of anthropological vulnerability and to concentrate solely on special vulnerabilities [64]. But this negative perception of vulnerability is wrong. The basic idea of the philosophical perspective on vulnerability is that anthropological vulnerability demonstrates what we are, and not what we do. It is therefore not a negative condition that can be reduced or eliminated through rational decisions of self-interested individuals. It is wrong to assume that the neoliberal social and political order established as a result of these rational decisions will create security that 'abrogates' vulnerability. On the contrary, if vulnerability constitutes human subjectivity, it cannot be eliminated since it is the precondition of ethical responsibility. Ethics is prior to politics; ethical responsibility does not depend on social agreement or a secure political order but is prior to them [65]. This perspective also illustrates that it is equally wrong to assume that anthropological vulnerability does not generate responses. Instead, it leads to respect and care, responsibility and empathy. Because it is a shared condition as a consequence of being embodied social beings, it necessitates that humans cooperate and create social institutions in order to protect themselves from the threats of everyday life. Acknowledging vulnerability and therefore expressing it can be strengthening [66]. It will engage people to take care for and assist each other. They feel connected to other humans because they are all equally vulnerable and thus recognize the situation of the other. This incites action. It encourages efforts to equally protect all human beings against threats, but it also opens up individual human beings towards each other. The work of Arnold Gehlen, for example, provides a lengthy argument for the productivity of anthropological vulnerability. If the human constitution is inherently deficient, how can we explain that such deficient beings have survived? Humanity has survived according to Gehlen because it is open to the world, transforming deficiencies into opportunities for survival and specifically creating a 'second nature', i.e. culture [67]. Anthropological vulnerability is therefore intrinsically linked with human freedom, plasticity, and development. Without this positive impulse, vulnerability would be a pathology, an inherent defect within the human constitution, an original fault that will always haunt human beings. Anthropological vulnerability is a relational concept; it connects human beings and is associated with solidarity.

From the political perspective, special vulnerabilities are in principle alterable and removable since they are produced by social, economic, and institutional conditions that are themselves the result of human activity. These vulnerabilities are not necessarily negative. Identifying subjects and populations as vulnerable focuses attention on structural violence and injustice. Recognizing and analyzing vulnerability, as Anita Silvers has argued, can have positive effects since it can protect against stereotyping and discrimination [68]. The language of vulnerability itself can generate a discourse that is more positive than the usual one of protection, reduction, and elimination. Furthermore, protective responses are less appropriate when vulnerability is related to social and economic conditions. They can easily become unduly paternalistic since there is nothing wrong with the decision-making capacity of the involved persons. The vulnerable do not lack autonomy or rationality but institutional support. According to this political interpretation of vulnerability, a different kind of empowerment is needed, an affirmative action based on dignity, respect, and social responsibility. More is required than protection. Vulnerability discourse appeals to positive duties such as benefit-sharing, solidarity in health care, post-trial access in research, and responsiveness to the needs of populations or communities in research [69]. But even authors who share this view often continue to use the language of protection. For example, Kottow and Solbakk call for two different regimes of protection: equal protection for every human being as essentially vulnerable, and specific positive action (remedial treatment and repair) for susceptible individuals [70]. This language still assumes the significance of individual autonomy. It focuses on the vulnerable as persons who cannot protect themselves, rather than interpreting vulnerability as a consequence of unequal and unjust conditions that cannot be addressed through protective mechanisms but call for a different type of intervention.

Conclusion

The two faces of vulnerability (negative as well as positive) are expressed in the formulation of ethical principles in the Universal Declaration on Bioethics and Human Rights, referring to *respect* for human vulnerability [71]. Before one can argue that the vulnerability of being human should be eliminated or accepted, one needs to respect it. The anthropological vulnerability that confronts us in the face of the other escapes comprehension. The ethical responsibility that is engendered through this experience can only be recognized, respected, and never fully understood. But anthropological vulnerability as the precondition of moral responsibility does not imply that we have to accept special vulnerability. Concomitantly with respect, we have the obligation to overcome special vulnerability if we have the opportunity and possibility. Medicine is the discipline that is distinctively focused on taking care of vulnerable subjects, not merely protecting them but also ameliorating and reducing their vulnerability. Medical research in particular is continuously searching for vulnerability reducing or eliminating means and strategies. As an ethical principle, respecting vulnerability opens up a range of opportunities that encourage solidarity, responsibility, assistance, mutual support, and care. The same respect also requires acknowledging that the fragility inherent in the human condition can be reinforced and multiplied by living in unjust societies and by increasing inequalities. This implies that more is required than the conventional focus on protecting the vulnerable individual.

Notes

1 See Chapter 2, 28–29.
2 See, Eckenwiler *et al.*, 'Hopes for Helsinki: reconsidering "vulnerability"'. Petryna, *When experiments travel. Clinical trials and the global search for human subjects.*
3 See Blacksher and Stone, 'Introduction to "vulnerability" issues'; Jaggar, 'Vulnerable women and neo-liberal globalization: debt burdens undermine women's health in the global South'.
4 Garrafa *et al.*, 'Between the needy and the greedy: the quest for a just and fair ethics of clinical research', 501, 504.

5   See, for example, Smith, 'The problematization of medical tourism: A critique of neoliberalism', and NaRanong and NaRanong, 'The effects of medical tourism: Thailand's experience'.

6   Scholte, *Globalization. A critical introduction*, 14.

7   Dower, 'Globalization', 2174. Steger defines globalization as a 'multidimensional set of social processes that create, multiply, stretch, and intensify worldwide social interdependencies and exchanges while at the same time fostering in people a growing awareness of deepening connections between the local and the distant' (Steger, *Globalization. A very short introduction*, 13). Scholte describes globalization as 'a transformation of social geography marked by the growth of supraterritorial spaces' (Scholte, Ibidem, 8).

8   Steger, *Globalization. A very short introduction*, 37 ff.

9   See, among others, Braedley and Luxton (eds.). *Neoliberalism and everyday life*; Cohen, *Homo economicus, prophète (égaré) des temps nouveaux*; Harvey, *A brief history of neoliberalism*; Kirby, *Vulnerability and Violence*; Kirby, 'Vulnerability and globalization'; Peck, *Constructions of neoliberal reason*; Scholte, *Globalization. A critical introduction*; Sobrino and Wilfred, *Globalization and its victims*; Steger, *Globalization. A very short introduction*; Supiot, *L'esprit de Philadelphie. La justice sociale face au marché total*.

10 Steger, *Globalization. A very short introduction*, 53.

11 Steger, *Globalization. A very short introduction*, 94. For a more extensive discussion, see Steger, *Globalism: The new market ideology*; Steger, *Globalism. Market ideology meets terrorism*; Steger, *Globalisms. The great ideological struggle of the twenty-first century*. In the first edition of his book (2002), Steger discusses only one globalism, namely neoliberalism. The third edition (2009) distinguished three globalisms: neoliberal, justice, and jihadist globalism.

12 For an elaboration of these ideological claims, see Harvey, *A brief history of neoliberalism*. Peck analyzes the historical evolution of the thought of two key leaders of neoliberalism, the political philosopher Friedrich von Hayek and the economist Milton Friedman (Peck, *Constructions of neoliberal reason*).

13 For the core claims of globalism, see Steger, *Globalism. Market ideology meets terrorism*, 47 ff.

14 As Thomas explains, it is 'the changed conditions of *mutual vulnerability* that increasingly characterize the globalized world' (Thomas, 'Globalization and human security', 119). See also Nef, *Human security and mutual vulnerability*.

15 Harvey, *A brief history of neoliberalism*, 168.

16 Scholte, *Globalization. A critical introduction*, 207 ff.

17 Bresson, *Sociologie de la précarité*, 82 ff.

18 See Kirby, *Vulnerability and Violence*; Kirby, 'Vulnerability and globalization'.

19 United Nations Department of Economic and Social Affairs, *Report on the world social situation, 2003. Social vulnerability: Sources and challenges*, 8.

20 UNDP, *Human development report 1999*, 90.

21 Watts and Bohle, 'The space of vulnerability: The causal structure of hunger and famine', 52 ff.

22 'Social processes and relations lead to unequal exposure to risks and the resulting inequalities must be treated as largely an expression and product of power relations at the national and global levels' (Beck, *World risk society*, 178). See also Beck, *Risk society: Towards a new modernity*. Nancy Fraser is talking about 'the insecurity society' (Fraser, *Scales of justice*, 111).

23 Bauman, *Globalization. The human consequences*, 5 and 119.

24 See, Bresson, *Sociologie de la précarité*; Scholte, *Globalization. A critical introduction*; Sobrino and Wilfred (eds.). *Globalization and its victims*.

25 Scholte, *Globalization. A critical introduction,* 259. Neoliberal policies '…usually distributed costs and benefits in ways that favour the already privileged and further marginalize the already disadvantaged' (Scholte, *Globalization. A critical introduction,* 236). 'Globalisation's rules favour the already rich (both countries and people within them) because they have greater resources and power to influence the design of those rules' (Labonté and Schrecker [eds.]). *Towards health-equitable globalization: Rights, regulation, and redistribution,* 10. Braedley and Luxton point out that 'neoliberalism is not advancing social justice and equality, but is, instead, reinscribing, intensifying, and creating injustices and inequality' (Braedley and Luxton [eds.]. *Neoliberalism and everyday life,* 6). See also Stiglitz, *The price of inequality.*

26 Labonté, Ronald and Ted Schrecker (eds.). *Towards health-equitable globalization: Rights, regulation, and redistribution,* 10, 34–42; WHO, *Everybody's business: Strengthening health systems to improve health outcomes,* 1, 9; Chapman, 'Globalization, human rights, and the social determinants of health'. For an extensive analysis of the positive and negative impacts of globalization on health and disease, see Bennett and Tomossy (eds.). *Globalization and health. Challenges for health law and bioethics,* and Cockerham and Cockerham, *Health and globalization.*

27 In India with one of the most privatized healthcare systems in the world, 'almost a quarter of the local population go undertreated for illness due to indebtedness…' (Smith, 'The problematization of medical tourism: A critique of neoliberalism', 6). See also NaRanong and NaRanong, 'The effects of medical tourism: Thailand's experience'.

28 Stiglitz, *The price of inequality,* 21 ff. See also Muller, 'Capitalism and inequality: What the Right and the Left get wrong'.

29 Bresson refers to the demarcation in France between town (the center) and 'banlieues sensibles'(the peripheries that are always regarded as places of trouble) (Bresson, *Sociologie de la précarité,* 65-66, 80, 95 ff.). Bankoff argues that vulnerable populations are created by social systems, not because they are exposed to hazard but because they are marginalized as weak and passive; large regions of the world then become imagined as dangerous (see, Bankoff, 'Rendering the world unsafe: "Vulnerability" as Western discourse', 25, 29).

30 Bauman, *Globalization. The human consequences,* 3; Sobrino and Wilfred (eds.). *Globalization and its victims,* 12.

31 Cox, *The political economy of a plural world: Critical reflections on power, morals and civilization,* 85 ff.

32 Von Wright is quoted by Bauman (see Bauman, *Globalization. The human consequences,* 57).

33 Kirby, *Vulnerability and Violence. The impact of globalisation,* 94.

34 Cohen, *Homo economicus, prophète (égaré) des temps nouveaux,* 14.

35 See, Kirby, *Vulnerability and Violence;* Kirby, 'Vulnerability and globalization'.

36 Cohen, *Homo economicus, prophète (égaré) des temps nouveaux,* 147.

37 Hinkelammert, 'Globalization as cover-up: An ideology to disguise and justify current wrongs', 27.

38 Alesina, Glaeser and Sacerdote. *Why doesn't the US have a European-style welfare system?* 30; Cohen, *Homo economicus, prophète (égaré) des temps nouveaux,* 83. See also Rougeau, 'Enter the poor. American welfare reform, solidarity, and the capability of human flourishing'.

39 Bresson, *Sociologie de la précarité,* 106.

40   Pinto, 'Beyond the state: The making of disability and gender under neoliberalism in Portugal', 121 ff.

41   Beck, World risk society, 62.

42   Luxton, 'Doing neoliberalism: Perverse individualism in personal life', 175. David Harvey summarizes the effect of exclusion as follows: 'For those left or cast outside the market system…there is little to be expected from neoliberalization except poverty, hunger, disease, and despair' (Harvey, A brief history of neoliberalism, 185).

43   Stiglitz, The price of inequality, 117.

44   Harvey considers neoliberalism first of all as a project to restore class power. One of the major effects of global policies has been the steep increase of income of the top one percent of income earners (Harvey, A brief history of neoliberalism, 16, 203).

45   Steger points out that the 2000s disclosed what was hidden in the 1990s behind the ideological veil of the 'self-regulating market': American Empire. Hence the new phase of imperial globalism (Steger, Globalism. Market ideology meets terrorism, 150).

46   See, Chamayou, Théorie du drone.

47   Alain Supiot refers to the architecture of the 'Grote Markt' of Brussels to show that the free market has historically been encompassed with institutions that safeguard that the economy functions well. The central market square is surrounded by the town hall and the guildhalls. See: Supiot, L'esprit de Philadelphie. La justice sociale face au marché total, 91 ff.

48   '…from its inception in the Thatcher and Reagan years, the globalist enterprise required frequent and extensive use of state power in order to dismantle the old welfare structures and create new laissez-faire policies' (Steger, Globalism. Market ideology meets terrorism, 150).

49   Frodeman, Bruggle, and Holbrook argue that even philosophy cannot escape from the framework of neoliberalism (see Frodeman, Briggle, and Holbrook. 'Philosophy in the age of neoliberalism').

50   Hibou, La bureaucratisation du monde à l'ère néolibérale, 25, 69, 80 ff.

51   Hibou, La bureaucratisation du monde à l'ère néolibérale, 109; '… un système de contrôle formé de codes, de procédures et de normes'(Hibou, La bureaucratisation du monde à l'ère néolibérale, 112). She refers to the concept of 'biopolitics' introduced by philosopher Michel Foucault (seeFoucault, The birth of biopolitics). See also Lemke, Biopolitics. An advanced introduction.

52   See Birch, 'Neoliberalising bioethics: Bias, enhancement and economistic ethics'; Savard, 'Personalised medicine: A critique on the future of health care'.

53   Savard, 'Personalised medicine: A critique on the future of health care', 199.

54   Sunder Rajan, Biocapital. The construction of postgenomic life, 144.

55   Cooper, Life as surplus. Biotechnology and capitalism in the neoliberal era, 45 ff.

56   Karlsen, Solbakk and Holm, 'Ethical endgames: Broad consent for narrow interests; open consent for closed mind', 575.

57   Birch, 'Neoliberalising bioethics: Bias, enhancement and economistic ethics', 6, 9.

58   Savard, 'Personalised medicine: A critique on the future of health care', 200.

59   Chamayou, Théorie du drone, 58 ff.

60   Murray, 'Do not disturb: "vulnerable populations" in federal government policy discourses and practices', 54 ff.

61   Brundtland, 'Global health and international security', 417.

62   Bezner Kerr and Mkandawire, 'Imaginative geographies of gender and HIV/AIDS: moving beyond neoliberalism', 462 ff.

63 Tomm-Bonde, 'The naïve nurse; revisiting vulnerability for nursing', 2.
64 Luna for example argues that vulnerability is 'inessential'; it is alterable (Luna, 'Elucidating the concept of vulnerability: Layers not labels', 129).
65 See Chapter 5, 115–116. Benaroyo, 'The notion of vulnerability in the philosophy of Emmanuel Levinas and its significance for medical ethics and aesthetics'; Nortvedt, 'Subjectivity and vulnerability: reflections on the foundation of ethical sensibility'. A document of the Church of Norway underlines this as follows: 'The vulnerability and defencelessness of humankind are the preconditions for its capacity for openness and solidarity. Vulnerability also represents a unique capacity for susceptibility and compassion that enables people to recognise and fulfil their ethical responsibility for their fellow human beings, their community and their surroundings... Recognition of our own vulnerability can encourage a desire for cooperation instead of conflict' (Church of Norway. *Vulnerability and security. Current challenges in security policy from an ethical and theological perspective,* 4–5). The same view that vulnerability should not be viewed as weakness but as strength is endorsed by the Evangelical Lutheran Church in America (Rolfsen, 'Vulnerability and the role of churches').
66 Ells, 'Respect for people in situations of vulnerability: A new principle for health-care professionals and health-care organizations', 184.
67 Gehlen, *Man. His nature and place in the world*, 28, 29.
68 Silvers, 'Historical vulnerability and special scrutiny: Precautions against discrimination in medical research', 57.
69 See, for example, Jotkowitz, 'Vulnerability from a global medicine perspective'; Zion, Gillam and Loff. 'The Declaration of Helsinki, CIOMS and the ethics of research on vulnerable populations'.
70 See: Kottow, 'The vulnerable and the susceptible'; Kottow, 'Vulnerability: What kind of principle is it?'; Solbakk, 'Vulnerability: A futile or useful principle in healthcare ethics?'.
71 UNESCO, *Universal Declaration on Bioethics and Human Rights*; Patrão Neves, 'Article 8: Respect for human vulnerability and personal integrity'. It is regrettable that Article 8 of the Universal Declaration on Bioethics and Human Rights continues to use the language of protection by stating that 'Individuals and groups of special vulnerability should be protected...' The implication of the article goes far beyond the need for protection, and in fact refers to most of the other ethical principles listed in the same document.

8

THEORETICAL IMPLICATIONS
OF VULNERABILITY

The previous chapters argued that vulnerability has three components: exposure, sensitivity, and adaptive capacity. Philosophical perspectives on vulnerability elaborate the component of sensitivity. Vulnerability is regarded as a species-related rather than individual characteristic. It is the shared condition of all humans, and as such the basis for activities of care, respect, cooperation, and solidarity. Human beings are vulnerable because they are connected and mutually dependent. The relatedness explains why human vulnerability is the condition for autonomous agency, instead of the result of failing autonomy. Since human beings are vulnerable they are dependent on the social conditions to flourish and to develop individual capabilities. Developing and sustaining those conditions is therefore necessary to compensate for the anthropological vulnerability of the human species. However, in certain circumstances some individuals, groups, and populations are more vulnerable than others. Vulnerability can be exacerbated under the influence of social, economic, cultural, and political determinants that increase exposure to threats and decrease the capacity to adapt. Special vulnerabilities are the focus of political perspectives that examine the production of vulnerability for specific individuals and groups of individuals.

In both philosophical and political perspectives vulnerability is understood as a social phenomenon. Although it is manifested in individual subjects who are in need of healthcare or who are recruited for research projects, vulnerability is in most cases not an individual affair. Even when vulnerability apparently is the result of limitations of autonomous decision-making capacity, as in the conventional cases mentioned in the CIOMS guidelines, the fact that, for example, children and people with intellectual disabilities are vulnerable subjects is not merely the consequence of their individual status but the result of challenges presented to them in particular circumstances. When researchers want to involve them in clinical trials, or physicians propose medical interventions, then these subjects are perceived as vulnerable because they can possibly be harmed due to specific exposure and because insufficient mechanisms are available for them to adapt to or cope with harmful exposure. Interpreting vulnerability as the inability to protect oneself therefore isolates the individual from the context that highlights vulnerability and contributes to the exacerbation of vulnerability; it misconstrues the conditions in which individuals are threatened and not able to adapt.

This chapter will examine the theoretical implications of these perspectives on vulnerability. In order to comprehend vulnerability as an anthropological and social phenomenon, contemporary bioethics will need to expand its theoretical framework beyond the primary focus on individual agency. It will be shown that several theoretical approaches are in principle available for developing such a broader framework. They will allow bioethical debate to have a wider and richer scope – one that can scrutinize the current sources of vulnerability discourse – and that is more appropriate to analyze and criticize the processes of globalization that increasingly produce vulnerability. Four theoretical frameworks will be analyzed that can assist contemporary bioethics to better address vulnerability in its multiple dimensions: human rights, social justice, capabilities, and global care ethics.

Human rights

Michael Kottow has argued that vulnerability requires equal protection of every human being. The framework for this protection is provided by international human rights. However, this will not be sufficient for special vulnerabilities [1]. The association of anthropological vulnerability and human rights is specifically elaborated by Bryan Turner. Because vulnerability is the condition shared among all human beings, it is the basis of human rights [2]. This grounding in the same condition of anthropological vulnerability also explains the universality of these rights. Human rights are an essential component in the social institutions that human beings need to construct in order to compensate for their frailty. They should ensure that the basic needs of humans can be met. Vulnerability, according to Turner, not only explains the emergence of human rights but also their purpose; they compensate for 'the inequalities of the natural lottery' [3]. But Turner differs from Kottow in that he furthermore relates human rights to special vulnerabilities. The social institutions that humans create to protect themselves are not stable and static since the socio-cultural context might change, especially through processes of globalization. This may lead to 'precariousness' of the context, producing an environment that leads to special vulnerabilities. Human rights do not just provide general protection but should also be directed at these special vulnerabilities through decreasing social inequalities and safeguarding socio-cultural conditions for human flourishing. Vulnerability therefore explains the origin of human rights, its universality as well as its object [4].

International human rights discourse has expanded, especially since the 1990s, in the area of medicine and health. The pioneering work of Jonathan Mann is generally recognized [5]. He has advocated that health and human rights are complementary approaches. Public health has an impact on human rights, while human rights violations impact health so that both are linked [6]. For Mann, the emphasis on vulnerability shifts the focus of HIV/AIDS prevention strategies from individual risk reduction to social and contextual determinants. The conceptual framework of human rights can then help to identify and analyze 'the societal root causes of vulnerability to HIV'. This framework not only provides a vocabulary to describe the commonalities in the specific situations of vulnerable people, but also provides directions about how to change these social conditions in order to promote health [7].

Human rights and bioethics

Human rights have not only been connected to vulnerability but also with the development of global bioethics [8]. International and intergovernmental organizations such as the World Medical Association, the Council for International Organizations of Medical Sciences, the World Health Organization, and the United Nations Educational Scientific and Cultural Organization have taken human rights as foundational for the documents and guidelines they have issued. Their approach locates vulnerability clearly within the encompassing framework of human rights. The Universal Declaration on Bioethics and Human Rights goes one step further. Besides taking human rights as the starting point and context of bioethics, it considers human rights itself as a basic bioethical principle that needs to be balanced with other bioethical principles. It furthermore regards human rights as the ultimate authority and final constraint in the interpretation and application of the other principles. For example, the principle of respect for cultural diversity cannot be justified if it violates human rights [9].

Human rights discourse seems particularly appropriate in the context of globalization. It is regarded as an alternative to neoliberal policies. Since many of the bioethical issues and problems today cross borders, human rights present a universal framework for analyzing and addressing them. Human rights are recognized and used by individuals, groups, agencies, organizations, and movements around the world. They are particularly relevant to address vulnerability for two reasons. One is that human rights language emphasizes the equal value of human beings. A human rights based approach is specifically targeting inequalities, power differences, marginalization, and discriminatory practices. The consequence is that this approach leads to a focus on examining how the situation of the vulnerable can be improved and how the most disadvantaged subjects and populations can be protected. As Michael Ignatieff puts it: 'Human rights have gone global not because it serves the interests of the powerful but primarily because it has advanced the interests of the powerless' [10]. The other reason is that, especially in connection to public health, human rights address basic underlying conditions such as safe and potable water, nutritious food, sanitation, housing, access to healthcare and medicines. This is the point reiterated by Mann: within a broader concept of health the focus of activities should be on the social context of individual and population well-being, and not only on healthcare [11]. The production of vulnerability is associated with lack of respect for human rights [12]. This perspective is endorsed by Farmer. For him, respecting human rights in healthcare means the promotion of equity, thus first of all the advancement of social and economic rights [13].

Scope of human rights

The advantage of human rights discourse is that it generates a broad scope of activities [14]. Emphasis is no longer merely on protection but also on promotion and fulfillment of rights [15]. Health is considered as a public good that needs to be encouraged, not a commodity that can be exchanged. Human rights discourse furthermore goes beyond academic research. It inspires practical normative actions that are not limited to lawmaking and that are not merely interesting domestic issues. Of course, this presupposes more than a legalistic understanding of human rights. This is exactly what scholars like Amartya Sen argue: human rights are ethical rights: they are 'meant to be significant ethical claims', they express a commitment to social ethics [16]. They also focus on the development and implementation of policies, not only addressing states but also non-state actors. They finally imply advocacy involving civil society organizations and initiatives.

Individual focus?

Obviously, human rights discourse encounters many theoretical and practical challenges [17]. One of the problems is that human rights are often regarded as the rights of individuals. French philosopher Corine Pelluchon, who published several books on ethics and vulnerability, is critical about human rights discourse since it is primarily conceived as instrument of individual liberty [18]. This is not an isolated opinion. Several authors have pointed out that especially American principlism regards human rights as individual negative rights [19]. However, this might be a matter of evolution. The human rights tradition itself has evolved and has a dual focus nowadays: an individualistic perspective as rights of individuals to be free from government interference, and a societal perspective viewing human rights as involving an obligation to act for common welfare [20]. While the initial emphasis in the area of healthcare has been on individual rights, and in particular patients' rights, attention has now shifted towards improving social conditions and addressing social determinants of health, thus towards social and economic human rights [21]. This is reflected in the increasing interest in the explanation, interpretation, and monitoring of the International Covenant on Economic, Social and Cultural Rights (ICESCR), an international treaty formulating economic, social and cultural rights, including the right to health. This right is interpreted, at least by its monitoring body, as an inclusive right that not only refers to access to health care but also to 'underlying determinants of health' [22]. Core obligations are formulated that substantially address the conditions necessary for a healthy life and aim at influencing the social determinants of health. Again, the focus of comments and interpretations is on the vulnerable. Priority should be given to addressing the situation of the most vulnerable populations [23]. The right to health for example is interpreted as the obligation 'to ensure the right of access to health facilities, goods and services on a non-discriminatory basis, especially for vulnerable or marginalized groups' [24]. Human rights discourse in general is therefore not only focused on protection of individual physical integrity (against violence and torture) but also on satisfaction of basic human needs (education, health, food, work, housing) as well as acceptance of differences in gender, ethnicity, and culture. The relatively new emphasis of human rights on social conditions is consistent with the notion of anthropological vulnerability. This notion is at the origin of human rights because it expresses a human species-related rather than an individual characteristic of human beings. Existing in a shared predicament, humans are not merely individual consumers or right-bearers; they are embedded, dependent, and relational, and therefore appeal to solidarity. It is interesting that the significance of solidarity for human rights discourse is used for the first time in the African Charter on Human and Peoples' Rights, adopted in Banjul in 1981 [25]. This document declares that the human person is not an individual but always connected to others. Alain Supiot has pointed out that this is an important view in African culture, highlighting that the word 'poor' in various African languages refers to a person who has few people around him or her, who can in other words not count on the solidarity of others [26]. Later, solidarity is also included in the Charter of Fundamental Rights of the European Union devoting a separate chapter to this notion [27]. This charter not only considers human rights as rights to be protected but also as rights to act and provide information, consultation, security and health so that conditions will be fostered and reinforced for human flourishing.

Social engagement?

A second challenge for the further development and application of human rights discourse in bioethics has to do with the relationship between human rights and bioethics itself. Significant theoretical, practical, and institutional differences make this relationship, as diagnosed by Richard Ashcroft, 'troubled' [28]. The interaction between the two discourses is rather recent. It is not certain in what direction the interaction will develop; will there be two complementary discourses, will there be a merger, or will one be subsumed into the other? It is clear that more research and reflection will be required and that a finished product will not be in the making soon. But one of Ashcroft's observations is actually worrisome for bioethics. He states that for professional bodies, governments, healthcare systems, research institutes, pharmaceutical companies, and universities, bioethics is more attractive than human rights language because of 'the lower ideological temperature' and because there is no risk for a 'social movement' [29]. This means that for many stakeholders bioethics is rather innocent and risk free compared to human rights discourse. Several bioethicists would possibly regard this weakness as the strength of their discipline since it is a merely academic enterprise [30]. Ashcroft is right that this would be the wrong conclusion; the relationship deserves deeper exploration. Evidently, a balance needs to be found between moral analysis and social change. Scientific inquiry and activist commitment are not necessarily in opposition [31]. But even if bioethical discourse will be more theoretical, moral analysis can contribute to develop alternative visions instead of the current values promoted by neoliberalism. Reflecting seriously on the notion of vulnerability and not reducing it to 'failed autonomy' implies that bioethics has the possibility to inspire society with ideas and ideals to counter the effects of neoliberal globalism on health and healthcare.

The use of human rights

Human rights discourse can provide a much needed theoretical injection to bioethics. It can encourage bioethical reflection to put more emphasis on the relational dimension of human beings, as long advocated by Thomasma, and on the concept of solidarity, as defended by Baker [32]. Bioethics can then move from a focus on individual to social perspectives, underlining ethical notions of social responsibility and obligations [33]. This will produce a better understanding of vulnerability. Human rights can also provide practical leverage to bioethical discourse, criticizing and influencing policies for vulnerable populations, for example policies to prevent people with HIV to enter countries as immigrants or to prohibit the sale of effective generic medication in developing countries [34].

Social justice

Human rights-based approaches are criticized by scholars who emphasize that social injustice is at the basis of structural violence. Though human rights provide equal norms, they focus more on the personal than structural level. Human rights do not change power structures: 'they refer to distribution of resources, not to power over the distribution of resources' [35]. The consequence of this criticism is that a proper analysis of vulnerability should proceed from the perspective of social justice. This requires a change in the orientation of bioethics. Especially scholars from the Global South have argued for such a change. Marcio Fabri dos Anjos for example states: 'Latin America's disastrous experience with a social order that does not attach priority to the lives and health of all sectors of its population illustrates the real link between unjust social structures and the need to recast medical ethical issues as concerns about social justice' [36]. In her presidential address to the International Association of Bioethics, Florencia Luna repeated the same message, almost ten years later: bioethics should have a 'deep concern for injustice, global inequalities, and the imbalance of power in the world' [37]. Instead of articulating and reinforcing rights of individual persons, the unjust context in which individuals are continuously rendered vulnerable should be scrutinized. This will allow bioethics to relate vulnerability not merely with deficient autonomy but with the perspective of justice.

The need for a justice perspective

Three arguments are used to direct attention towards social justice in connection to vulnerability. The first is that today it is more often recognized that health is linked to social conditions and structures. As discussed in previous chapters, social determinants of health contribute in many cases to making populations vulnerable to disease, disabilities, and suffering. A broader concept of health requires a broader concept of justice that considers the social and economic conditions that make some people more vulnerable than others. Usually discussions of justice in healthcare are centered on issues of access to care and distribution of resources for those who are already harmed. But, as Nancy Jecker argues, instead of concentrating on the endpoint of healthcare needs, justice concerns should address how those needs arose [38]. A more equal distribution of health resources or better access to health services will not be sufficient as long as the sources producing vulnerability are not addressed.

The second argument is related to globalization. Neoliberal policies benefit the private healthcare sector, particularly in developing countries while public health services have become weak and less accessible [39]. Giving advantage to the stronger parties, the benefits and burdens of processes of globalization are not equally distributed. The results are global injustices that produce widespread vulnerabilities [40]. Such special vulnerabilities cannot be addressed at a micro-level since they are the consequences of difference in power at the macro-level of societies and the world as a whole. It is also argued that there is a global responsibility to make processes of globalization and global structures more just because the injustices that they embody and the vulnerability they engender are not natural phenomena but the product of human efforts and activities so that in principle they can be redressed [41].

The third argument to focus on social justice is related to the political perspective on vulnerability. Vulnerable means having a weak position, being unable to secure a fair share of benefits. In resource-poor countries it is easier to recruit people to participate in research, especially if they have no access to healthcare or medication. At the same time people often do not receive benefits from research [42]. Interpreting vulnerability not as individual weakness but as the consequence of an exploitative order produced by neoliberal globalism, will focus attention on structural injustice; this is not merely a matter of unfair distribution of harms and benefits but concerns the social framework that differentiates categories of people as powerless.

The problems with social justice

Applying the notion of social justice in the current debate on vulnerability is problematic for two reasons. First, it is argued that the scope of justice is domestic; social justice only applies within the context of the modern nation-state [43]. Its focus is on what citizens owe to each other in terms of resources and respect. Social justice in this domestic context is important since just social conditions will reduce inequalities in health. Associating vulnerability with processes of globalization is fine but in this view it can only be addressed within the territory of the nation-state. Social justice cannot be extended to the global domain. The pursuit of global justice can even be dangerous according to some authors since it requires powerful global agencies that will often not promote justice but will serve the interests of the powerful who create them [44]. If vulnerability is primarily related to differences in power, emphasizing justice at a global level is not the right answer. Pursuing global justice will only reinforce and exacerbate special global vulnerabilities. The second problem with the notion of social justice is that neoliberal globalization has fundamentally transformed the idea of the social. Several studies have shown how the term 'social justice' emerged in the late eighteenth century in response to *laissez-faire* liberalism during the industrial revolution in Europe, shaping a new understanding of the social as the collective domain of the community or society in which the state could redistribute resources to remedy systemic inequalities [45]. The emphasis on social justice has not only redefined the relations between state and market, but it has also promoted a vision of the citizen as social being and the state as protector of social rights and provider of social security. However, neoliberal policies have significantly converted the context of social justice. The contemporary state follows market logics, more often exploring what is efficient rather than what is fair or right. Neoliberal rationality furthermore pervades the conduct of everyday life. The effect is that the social itself is redefined in economic terms. Citizens have become consumers and clients. Social differences are the results of the individual choices of rational actors. Social justice demands that resources are awarded on the basis of effort and individual responsibility. Within the neoliberal context of existential insecurity and generalized vulnerability, the language of social justice can no longer be used to argue in favor of social responses and policies to address systematic disadvantages and inequalities [46].

Scope of social justice

Analyzing vulnerability with the theoretical discourse of social justice therefore requires addressing these problems of scope and relevance. The increasing interest for global justice indicates that for many scholars the scope of social justice can no longer be restricted to territorial borders. Social processes that are disadvantaging individuals, groups, and populations do not stop at the state borders. These borders do exist and continue to exist but they are becoming more and more 'porous' [47]. Empirical reality makes the idea of bounded justice increasingly obsolete. In the era of globalization it is not possible to separate the domestic and global institutional orders [48]. But there are also substantial philosophical arguments in various forms of cosmopolitanism advocating ideas of global citizenship, global moral community, and global responsibilities and obligations. They share the conviction that all human beings live in one 'moral community' and that they have global obligations [49]. Ultimately, national boundaries do not have moral significance. According to Peter Singer we need to take a 'global ethical viewpoint' because 'we are facing issues that affect the entire planet' and that require global solutions [50]. He strongly argues that global institutions should be built and strengthened so that greater global governance will develop. Since we are connected with vulnerable subjects and populations around the world, the implication is that we have duties to assist them to improve their situation, not as a matter of charity but as an obligation of global justice.

Relevance of social justice

The criticism that social justice is no longer relevant in the era of globalization has energized efforts to reinterpret the notion. The issue of relevance arises because social justice is considered as interchangeable with distributive justice. According to the analysis of Iris Young, social justice is restricted to 'the morally proper distribution of benefits and burdens among society's members' [51]. Emphasis is on the allocation of material and non-material goods. Applying social justice to vulnerability means that if vulnerability is the result of the lack of power of people to protect themselves or the manifestation of social, cultural, and political inequalities, there are two ways to respond: protection so that it becomes more difficult for the powerful to exploit the vulnerable, and redistribution so that material goods but also power are more balanced so that people become less vulnerable. The problem with this so-called 'distributive paradigm' is twofold [52]. Focusing on distribution ignores the background conditions that determinate distributive patterns. This is also the concern of Jecker, mentioned above, that distributive justice focuses on the endpoints not on the conditions that produce inequalities. Additionally, the paradigm considers citizens as consumers of goods; in disregarding the social and institutional context it assumes 'social atomism' [53]. This reduces issues of justice to preferences and assets of individuals while justice as social justice should refer to social and institutional conditions [54]. The effect of these two problems is that social justice does not concentrate on the systematic constraints that are operating in distributive processes. The discourse of justice can therefore only properly address vulnerability if it goes beyond the level of individual subjects and distribution of goods. This is illustrated in the example of senior citizens whose vulnerability is increased as access to healthcare is more and more difficult to secure since pensions are reduced and care increasingly privatized.

Of course, individuals should be involved in decision-making concerning their care, and vulnerable subjects should be protected. Furthermore, society should guarantee that material and non-material goods are fairly distributed among the elderly and other members of society. But the scope of justice should expand to background conditions that exacerbate vulnerability since the elderly are excluded from labor, often stereotyped in culture, and most frequently not involved in social and political decision-making processes. Vulnerability due to age moreover illustrates that cultural recognition is an important remedy for injustices, just as in cases of vulnerability due to gender, race, or ethnicity [55]. The same expansion of the justice discourse will be necessary to address the vulnerability of the homeless, to take another example. People who lost their jobs and homes due to the economic crisis, and now are living on the street, perfectly illustrate the statement of Bauman: 'All of us are doomed to the life of choices, but not all of us have the means to be choosers' [56].

A broader concept of justice

Addressing vulnerability in these examples is not, or at least not merely, a matter of just distribution or redistribution of resources. Justice here requires a critical analysis of the background conditions that produce vulnerable subjects. It also requires 'transformative remedies' that restructure the underlying processes [57]. Young particularly criticizes ways in which vulnerable subjects are excluded from decision-making power and procedures. In fact, they live in oppressive conditions ('structural phenomena of domination') that do not provide them with the means to exercise their capacities [58]. As long as social justice does not scrutinize these conditions, it will contribute to the continuation of oppression and domination, and thus structural injustice, because it reinforces unjust social structures and relations [59]. Utilizing justice discourse to address the different faces of vulnerability will require more than protecting personal autonomy and compensating for the lack of it; it also demands providing or liberating the means to exercise autonomy, expanding the range of options, and rebalancing the power differentials that determine the background conditions [60]. Similar reasoning is applied by Alex London in the area of international medical research. He argues that a minimalist or distributive view of justice as mutual advantage in which harms and benefits are fairly distributed is not sufficient to reduce vulnerability. International research cannot be abstracted from the conditions in the host communities. A broader concept of justice requires that research will assist the basic social structures of these communities to address the needs of its members [61]. In other words, justice demands that vulnerability is reduced through remedying structures that produce vulnerability. If, as argued earlier, vulnerability in many cases is a symptom of social inequalities, justice implies more than redistributing available resources. Involving vulnerable groups in research is just if it strengthens communities to meet the basic health needs of its members.

The broader concept of social justice as global justice is elaborated and specified in various directions that are relevant for addressing vulnerability. Emphasis can be put on respectively, obligations, responsibilities, institution-building, representation, and capabilities [62].

Obligations

Rather than emphasizing rights of vulnerable persons, the argument can be made that social inequalities and differences in power that exacerbate vulnerability generate obligations for the powerful towards the vulnerable. This approach is taken by several scholars. Goodin, for example, considers vulnerability as the source of special obligations, in particular the obligation to prevent people from becoming vulnerable, and to protect those who already are vulnerable [63]. Rogers and colleagues have argued that there is a wider set of obligations towards the vulnerable than protection [64]. Although James Dwyer is skeptical about a cosmopolitan perspective, global justice in his view implies three duties: a duty not to harm, a duty to reconstruct (creating more just international institutions), and a duty to assist (help societies to create and maintain just institutions) [65]. In Thomas Pogge's cosmopolitan perspective, vulnerability is manifested in the fact that each year 18 million people die from diseases that are preventable, curable, or treatable. This global injustice requires institutional rearrangements notably with regard to access to medical interventions. Since the global institutional order is a human creation, its maintenance is a violation of the negative duty not to harm. If widespread vulnerability is produced and actual harm done, there is an obligation to rectify matters [66].

Responsibilities

Another approach is to articulate responsibilities. As discussed earlier, both philosophical and political perspectives on vulnerability criticize the assumption that vulnerability is an individual deficit and a matter of personal responsibility. It is often the result of structural injustice. In these circumstances it is difficult to identify who is responsible for the potential harm that may occur. The model of individual responsibility will not suffice. A different model of responsibility has therefore been proposed: the 'social connection model' [67]. It is in fact based on the idea of anthropological vulnerability: humans are social beings who are always in interdependent relationships with other humans. They are all involved in contributing to and sustaining unjust background conditions. Harm may be produced and people hurt but there is not one individual to blame. This kind of responsibility requires collective action. It is a political rather than personal responsibility since it can only be discharged if people cooperate [68]. Compared to the emphasis on obligations, political responsibility is broader and at the same time more restricted. It is broader because not only institutional arrangements contribute to structural injustice but all types of social processes, for example in communication and consumption. It is more restricted since responsibility will vary according to the social positions of agents, depending on power or influence, privilege and benefit, interests, and collective ability.

Institution-building

A third approach underlines the relevance of institutions for global justice. If justice is no longer bounded by states, there is a need for global institutions to take care of the vulnerable. This 'institutional cosmopolitanism' is advocated by Onora O'Neill. It does not imply a world government but rather the construction of robust institutions that provide security for vulnerable populations. Vulnerability is a relative notion. People are vulnerable in relation to others who are stronger and more powerful, and who are prepared to use these advantages to their own benefit. For this reason, it can be argued that the demands of justice are stronger in regard to the weak than to the powerful [69]. Allocating rights to the vulnerable will not help because the weak are not in the position to make sure that these rights are respected. In order to reduce differentials of power it is better to focus on obligations of the strong [70]. For this reason non-territorial institutions for global justice must be built that involve supporting the capacities of the vulnerable and 'disciplining the action of the powerful' [71]. It might be a return to what Alain Supiot has called the 'spirit of Philadelphia', referring to the Declaration of Philadelphia (in May 1944) that constituted the International Labor Organization as the first specialized agency of the United Nations and that emphasized social justice as one of the key foundations for the security of the international order [72].

Representation

A broader view of justice can also articulate the issue of representation. The vulnerable are often misrepresented. Their voice is not heard in the international debate and therefore they are not able to participate in social interaction and to protect their interests. Representation is the political dimension of justice, besides distribution (the economic dimension) and recognition (the cultural dimension) [73]. This political dimension is particularly important in a globalized order since it determines who is included in social arrangements in which everybody can participate on an equal basis in social life. If social justice is limited to the territorial framework, structural injustices that produces vulnerability cannot be addressed; the global forces that create and reinforce vulnerabilities are out of the reach of justice [74]. Therefore, it is important to redefine the frame of justice. It is not simply the case that vulnerable subjects and populations are not included in public deliberations and democratic decision-making. The problem is deeper. Since it is the powerful that set the rules of social and economic life, it is not even possible to analyze and critique the forces that produce vulnerability. The reason is that justice claims of the vulnerable in this framing can be addressed only from the perspective of distribution and recognition. Deconstructing the process of framing itself and redrawing the boundaries of who is included in social life is the way to redress the injustice of misrepresentation and to eliminate the 'political voicelessness' of the vulnerable [75]. As discussed in Chapter 5, this is exactly the purpose of the notion of anthropological vulnerability: if the human species is basically vulnerable, social exclusion is impossible.

The challenge of global justice

Anthropological and special vulnerability both refer to conditions beyond the level of individual agency. Therefore, analyzing and using the notion of vulnerability requires the perspective of justice. However, as argued in this section, such perspective should not only take into account how resources are allocated. This implies that the notion of justice in this context should not be merely distributive justice. Justice requires critical examination of the power differences and inequal structures within which resources, and thus vulnerability, are distributed. This focus on the social context is contested. One argument is that social justice is only possible within the boundaries of the nation-state. The other argument points out that the entire idea of social justice has become irrelevant because of neoliberal globalization. This section has shown that these arguments do not apply in the current stage of globalization since social justice is increasingly expanded into the notion of global justice. The idea of social justice itself is no longer confined to the individual state so that social justice has become global justice. Going beyond the usual confines of distributive justice global justice focuses attention on the social, political, and economic background conditions that exacerbate vulnerability. Global justice presents a discourse that criticizes the common assumptions of neoliberal ideology. Global justice furthermore inspires the bioethical debate to search for approaches of vulnerability that do not merely concentrate on protecting vulnerable subjects or remediating deficient autonomy. The specific implications of this broader concept of justice can differ. Emphasis can be put on the moral obligations towards the vulnerable, on global responsibilities, construction of global institutions, or issues of representation of the vulnerable. One particularly interesting theory of global justice elaborates human capabilities as a way to address vulnerability.

Capabilities

Michael Higgins, the President of Ireland, in September 2013 criticized neoliberal economic policies and called for a change in public consciousness, placing human flourishing at the center of political action. The basic question he asked was: 'What kinds of human capabilities do particular societies value, encourage, genuinely enable, or block' [76]? Higgins' question points to the capabilities approach as another theoretical framework used to address vulnerability. This approach has been developed by Amartya Sen and Martha Nussbaum as a framework for the analysis of poverty, inequality, and human development. Sen argues that development and justice should not be understood in economic terms, i.e. in relation to wealth, income, or resources. Resources are only means to opportunities for a good life [77]. Well-being or human flourishing are not the same as income, satisfaction, or happiness. Development means expanding human capabilities and not more economic growth. The focus should therefore be on what people are able to achieve, how they are able to function with the means they have. Instead of focusing on commodities, the central question is how people can convert commodities into achievements. The abilities of individuals to do this vary; they also do not have the same needs and values. The availability of resources is therefore not determinative for human well-being, but it is the opportunity to choose to do what people value [78]. According to Sen, a better understanding of development associates human flourishing with freedom and human capabilities. In this connection, he makes the well-known distinction between functionings (i.e. the actual achievements of a person; 'what he or she manages to do or to be') and capabilities (the ability to achieve a functioning, the freedom or opportunity to achieve) [79]. Capabilities are more important than functionings. Available commodities (for example food or bikes) are used to achieve a functioning (e.g. being adequately nourished or biking). Capabilities (for instance, the ability to avoid hunger or the capability to move freely) are the opportunities to function; they reflect freedom of choice [80]. To illustrate the difference, Sen frequently refers to the difference between a starving and a fasting person [81]. Both are hungry and have the same functioning but different capabilities. The starving person lacks the capability to eat; his functioning is not the result of free choice while it is for the fasting person.

Sen's framework has been broadened by other scholars, most notably philosopher Martha Nussbaum. The new view of development as expansion of human capabilities is elaborated in her many publications into a general theory of justice and human flourishing, beyond the domain of economic development. A good and just society is concerned with the distribution of capabilities. The quality of life is determined by the opportunity to function, more than by people's actual functioning [82]. For Nussbaum this implies that there is a minimum level of capabilities that should guarantee 'a life that is worthy of the dignity of the human being' [83]. Without certain functions human life itself will be in danger or human dignity will be violated so that human life will either come to an end or lose its specific human character. Because there is a basic social minimum each human capability has a threshold 'beneath which … truly human functioning is not available to citizens' [84]. Nussbaum also holds that certain capabilities are universal; they are important for every person regardless of who and where he or she is, and whatever the person chooses. As a result of these considerations she proposes a list of central human capabilities [85].

Capabilities and vulnerability

The capabilities approach has been applied in many areas beyond the initial scope of human development studies. In some of these applications an explicit link is made with the notion of vulnerability. The approach is for example used to discuss the condition of vulnerable migrants, especially overseas domestic workers, although the relation between capabilities and vulnerability is not elaborated [86]. Sometimes one specific capability is mentioned. In disaster research, for example it is argued that the impact of a disaster is more determined by vulnerability than by the hazards and damaging events themselves. But vulnerability can then be understood in two different ways, as 'liability' (the susceptibility to harm, in other words as the component of sensitivity in the technical definition of vulnerability) or as 'capability' (the ability to withstand harm or the adaptive capacity to resist and recover from threats) [87]. In an application of the capabilities approach to the vulnerable population of the elderly, no specific capabilities are mentioned [88]. An important advantage of the approach is highlighted: it regards older people as resourceful persons who contribute to their families and communities, rather than as passive, dependent, and vulnerable subjects. Promoting capacities to enhance quality of life and social involvement encourages policies that consider older people as agents, emphasizing their abilities instead of their deficiencies. Disability is another area of work. Applying the capabilities approach to the vulnerable populations of disabled people produces an understanding of disability as deprivation of either capabilities or functionings. The first is potential disability since it means that due to an impairment an individual is deprived of practical opportunities; the second is actual disability because the individual's functionings are restricted [89]. Nussbaum herself has extensively discussed disability but in rather general and policy-orientated ways [90]. Theories of justice assume equality in capacities; they further assume that social cooperation is based on mutual advantage and reciprocity. They therefore usually do not include people with disabilities. Nussbaum argues that theories should acknowledge that most human beings will experience need, dependency, and impairment, so that there is not a substantial difference with people with lifelong disabilities [91]. This statement refers to what previously has been called anthropological vulnerability, and it is apparently accepted at least in Nussbaum's capacities approach; there is also agreement that the person is a social animal. Nussbaum thus explains that though the capabilities approach is a liberal approach it certainly is a modified liberalism with a different view on human beings; they are essentially cooperative, and they have various motives to cooperate, not merely self-interest [92]. The implications of this view is that in order to sustain the capabilities of disabled people there is a need for a new approach to social cooperation, not only instituting atypical social arrangements (such as custodial care, special education, and redesigning of public space) but also policies for people who provide care [93]. A broad spectrum of activities is required because capabilities are nonfungible; one cannot simply concentrate on one capability but all central capabilities need to be addressed [94]. According to Nussbaum, disabilities should not be a reason to adapt or modify the central capabilities themselves or the threshold for each capability. We should try to improve the capabilities even if our efforts will not be successful for some people with serious disabilities [95].

It is not immediately obvious how vulnerability can be understood within the perspective of the capabilities approach. One of the relatively few times the term vulnerability is used in Sen's work is in his discussion of constitutive and instrumental roles of freedom [96]. The first refers to the intrinsic importance of freedom as the end of development, while the second refers to freedom as the principal means of development. According to Sen there are five types of freedom that contribute to the further expansion of human freedom in general: political freedoms, economic facilities, social opportunities, transparency guarantees, and protective security. The last freedom is necessary to prevent misery for people who are 'on the verge of vulnerability' [97]. It is tempting to assume that insufficiency or deprivation of each of these five instrumental freedoms can contribute to vulnerability, even more so because they are strongly interlinked, but this is not further elaborated. In a recent publication Sen discusses four sources of deprivation that arise because of difficulties in converting resources as means of living into actual opportunities for good living: personal heterogeneities, diversities in the physical environment, variations in social climate, and differences in relational perspectives [98]. These sources combine internal and external characteristics, such as age, gender, and disability on the one hand; and environmental, social, and cultural conditions on the other. They explain why capabilities can be restrained. Sen's analysis is focused on poverty and disability, but it can be used to explain vulnerability.

Vulnerability as limitation of capabilities

Although in the relevant literature the notion 'vulnerability' is not explained in relation to capabilities, a few observations can be made. Nussbaum argues in favor of a new form of liberalism that regards human beings as 'vulnerable temporal creatures, both capable and needy' [99]. Vulnerability therefore should be accepted as anthropological and characteristic of human beings. At the same time, Nussbaum considers women in developing countries as particularly vulnerable. Women are suffering from 'capability failure' as a result of poverty, violence, discrimination, or lack of education [100]. Special vulnerability is apparently interpreted as lack or impairment of capabilities. This is in line with Sen's view that the poor are suffering from 'capability deprivation' or 'capability inadequacy' [101]. If capabilities are opportunities that allow people to choose what they actually do or are, then vulnerability is primarily the restriction of those opportunities – in other words, limited freedom. For Nussbaum, human capabilities are the 'basic powers of choice', just like according to Sen a capability is 'the power to do something' [102]. In both cases, power apparently is an individual property. If this interpretation is applied on the notion of vulnerability, it will not produce a different perspective as provided in mainstream bioethics. Vulnerability should primarily be regarded as impairment of choice [103]. Although the terminology has changed, it would still assume that being vulnerable means lacking individual autonomy or freedom. This conclusion, however, is not correct.

The capabilities approach provides a different perspective. Regarding vulnerability as limitation of capabilities allows an important conceptual shift. Vulnerable subjects can have the same functionings as subjects who are not vulnerable; there will only be a difference if they will actually be damaged [104]. The main divergence is in the opportunities to do or to be. For vulnerable persons opportunities are constrained either because of individual characteristics or due to social, economic, and physical conditions. Being vulnerable therefore means having few opportunities to achieve states of being or to undertake activities. This interpretation of vulnerability can be connected to the distinction between different types of capabilities made by Nussbaum [105]. Failure of basic as well as internal capabilities can explain the internal conditions of vulnerability, or what has been called inherent vulnerability. Combined capabilities in Nussbaum's typology refer to external conditions. For their realization, not only development of internal capabilities is required but also an appropriate environment. They are opportunities for choice within a specific situation; if they are constrained or inadequate, they result in special vulnerability. It is also attractive to explain vulnerability in relation to failure of the central capabilities listed by Nussbaum [106]. For example, the capability of being able to live to the end of a human life is constrained for poor children in resource-poor countries, and this explains their vulnerability. Or, the capability of being able to be treated as a dignified being is often limited for homeless people. The difficulty with such exercise is that frequently multiple capabilities are affected at the same time. Capabilities are not isolated items but are interconnected since they affect one another. Also theoretically it is argued that all capabilities are relevant as opportunities for a good life because individuals can value them differently [107]. It is not possible to select one core capability as more fundamental; they all need to be addressed, and cannot compensate for each other. Although vulnerability therefore cannot be analyzed in connection to one or more specific capabilities, the framework of capabilities at least encourages an analysis of specific ways to expand specific capabilities. It generates a multidimensional examination of what exactly is constraining capabilities, for instance the difficulties in converting resources or commodities into capabilities or the limitations of possibilities for choosing specific functionings. One observation therefore is that the interpretation of vulnerability as capability constraint opens up specific remediation possibilities in order to enhance, expand, or recover capabilities.

Structural capability constraints

The second observation is that the relevant capabilities are not merely individual powers. They are the result of the interplay between internal conditions and the external environment, so that agent as well as structure needs to be considered [108]. For example, it is appealing to argue that vulnerability in the capability approach is primarily associated with the adaptive capacity. This is a clear capability dimension of the functional notion of vulnerability. Reducing vulnerability then implies enhancing the capabilities to anticipate damage, to cope and adapt, and to be more resilient [109]. Such interpretation, however, will focus on individual capacities, considering vulnerability as a negative capability, more or less equating being vulnerable with being capable to be damaged and not being able to protect oneself. But this individualistic approach neglects that lack of capacities can also result from social contexts that produce unequal exposure to threats. Poverty and social inequality diminish the abilities of persons to achieve their potential; they therefore result in restricted, deprived, or inadequate capacities explaining why some populations have special vulnerability. The capability approach therefore not only focuses on the adaptive capacity of individuals but also on the components of exposure and sensitivity that create or exacerbate vulnerability for groups and populations. In this perspective, even if vulnerability is regarded as capability constraint, and thus as limitation of opportunities for choice, it is most often not the result of the deterioration of internal capabilities alone but of the impact of external conditions. Nussbaum's recent articulation of the idea of 'capability security' underlines the same conclusion. It is not sufficient that people have capabilities, but they need to be sure that these capabilities will exist in the future. Each capability should be protected from the insecurities of the market and power politics [110]. It is important to secure the structural context in which capabilities can be exercised. Capabilities are not only interconnected but also dependent on the conditions in which people live; these conditions can expand or constrain capabilities.

The sociality of capabilities

The third, related observation is that the capabilities approach is often criticized for being individualistic [111]. The objection is that viewing capabilities as each person's actual freedom to achieve does not sufficiently recognize social power and significant power differences in human existence. Sen's recent book on justice for example does not pay much attention to explaining the role of power dynamics or critically analyzing the strategies of transnational companies such as the pharmaceutical industry [112]. Focusing on the capabilities of individuals also tends to underestimate the role of social context and structure that create and sustain individual capabilities. The main weakness of the capabilities approach is that 'it takes freedom as the ultimate value of human life' [113]. The core of this critique is that there is insufficient appreciation of the social nature of human beings. In a certain way this refers to similar criticism of the emphasis on individual autonomy in bioethics and the individualistic interpretation of vulnerability discussed in previous chapters. If vulnerability is regarded as capability constraint for individual persons, it isolates the individual, since it is not taken into account that individual freedoms are constituted by social arrangements, and that capabilities are affected by being a member of groups or communities [114]. Capabilities are collective in the sense that their exercise is possible because people interact and cooperate within a shared social context. Whether or not an individual will achieve particular functionings is not only the outcome of individual choice but will depend on the actions of other individuals [115]. However, as discussed earlier, the capabilities approach is not merely individualistic and this makes it an interesting perspective for interpreting vulnerability. But there is an ambiguity that facilitates the above critique. Nussbaum for example gives much attention to the role of social conditions; she shows how the inequality of social and political circumstances produces inequality of human capabilities, especially for women confronted with poverty, violence, or discrimination. The result is failure of capabilities, thus vulnerability [116]. She also argues that the individual is not alone. Human beings shape their own life but 'in cooperation and reciprocity with others' [117]. The social dimension of human existence has implications for policy. The aim of government is to deliver 'the *social basis*' of human capabilities [118]. Human institutions should be shaped in such a way that they enable the flourishing of every citizen by making sure that the threshold levels of each capability can be reached. On the other hand, Nussbaum argues that ultimately, promoting the capacities of each person is the political goal. The social context is secondary to this goal. The context is necessary for the training of internal capabilities and the formation of individual agency. Social arrangements are therefore instrumental; they should expand capabilities, thus the freedom of individuals to achieve valuable doings and beings. Of course, the aim of policies according to the capabilities approach is focused on human beings rather than on wealth, income, growth, or material resources, but it still reflects a reduced view of the social nature of human beings. The work of Sen has similar ambiguities. 'Societal arrangements' are examined because they contribute to the enhancement of individual freedoms [119]. He acknowledges that opportunities 'depend crucially on what institutions exist and how they function' [120]. At the same time, his conception of the social context it rather thin; it does not take into account the history and community within which freedom is exercised [121]. In the end it is the capabilities of individuals that need enhancement. Sen's philosophy is agent-oriented: 'With adequate social opportunities, individuals can effectively shape their own destiny and help each other' [122]. Although Sen puts a lot of trust in public reasoning, he does not address the question of how within a context of structural inequality and injustice, vulnerable populations can advance reasons for valuing certain capabilities, particularly when there is little room for free decisions. Nonetheless, Sen emphatically replies to this criticism that human beings are 'societal creatures' and that individual capabilities are socially dependent [123]. The capabilities approach recognizes that there are structural barriers to the development and exercise of human capabilities, but Sen concedes that he should have said more about the obstructive effects of market processes.

Advantages of the capabilities approach

Interpreting and elaborating vulnerability from the perspective of the human capabilities approach has four advantages. First, emphasis is put on people instead of material resources or commodities. This is an important corrective upon the dominant neoliberal framework of policy-making. Vulnerability therefore is not related to people's actual doings or beings but to what they are able to do or be; it is not a determinant of individual persons but relates to the choices and freedoms that are available to them. Second, the approach encourages the empowering of people as active agents. Vulnerable persons are not victims or passive recipients of protection but they should be enabled to develop their capacities [124]. Third, what capabilities are relevant to enhance is determined in a process of public reasoning. The importance of capabilities is dependent on the circumstances as well as the purpose for which they are necessary. For example, in research different capabilities will be more important than in long-term care. Fourth, practical attention is directed to external conditions. Creating possibilities of choice requires an appropriate environment in which vulnerability can be respected [125]. This is the task of government and legislation: supplying the external conditions for human flourishing since they enable central human capacities [126]. The adequate response to vulnerability therefore is not protection but the provision of opportunities. This is the consequence of an emphasis on freedom as positive liberty. While neoliberal policies are mainly concerned with negative liberty, the capabilities approach stresses the promotion of people's capabilities through the creation of opportunities, and thus freedom to act or to be according to what they value. This freedom should be actively supported by affirmative government action [127]. Not interfering with individual liberty will not be sufficient for enhancing capabilities [128].

Challenges

The capabilities approach has had an enormous global impact on social policy, economics, and development studies, particularly through the Human Development Reports since 1990, that challenged the neoliberal ideology of economic growth and human progress [129]. At the same time, it is confronted with practical challenges. It has a strong focus on global poverty and inequality, but how can the idea that all people 'deserve core life opportunities as a matter of basic justice' as Nussbaum argues, be put into practice [130]? As argued earlier, action is not simply individual choice, and expanding capabilities depends on the opportunities provided. There is therefore a need for an institutional solution, but it is not clear how duties should be assigned and to whom exactly. How can legal and political structures be created that secure a minimum level of central capabilities [131]?

Within bioethics, a systematic connection between the capabilities approach and the notion of vulnerability has not been made. Alex London has introduced the 'human development approach' to reframe issues of justice in international research [132]. He criticizes the view that justice is concerned with distribution of commodities (access to the outcomes of research) and mutually advantageous exchanges, advocating instead a concern with social structures that will allow the development of basic human capacities of members of the host population. He emphasizes the capabilities of social institutions rather than those of individuals. A recent application of the capabilities approach, taking a lead from Sen, distinguishes health (as achievement) from the capability to be healthy [133]. Since the capability to be healthy is determined by social conditions, not only healthcare institutions are required but particularly broader social arrangements that determine health, well-being, and life expectancy [134]. Health justice therefore implies more than just health policies. Furthermore, following Nussbaum's ideas, the capability to be healthy is grounded in the conception of human dignity, and it should be regarded as pre-political moral entitlement of every human being, referring to the basic right to health [135]. In this view it is imperative to create and reinforce the social context that is required for the capability of being health. If this capability is not present, individuals are not free to make choices. The human right to health should be interpreted as the moral entitlement to this capability. This entitlement has a global nature since the determinants of the health of individuals and populations are crossing borders. The interconnections of individuals and societies demonstrate the shared vulnerability of human beings and encourage cooperation across societies. The challenge is to improve the material and social conditions that constrain the capability of individuals to be healthy. Other applications of the capabilities approach, however, go in another direction, pointing out that social action should be limited to healthcare policy which is more narrowly defined and does not include social mechanisms that influence health [136]. Broad 'non-health policies' should not be attempted since there is not sufficient understanding of the social determinants of health [137]. Injustices caused by social conditions are therefore not addressed in health policies. Fundamentally, in this view, at stake in health justice is individual health capabilities.

The central role of health capabilities in theories of justice, whether conceived more broadly or narrowly is criticized by Powers and Faden. Although their well-being theory is inspired by the capabilities approach, the authors argue that the priority given to opportunities available to people rather than actual functionings is misplaced [138]. Not only is the language ambiguous since for vulnerable categories such as children actual functioning is primordial (because it is the only way to develop capabilities) but also since for most people it is health, not the capability to be healthy that matters most for their well-being. The focus on capabilities furthermore misdirects the central aim of justice, which is the achievement of well-being rather than the freedom to achieve it. Inequality of health is the most serious problem of global justice. To face this problem it is not sufficient to argue for equal opportunities but what is necessary is equality of outcomes. Terminological differences notwithstanding, Powers and Faden join the capabilities approach in focusing on the creation of opportunities. Appropriate background conditions and social structures should be designed so that the relevant dimensions of well-being can be achieved for all human beings [139]. This must be the primary approach to reducing and eliminating special vulnerabilities.

Global care

The view that vulnerability is a characteristic of being human has particularly been espoused by scholars in feminist and care ethics. As discussed in Chapter 5, Fineman, Butler, and Kittay, for example, argue that vulnerability arises from the relatedness and interdependency of humans. Anthropological vulnerability is a shared condition typical for the human species and also more fundamental than individual autonomy since it defines the possibility to develop human agency and activity. It is furthermore a positive predicament because it implies openness to the world and necessitates cooperation among people. The philosophical perspective on vulnerability reorients thinking and acting towards the condition of other people. This is what links this perspective with the political one. The same scholars argue that due to social and political conditions vulnerability can be exacerbated so that some people and some populations are especially vulnerable. This is because, as Judith Butler has pointed out, vulnerability is not equally allocated. Human beings are connected and dependent; this precisely entails inequality of power and thus the possibility of violence, domination, exploitation, marginalization, and discrimination.

Although care ethics has frequently addressed issues of vulnerability, it is only recently that it examines the notion as a phenomenon of globalization. Ethics of care, arguably, develops through different stages, with circles of care expanding from the private, to the public, to the global domain [140]. Theories of care are now becoming more encompassing and are being applied to broader settings. Initially, care ethics was concerned with the private sphere where we have specific relationships with particular others and where care is provided and received in concrete arrangements. In this stage it was assumed that ethical considerations of justice and care have a different focus; justice is concerned with institutions, care with character [141]. Recently, scholars such as Held, Robinson, and Tronto have broadened the scope of care ethics, taking into account the social, political, and the global realm [142]. Care is an excellent notion to accomplish a wider perspective since it 'embeds personal practices within the context of social structures and social relations' [143].

The impact of globalization

Processes of globalization have clearly impacted care as a practice, as a set of activities. Not only are care patterns within countries changing under the influence of neoliberal policies, but they are also transferred from wealthy to poor countries. The phenomenon of 'care drain' shows how immigrant care workers from developing countries leave their own dependants alone to take care of families in more developed countries. This creates 'global care chains' reasserting existing socio-economic inequalities and power differences [144]. Phenomena like these also make clear that women are disproportionally disadvantaged by globalization. Care for vulnerable subjects as well as access to care resources are unequally distributed [145]. Care is usually provided as unpaid labor of women from poor countries or minorities. While states have decreased social services and support for care work, the burden for women has increased since usually they are the ones taking care for vulnerable subjects such as children, disabled or diseased family members, and the elderly [146].

Care as a normative framework is equally challenged. Given the global scale of processes, the basic question is how the fundamental notions of care ethics can be applied in such a broad context. How can the focus on concrete individuals in specific settings be maintained when the needs of distant others are taken into account? Or, how are the central role of dependence and relationship articulated in a global context? How is the emphasis on responsibility and cooperation understood when confronted with global phenomena such as poverty, violence, and structural injustice? The challenge therefore is to develop a cosmopolitan account of care that goes beyond the level of personal attachments to that of social relationships across borders [147].

Globalizing the discourse of care

Political scientist Joan Tronto proposed to move care ethics from the private to the public arena, highlighting its political implications. Care ethicists should contest the boundary between moral and political life since it ensures that vulnerability is not regarded as a social and political phenomenon [148]. Defining certain categories of people as vulnerable, and thus in need of care, merely reinforces the discourses of power [149]. They should also reject the current boundary between private and public domains since it reinforces the view that providing and receiving care is a private and domestic activity while vulnerability is contained in the same domain, and therefore beyond political concern. According to Tronto, care should be used as an ethical notion to assess the quality of societies rather than only that of persons [150]. Moving care ethics into the public and political arena is a first step towards globalizing the discourse of care ethics. The next step is to rethink and reinterpret its basic notions at the global level.

Relatedness

The fundamental idea of human relatedness receives a broader meaning with a new vision of relational autonomy as a global self, 'embedded in a web of interdependence and interconnection' [151]. Globalization does not produce relatedness but it intensifies it, facilitating connections across borders. The discourse of care ethics makes it easier to understand the significance of connectedness, but it also takes it as an ethical quality. Unlike the capabilities approach, care ethics fundamentally is a relational ethics. It proceeds from a relational ontology; it does not presuppose individualized moral agents but conceives persons as necessarily constituted through relationships [152]. It is therefore not simply the case that individuals are connected to each other, and even more so as a result of globalization, but there is a fundamental significance: the individual self is not understandable without relatedness to others. The basic condition of relatedness moreover implies a fundamentally different appreciation of the environment of human existence. The social, economic, cultural, and political conditions in which humans live are not just the context in which interactions between individual actors take place; human beings depend on these conditions to flourish; they provide the possibility to develop and nurture autonomous agency [153].

Dependency

A similar broadening is applied to the notion of dependency. Processes of globalization demonstrate the increasing interdependency of human beings even in distal relations. Phenomena as global care chains, human trafficking, and organ trade also show that such interdependencies can intensify structural injustice and systems of exclusion and oppression. The view of care ethics that all human beings are dependent implies that care concerns everybody; each person needs care and is capable to provide it. Inequalities in care should therefore be scrutinized and the background conditions of care practices ethically assessed. The dependencies manifested at global level underline that care is important for the public sphere. Care is not a commodity that can be delivered and transacted. Since it is associated with the basic feature of human dependency, care is a general concern; it is the cement of society, the indispensable ingredient of social cohesion, and even the cornerstone of living together [154]. Care ethicists, however, are aware that the capacity to care is not the same for everybody, determined as it is by contexts that enable people to care. They are also aware that human dependency and thus dependency care is mostly invisible since it is often located and restricted to the private domain [155]. It is therefore imperative to critically examine the mechanisms that leave the context and public sphere out of the bioethical debate. This perspective furthermore requires rethinking of the notion of freedom. It should no longer be conceived as individual independence but as interdependence; in other words, freedom should be regarded as 'an expansion of self in relationship' [156]. The notion of dependency entails that we are not free if others with whom we are connected are not free.

Responsibility

The third revisited notion is responsibility. Caring relations with distant others generate responsibility. The perspective of care, argues Tronto, introduces a different view of public life [157]. It promotes a vision of interdependent actors with collective responsibility, particularly for vulnerable groups. Since all human beings are dependent on care, this responsibility should not only be assumed by individuals but also by states, institutions, and organizations.

Cooperation and solidarity

The notions of cooperation and solidarity will play an important role in globalized care ethics. The starting point for care ethics is radically different from the neo-liberal perspective that considers cooperative interaction as based on self-interest. For care ethics, the person who needs care is crucial; ethics means 'reaching out to something other than the self' [158]. Rather than assuming opposed interests, human cooperation is founded on the recognition that interests are mutually dependent [159].

Contextuality

Finally, the notion of contextuality is a basic notion in care ethics. Ethical analysis is not applying abstract principles or universal concepts such as rights or justice but is attending to concrete subjects in specific conditions that produce vulnerability. Within a global perspective the same contextualizing approach should be used. Global care ethics must focus on experiences of real people in concrete situations. Like the capabilities approach it is directed towards what people are actually able to do and to be, but unlike this approach care ethics does not claim universalism by identifying central capabilities. Its contextual focus also provides care ethics an advantage over the ethics of justice in understanding the particularities of different situations [160]. This focus does not mean that individuals are the primary object of analysis. The human condition can only be understood within a wider context of social, economic, political and cultural factors, which means explaining differences on the basis of gender, race, ethnicity, class, and geography [161]. In the analysis of structural inequalities, more attention will furthermore be given to the different starting points and disadvantages that mark the existence of specific groups and populations [162]. Focusing on contextuality allows global care ethics to be more sensitive and open to cultural differences.

Advantages and challenges

Rethinking its conceptual framework has assisted care ethics to develop a more global approach to vulnerability. As a moral orientation it provides new perspectives on human existence and social life that can better explain the global phenomenon of increasing vulnerability. Furthermore, it introduces into the ethical discourse a new vocabulary that can serve as inspiration and foundation for policy change and political action since values are reintroduced that have been eroded in processes of globalization driven by a neoliberal ideology. This reorientation has practical implications for society, international relations, policies of globalization, and global agencies and institutions.

Society

Society, in the perspective of global care ethics, is not based on a contract between free individuals. On the contrary, it is founded on the interdependency created by the shared vulnerability of human beings. They are all in a state of weakness and lack of freedom, and therefore need care. At the same time, not all are symmetrically situated; some individuals are more dependent than others [163]. The core of society therefore is care. Moreover, dependency is not a negative obstacle but an opportunity for human flourishing. This means that providing and receiving care should be a collective activity. Vulnerability requires solidarity. Instead of considering who is providing care (individuals, families, agencies, or the state) it should be emphasized that care is an ongoing social process, requiring the continuous engagement of citizens. Care arrangements reflect the quality of a society, particularly a democratic society. Tronto has made the important observation that 'the practice of care describes the qualities necessary for democratic citizens to live together in a pluralistic society…' [164].

International relations

International relations between nation-states should be understood as global networks of mutual dependency and thus vulnerability [165]. The example of Tuvalu in Chapter 1 highlights this global connectedness and inequality. Care ethicists argue that processes of globalization must be approached in a different way [166]. Rather than assuming that a cosmopolitan universe will emerge with world citizenship governed by universal human rights or abstract principles of justice, the experience of global relationships should be analyzed in practical and contextual ways, examining how social, economic, and political arrangements are affecting specific populations. What Robinson has called the 'increasing sociopolitical intertwinement brought about through globalization' demands a detailed examination of the elements that contribute to the structures of injustice and inequality at the global level [167]. Analyzing the phenomenon of 'care drain' for example shows that the problem is not simply the inequality of power and knowledge between more and less developed countries. There is also the gender dimension of inequalities, reflected in the everyday life of people. Because most of the unpaid carework and low-cost labor is done by women, the global inequalities can continue to exist [168]. The same conclusion is drawn by Tronto: as long as care is regarded as a private activity and provided by unequal citizens, the inequalities and exclusion of vulnerable groups will only be deepened [169]. Solving the care crisis in developed countries by importing care workers from other countries will produce other care crises in poor-resource countries. An inclusive approach is needed, not merely focused on care receivers but also on care providers, taking into account the inequalities of power and wealth, and the differences in concrete lives of real people.

Globalization policies

Global care ethics is highly critical of neoliberal policies of globalization. Following the capabilities approach it argues that human flourishing does not depend on commodities and cost-benefit ratios. Global policies must be challenged by the priority of care as a collective need. Neoliberal discourse promotes a view of care as individual choice on the market place; care is regarded as private activity, the result of individual action. It is transformed into a commodity, a 'service' that can be purchased and commanded by the more powerful [170]. This view is devastating since care is no longer taken as a public value and as an ongoing social process that contributes to social cohesion. The neoliberal view contributes to the marginalizing and ignoring of care in public policies. It reinforces the idea that care reflects weakness and 'uncontrollable vulnerability' [171]. The question how care ethics can help to produce an ethical globalization as a real alternative to neoliberalism is explicitly explored by Fiona Robinson [172]. She argues that in debates on globalization the focus should shift from autonomous individual actors to connected and interdependent communities, groups, families, and households. States should assume a different role; more is required than respecting rights, but the social cohesion that is affected by neoliberal policies should be restored. Other values than market values should be promoted, so that a caring society can emerge [173]. Furthermore, transnational corporations should be regarded as communities with specific responsibilities.

Global institutions

The role of global agencies and institutions should be redefined in the perspective of global care ethics. As Robinson indicated, transnational corporations should act as socially responsible actors. They have specific global responsibilities, articulated in the idea of corporate social responsibility. From the point of view of care ethics, states, corporations and international organizations and institutions are not 'atomized moral agents' but 'social-moral communities' deeply embedded in networks of relationships across the world [174]. The challenge, especially important for vulnerable people, is to create caring institutions. At a global level, institutions need to reflect care practices. As Tronto argues, providing and receiving care requires infrastructure, institutional practices, policing regulations, and communication networks [175]. It is clear that at the global level the work of creating caring institutions, or transforming existing institutions into caring ones, has hardly begun.

Conclusion

The aim of this chapter has been to show how contemporary bioethics can mobilize four theoretical frameworks to address vulnerability: human rights, social justice, capabilities, and global care ethics. International human rights discourse is increasingly connected with the globalization of bioethics. It not only concerns similar global problems as bioethics but also provides a robust framework to approach those problems in terms of rights that apply to everyone regardless of context and conditions. Human rights are therefore particularly appropriate to scrutinize global conditions that produce vulnerability. First, human rights are based on the notion of anthropological vulnerability. Human beings are individual exemplars of the human species that are equally vulnerable and are entitled to the same fundamental rights, independent of where they are and in what conditions. Second, human rights can be used to criticize social and cultural conditions that produce special vulnerability within the context of contemporary globalisms. Since human rights discourse emphasizes a universal approach based on human dignity and equality, it is critical of the neoliberal ideology that is driving processes of globalization and that produces many of today's bioethical problems. It is therefore directed at remediating and reducing the basic mechanisms (such as power imbalances and inequalities) that are responsible for creating and exacerbating special vulnerability. However, the human rights-based approaches are very general and require specification and articulation in order to be implemented in global bioethics. One challenge is that human rights are often primarily regarded as individual rights. Interpreted as such, they will not be able to provide an ethical perspective that transcends the primacy of individual interests. However, the increasing emphasis in the international legal and political community on socio-economic rights, particularly the right to health, demonstrate that human rights include a social focus, formulating societal obligations to address basic human needs and to provide conditions for dignified human existence. The second challenge is that human rights discourse combines theoretical inquiry and practical activism such as advocacy and social engagement. This combination is often required. There is an abundance of human rights documents. Rather than formulating new human rights the major problem is their implementation and application. Since human rights are available for everyone and apply to everyone, the discourse is often a critical instrument for citizens and civil society rather than legal experts or political decision-makers. Bioethics on the other hand is used to separating moral analysis and moral activism. Much of the bioethical discourse is promulgated by experts. Authoritative documents formulating a global ethical framework are scarce. There are only a handful of nongovernmental organizations and civil society movements in the area of bioethics. At a global level, bioethics seems to be a rather academic and isolated exercise without much practical clout.

These two challenges need to be addressed in the further development of global bioethics. For the analysis of vulnerability, human rights discourse has two advantages. First, it provides a normative framework that escapes the neoliberal ideology, articulating that health is not a commodity but a right. It therefore addresses the sources of special vulnerability in today's societies. Second, it delineates a basic foundation for ethical analysis, articulating that every human being is entitled to the same rights since they all share the same vulnerability. Human flourishing requires that basic needs are satisfied through respecting fundamental rights. Since these rights are universal in scope, they define a minimum that should be respected at the global level. Human rights discourse will redirect the main bioethical concerns from the individual level of autonomous decision-making and protecting vulnerable subjects to the social level of processes and mechanisms that make individuals more and more vulnerable in the first place. Since the discourse includes individual (civil and political) and socio-economic rights, the focus on social conditions that produce vulnerability is not antagonistic with respect for individual autonomy.

However, articulating human rights will not be sufficient to address vulnerability. Human rights underline individual entitlements and state obligations but vulnerability is often produced by power imbalances and socio-economic inequalities. What is needed is the perspective of social justice. This chapter has argued that theories of justice may do a better job in examining the global contexts that make individual subjects particularly vulnerable. A broader concept of global justice is developing that transcends the common distributive paradigm of justice. It moves beyond the concentration on scarce medical technologies, individual decision-making capacity and efficient allocation of resources for health services to critically question unjust structures and relationships. Global justice implies therefore not merely protection or fair and equal treatment of vulnerable subjects but a rebalancing of the background conditions that determine their vulnerability. Theories of justice have several theoretical instruments to address vulnerability: formulating moral obligations towards the vulnerable, specifying global moral responsibilities, arguing for global institutions to care for the vulnerable, and giving voice and better representation to the vulnerable in ethical and policy debates. The human capabilities approach is another specific theory of global justice with particular relevancy for the notion of vulnerability. It specifically intends to overcome a restricted economic focus on human flourishing by asking what people are able to achieve. The emphasis therefore is on the opportunities to function as a human being. There is a minimum level of such capabilities which is in fact defined by human rights discourse. Vulnerability in this perspective is understood as inadequate or failed capability. It refers to restricted opportunities to achieve. This is not merely limited individual choice but often due to inadequate socio-economic conditions. Analyzing the specific conditions that constrain capabilities may generate practical strategies and remedies to enhance capabilities and reduce vulnerability. Like human rights discourse, the capabilities approach is criticized for centering primarily on the individual subject. This chapter has argued that such critique is not correct. Although the human capabilities approach is sometimes ambiguous, it recognizes that individual freedom presupposes social arrangements. Expanding opportunities of people therefore requires creating and sustaining an appropriate environment for human flourishing. Positive action is therefore required to reduce and eliminate special vulnerability.

Another theoretical framework to address vulnerability is provided by care ethics. This framework insists on the interdependency of human beings (thus the notion of anthropological vulnerability). It has also pointed out the inequal allocation of vulnerability according to various contexts (thus the notion of special vulnerability). This theoretical backdrop notwithstanding, care ethics is only recently expanding into a global theory of care. It is obvious that globalization affects care practices worldwide. The challenge is to embed care practices that are often concrete, personal and inter-relational, within broader social structures. Broadening of core notions of care ethics such as relatedness, dependency, responsibility, cooperation, and contextuality can help to develop a global theory. These notions do not necessarily focus on discrete, individual subjects; they can also be applied to the structures and settings on which they depend and that make them vulnerable as groups and populations. Global care ethics, however, is less concerned with abstract and universal notions such as rights and justice. Its primary concern is with concrete subjects in concrete circumstances that produce special vulnerability. At the same time, this concern does not imply that vulnerability is individualized. Care cannot be reduced to a private activity. It is an ongoing social process that reflects the quality of social life. As an ethical notion is should pervade international relations, policies of globalization, and global institutions. Practices of care that are able to counter the production of vulnerability can only emerge if they are sustained by caring institutions.

All four theoretical frameworks discussed in this chapter point in the same direction. They not only recognize vulnerability and its sources in processes of globalization, but they also focus on the social, political, and economic conditions that produce and exacerbate special vulnerability around the world. Combining philosophical and political perspectives on vulnerability, they offer a wider range of possible actions than merely protecting vulnerable subjects. They do not disregard the ethical importance of individual autonomy but they embed it within concerns for the background conditions that sustain and nurture this autonomy. For an emergent global bioethics they therefore provide a broader theoretical perspective than the current discourse of mainstream bioethics.

Notes

1 Kottow, 'The vulnerable and the susceptible', 463. See also Kottow, 'Vulnerability: What kind of principle is it?'
2 There are several theories of justification of human rights. Gordon distinguishes between interest-based approaches (for example referring to needs or capabilities) and non-interest-based approaches. Turner's theory is classified in the second category. His theory explicitly relates human rights to vulnerability (see Gordon, 'Human rights in bioethics – Theoretical and applied', 285).

3 Turner, *Vulnerability and human rights*, 110.
4 Turner's theory has been criticized with the argument that medical and technological progress may diminish anthropological vulnerability and may also ameliorate specific vulnerabilities by improving global health and socio-cultural conditions. If vulnerability is decreasing, the future relevance of human rights might diminish. Turner correctly points out that anthropological vulnerability is constitutive for human beings so that this critique only applies for a post-human world. See: Turner and Dumas, 'Vulnerability, diversity and scarcity: on universal rights'.
5 Jonathan Mann, trained as epidemiologist, was founding director of the WHO Global program on AIDS from 1986 until 1990; he assisted in the launch of the journal *Health and Human Rights* as its first editor. See: Bayer, 'In memoriam, Jonathan Mann, 1947–1998'; D'Oronzio, 'The integration of health and human rights: An appreciation of Jonathan M. Mann'.
6 Mann, Gostin, Gruskin, Brennan, Lazzarini, and Fineberg, 'Health and human rights', 13, 19 ff. See also, Mann, 'Medicine and public health, ethics and human rights' and Mann, Gruskin, Grodin and Annas (eds.). *Health and human rights*. New York and London: Routledge, 1999.
7 Mann, 'Human rights and AIDS: The future of the pandemic', 218, 222.
8 See: Ashcroft, 'The troubled relationship between bioethics and human rights'; Andorno, 'Human dignity and human rights as a common ground for a global bioethics'; Baker, 'Bioethics and human rights: A historical perspective'; Faunce, 'Will international human rights subsume medical ethics? Intersections in the UNESCO *Universal Bioethics Declaration*'; Gordijn and Ten Have, 'Future perspectives'; Gordon, 'Human rights in bioethics – Theoretical and applied'; Knowles, 'The lingua franca of human rights and the rise of a global bioethic'; Thomasma, 'Bioethics and international human rights'.
9 Ten Have, 'Bioethics and Human Rights – Wherever the twain shall meet'.
10 Ignatieff, *Human rights as politics and idolatry*, 7.
11 Mann, Gostin, Gruskin, Brennan, Lazzarini, and Fineberg, 'Health and human rights', 9.
12 '…individual and population vulnerability to disease, disability and premature death is linked to the status of respect for human rights and dignity' (Mann, Gostin, Gruskin, Brennan, Lazzarini, and Fineberg, 'Health and human rights', 21).
13 'Rights violations are…symptoms of deeper pathologies of power and are linked intimately to the social conditions that so determine who will suffer abuse and who will be shielded from harm'. (Farmer, *Pathologies of power. Health, human rights, and the new war on the poor*, 7). See also Farmer, 'Challenging orthodoxies: The road ahead for health and human rights'.
14 Ashcroft makes a distinction between academic bioethics and policy bioethics; human rights can specifically inspire policy-making (Ashcroft, 'Could human rights supersede bioethics?', 643–44). See also: Gordijn and Ten Have, 'Future perspectives'.
15 See Mann, Gostin, Gruskin, Brennan, Lazzarini, and Fineberg, 'Health and human rights'.
16 Sen, 'The global reach of human rights', 94; see also Sen, *The idea of justice*, 362–3; human rights according to Sen should not be regarded as children of law but as parents of law; they are moral propositions that can serve as grounds for legislation. This view of human right moderates the distinction that is often made between 'hard' and 'soft' human rights instruments. Soft law documents such as declarations that are not binding should be considered as ethical commitments rather than weak legal statements.

That human rights entail a significant moral thesis is also the view of Thomas Pogge. Human rights provide a framework to assess and reform the global institutional order. (Pogge, 'The international significance of human rights', 53, 55).

17   For a recent overview, Gordon, 'Human rights in bioethics – Theoretical and applied'.

18   Pelluchon, *La raison du sensible. Entretiens autour de la bioéthique,* 86. See also: Pelluchon, *L'autonomie brisée. Bioéthique et philosophie;* Pelluchon, *Éléments pour une éthique de la vulnérabilité. Les hommes, les animaux, la nature.*

19   Baker, 'Bioethics and human rights: A historical perspective', 248 ff; Knowles, 'The lingua franca of human rights and the rise of a global bioethic', 256; Thomasma, 'Bioethics and international human rights', 296.

20   Gostin, *Public health law and ethics,* 258–9.

21   Chapman, 'Globalization, human rights, and the social determinants of health', 105; See also for the approach in the earlier days: Dujardin, 'Health and human rights: The challenge for developing countries'.

22   UN, Committee on Economic, Social and Cultural Rights. *General comment no. 14: The right to the highest attainable standard of health,* para. 11. See also: Chapman, 'Globalization, human rights, and the social determinants of health', 101.

23   The document of the committee has many references to the vulnerable: para. 12, 18, 35, 37, 40, 52, 62, 65 (UN, Committee on Economic, Social, and Cultural Rights. *General comment no. 14: The right to the highest attainable standard of health*).

24   UN, Committee on Economic, Social, and Cultural Rights. *General comment no. 14: The right to the highest attainable standard of health,* para 43.

25   African Charter on Human and Peoples' Rights, Article 29(4): 'The individual has a duty to preserve and strengthen social and national solidarity…' (African Charter on Human and Peoples' Rights).

26   Supiot, *L'esprit de Philadelphie. La justice sociale face au marché total,* 166.

27   Charter of Fundamental Rights of the European Union, Chapter IV: Solidarity.

28   See Ashcroft, 'The troubled relationship between bioethics and human rights'.

29   Ashcroft, 'The troubled relationship between bioethics and human rights', 45.

30   An extreme position is taken by South African philosopher David Benatar who argues that bioethics is a branch of moral philosophy; it is an academic exercise and should have nothing to do with activism because it undermines the academic enterprise (Benatar, 'Bioethics and health and human rights: a critical view', 19).

31   Farmer, 'Challenging orthodoxies: The road ahead for health and human rights', 13; see also Gordijn and Ten Have. 'Future perspectives'.

32   Thomasma, 'Bioethics and international human rights', 298; Baker, 'Bioethics and human rights: A historical perspective', 248 ff. See also Thomasma, 'Proposing a new agenda: Bioethics and international human rights'.

33   Knowles argues that this will have significant implications especially for American bioethics: 'Adopting the language of human rights means moving toward a more expansive understanding of the relationships between human health, medicine and the environment, socioeconomic and civil and political rights, and public health initiatives and human rights'. (Knowles, 'The lingua franca of human rights and the rise of a global bioethic', 260).

34   McInnis and Lee, *Global health and international relations,* 67 ff.

35   'They refer to what individuals can do or can have, not to who or what decides that they can do or have; they refer to distribution of resources, not to power over the distribution of resources'. (Galtung, 'Violence, peace, and peace research', 188).

36   Dos Anjos, 'Medical ethics in the developing world: A liberation theology perspective', 633.

37 Luna, 'Poverty and inequality: Challenges for the IAB: IAB presidential address', 452.

38 Jecker, 'A broader view of justice', 4.

39 Ganguli Mitra, 'Off-shoring clinical research: Exploitation and the reciprocity constraint', 4, 7.

40 Snyder, 'Exploitation and sweatshop labor: Perspectives and issues', 195.

41 'I am arguing that the citizens and governments of the wealthy societies, by imposing the present global economic order, significantly contribute to the persistence of severe poverty and thus share institutional moral responsibility for it' (Pogge, *World poverty and human rights. Cosmopolitan responsibilities and reforms,* 121).

42 The statement made in the WHO report in 1976 is still valid today: 'the benefits of recent discoveries in the medical field may still be limited to a few persons' (WHO, *Health aspects of human rights: with special reference to developments in biology and medicine,* 13).

43 The notion of social justice in this view assumes according to David Miller that there is 'a bounded society with a determinate membership, forming a universe of distribution' (Miller, *Principles of social justice,* 4). Miller's earlier description of social justice clarifies that it is relevant in connection to a society. Social justice seeks to regulate the distribution of advantages and disadvantages among the members of a society. But 'society' here can only refer to the nation-state because the political community at this level has specific features necessary for social justice that do not exist at the world level (Miller, *Social justice,* 18 ff.) On the distinction between social and global justice, see also Dwyer, 'Global health and justice', 464 ff.

44 This is the view of Chandran Kukathas who also argues that social justice is not a useful concept within domestic affairs (Kukathas, 'The mirage of global justice', 3, 11, 20, 25).

45 Raphael, *Concepts of justice,* 59, 192–4, 233–6; Miller, *Social justice,* 336 ff; Miller, *Principles of social justice,* 2–4, 18–20; Jackson, 'The conceptual history of social justice', 367; Brodie, 'Reforming social justice in neoliberal times', 97 ff.

46 Brodie, 'Reforming social justice in neoliberal times', 104.

47 'State boundaries … can no longer be seen as legitimate bounds of justice' according to Onora O'Neill (O'Neill, *Bounds of justice,* 6). See also O'Neill, 'Justice, gender and international boundaries', 451.

48 Pogge, *World poverty and human rights. Cosmopolitan responsibilities and reforms,* 39.

49 Dower, *World ethics. The new agenda,* 199.

50 Singer, *One world. The ethics of globalization,* 13, 173. See also Appiah, *Cosmopolitanism. Ethics in a world of strangers,* and Ten Have, 'Global bioethics and communitarianism'.

51 Young, *Justice and the politics of difference,* 15.

52 Young, *Justice and the politics of difference,* 16 ff.

53 Young, *Justice and the politics of difference,* 18.

54 As Young explains: '…social justice concerns the degree to which a society contains and supports the institutional conditions necessary for the realization of these values [that constitute the good life]' (Young, *Justice and the politics of difference,* 37).

55 This point is made by Nancy Fraser who distinguishes between socio-economic and cultural injustice, and thus redistributive and recognition remedies (Fraser, *Justice interruptus,* 14 ff).

56 Bauman, *Globalization. The human consequences,* 86.

57 Fraser, *Justice interruptus,* 23.

58 'Justice should refer not only to distribution, but also to the institutional conditions necessary for the development and exercise of individual capacities and collective communication and cooperation'. (Young, *Justice and the politics of difference,* 39).

59 Young, *Justice and the politics of difference*, 31.
60 So that vulnerable subjects will have what Snyder calls 'meaningful choices' (Snyder, 'Needs exploitation', 394).
61 London, 'Justice and the human development approach to international research', 32–33.
62 Different articulations often will be combined.
63 Goodin, *Protecting the vulnerable*, 206.
64 Rogers, Mackenzie and Dodds, 'Why bioethics needs a concept of vulnerability'. See also: Mackenzie, Rogers and Dodds, *Vulnerability. New essays in ethics and feminist philosophy*.
65 Dwyer, 'Global health and justice', 468 ff.
66 Pogge, *World poverty and human rights*, 205, 222 ff. Pratt, Zion and Loff object that Pogge does not provide a mechanism for allocating specific duties to specific actors (Pratt, Zion, and Loff, 'Evaluating the capacity of theories of justice to serve as a justice framework for international clinical research'. 35). For Pogge it is 'the governments, corporations, and citizens of the affluent countries who have this obligation because they are "greatly and disproportionately benefitting" from current global arrangements' (Pogge, *World poverty and human rights*, 264).
67 'The social connection model finds that all those who contribute by their actions to structural processes with some unjust outcomes share responsibility for the injustice' (Young, *Responsibility for justice*, 96). See also Chapter 6, 140–141.
68 Young, *Responsibility for justice*, 104 ff.
69 O'Neill, 'Justice, gender, and international boundaries', 456.
70 O'Neill, *Bounds of justice*, 136.
71 O'Neill, *Bounds of justice*, 140.
72 See Supiot, *L'esprit de Philadelphie. La justice sociale face au marché total*.
73 Fraser, *Scales of justice*, 16 ff. Previously, Fraser distinguished only two dimensions of justice, namely distribution and recognition. See Fraser, *Justice interruptus*.
74 Fraser, 'Reframing justice in a globalizing world', 9.
75 Fraser, *Scales of justice*, 59.
76 Higgins, 'Toward an ethical economy'.
77 'Commodity command is a *means* to the end of well-being, but can scarcely be the end in itself' (Sen, *Commodities and capabilities*, 19). The capability approach is a 'departure from concentrating on the *means* of living to the *actual opportunities* of living' (Sen, *The idea of justice*, 233).
78 Sen, *Commodities and capabilities*, 17 ff. See also Brooks, 'Capabilities'.
79 Sen, *Commodities and capabilities*, 7. See also Crocker, 'Functioning and capability: The foundations of Sen's and Nussbaum's development ethic'.
80 'Capability is thus a kind of freedom: the substantive freedom to achieve alternative functioning combinations (or, less formally put, the freedom to achieve various lifestyles)' (Sen, *Development as freedom,* 75).
81 Sen, *Inequality examined*, 52; Sen, *Development as freedom*, 76; Sen, *The idea of justice*, 237.
82 '… capabilities, not functionings, are the appropriate political goals' (Nussbaum, *Creating capabilities*, 25.
83 Nussbaum, *Women and human development*, 5.
84 Nussbaum, *Women and human development*, 6. This is what Nussbaum calls the 'principle of each person's capability'. It expresses the view that for each and every person it is true that 'beneath a certain level of capability, in each area, a person has not been enabled to live in a truly human way'. (Nussbaum, *Women and human development*, 74).

85 Nussbaum, *Women and human development*, 78-80; Nussbaum, *Creating capabilities*, 33-34. Sen often mentions valuable or basic capabilities. He has analyzed poverty in terms of 'insufficient basic capabilities' (Sen, *Inequality examined,* 151) or as 'deprivation of elementary capabilities' in relation to education, health, nourishment, and literacy (Sen, *Development as freedom,* 20, 87 ff.). However, he has never made a list of capabilities. In his view, the relevancy of capabilities depends on personal value judgments. The problem is not so much the listing but that the list is composed without public participation and discussion (Sen, 'Human rights and capabilities', 158). Nussbaum's list is the subject of extensive critique. See for example, Jaggar, 'Reasoning about well-being: Nussbaum's methods of justifying the capabilities'.

86 McEntire, 'Understanding and reducing vulnerability: from the approach of liabilities and capabilities', 295–7.

87 Lloyd-Sherlock, 'Nussbaum, capabilities, and older people', 1164-1165.

88 Briones, 'Rights with capabilities: A new paradigm for social justice in migrant activism', 136–138.

89 Mitra, 'The capability approach and disability', 241; Burchardt, 'Capabilities and disabilities: the capabilities framework and the social model of disability', 736 ff.

90 Nussbaum, *Frontiers of justice,* 96–223. See also Nussbaum, *Creating capabilities,* 149–152.

91 Nussbaum, *Frontiers of justice,* 87 ff.

92 Nussbaum, *Frontiers of justice,* 156 ff. Nussbaum underlines that 'The person leaves the state of nature ... not because it is more mutually advantageous to make a deal with others, but because she cannot imagine living well without shared ends and a shared life' (Nussbaum, *Frontiers of justice,* 158).

93 Nussbaum, *Frontiers of justice,* 99 ff.

94 '... lacks in one area cannot be made up simply by giving people a larger amount of another capability' (Nussbaum, *Frontiers of justice,* 166–7).

95 Nussbaum argues that public arrangements should provide the social basis of all capabilities (*Frontiers of justice,* 193). The norm in regard to people with disabilities should be that the person should be put in the position to choose functionings herself. Disabled persons should receive as many of the central capabilities as possible while the caregivers should receive all (Nussbaum, *Frontiers of justice,* 222).

96 Sen, *Development as freedom,* 36–40.

97 Sen, *Development as freedom,* 40.

98 Sen, *The idea of justice,* 255–256.

99 Nussbaum, *Frontiers of justice,* 221. Earlier in the same publication she acknowledges that 'we are needy temporal animal beings...' (Nussbaum, *Frontiers of justice,* 160). See also Goldstein, *Vulnérabilité et autonomie dans la pensée de Martha C. Nussbaum.*

100 Nussbaum, *Women and human development,* 6, 32. It is not clear however, how anthropological and special vulnerability are related. Because the various types of capability are linking internal and external determinants, it is demonstrated that capabilities have different origins but that at the same time they are nonetheless related. Given this view, one can also assume that anthropological and special vulnerability are associated to each other.

101 Sen, *Development as freedom,* 87 ff, 90.

102 Nussbaum, *Women and human development,* 298; Sen, *The idea of justice,* 19.

103 This is at least the suggestion of Ingrid Robeyns when she argues that young children and the mentally disabled 'might not be able to make complex choices' so that

it is more sensible to study achieved functionings rather than capabilities (Robeyns, 'The capability approach: a theoretical survey', 101).

104 This is also the description of Dubois and Rousseau who explain that 'vulnerability is the probability of falling to a lower state of well-being' (Dubois and Rousseau, 'Reinforcing households' capabilities as a way to reduce vulnerability and prevent poverty in equitable terms', 426). This means that the functioning (being in a state of well-being) is the same as for persons who are not vulnerable but at the same time this functioning is threatened to be decreased.

105 Nussbaum, *Women and human development*, 84 ff.

106 Nussbaum, *Creating capabilities*, 33–4.

107 Vallentyne, 'Debate: Capabilities versus opportunities for well-being', 363 ff.

108 'Individual freedom is quintessentially a social product', as Sen has famously summarized this view. He has also analyzed various example such as the discrimination that leads to sex-selection and infanticide, as well as the lower healthcare received by women and girls (Sen, *Development as freedom*, 31).

109 McEntire, 'Understanding and reducing vulnerability: from the approach of liabilities and capabilities', 296–7.

110 Nussbaum, *Creating capabilities*, 43.

111 See: Gasper and van Staveren. 'Development as freedom – and as what else?'; Deneulin, Nebel, and Sagovsky, *Transforming unjust structures. The capability approach*; Carpenter, 'The capabilities approach and critical social policy: Lessons from the majority world?'; Dean, 'Critiquing capabilities: The distractions of a beguiling concept'; Evans, 'Collective capabilities, culture, and Amartya Sen's *Development as freedom*'; Stewart and Deneulin, 'Amartya Sen's contribution to development thinking'.

112 Deneulin and McGregor, 'The capability approach and the politics of a social conception of well-being', 513. Vicente Navarro has argued earlier that Sen does not analyze the political context of development in his emphasis on freedom (Navarro, 'Development and quality of life: A critique of Amartya's Sen's *Development as freedom*', 465).

113 Deneulin and McGregor, 'The capability approach and the politics of a social conception of well-being', 509.

114 Frances Stewart has argued that the focus should be more on 'group capabilities' since group formation and affiliation is a source of power and a significant determinant of individual capabilities (Stewart, 'Groups and capabilities', 199).

115 Hartley Dean argues that in the capability approach the priority is individual liberty not social solidarity. The basic assumption is that human beings are not dependent on each other, and that they are not under any control in the public space. The individual essentially is an autonomous person who freely enters into relations with other individuals. The social structure is not considered relevant, though it may enable or constrain capabilities. (Dean, 'Critiquing capabilities: the distractions of a beguiling concept', 265 ff.).

116 Nussbaum, *Women and human development*, 3.

117 Nussbaum, *Women and human development*, 72. Nussbaum has also argued in favour of 'a political conception of the person that is more Aristotelian than Kantian, one that sees the person from the start as both capable and needy...' (Nussbaum, 'Capabilities as fundamental entitlements: Sen and social justice', 54).

118 Nussbaum, *Women and human development*, 81.

119 'Societal arrangements, involving many institutions (the state, the market, the legal system, political parties, the media, public interest groups, and public discussion forums, among

others) are investigated in terms of their contribution to enhancing and guaranteeing the substantive freedoms of individuals, seen as active agents of change, rather than as passive recipients of dispensed benefits'. (Sen, *Development as freedom,* xii-xiii).

120 Sen, *Development as freedom,* 142.

121 This point is elaborated by Séverine Deneulin who compares the application of the capabilities approach to the Dominican Republic and Costa Rica (Deneulin, 'Necessary thickening', 35–39). See also Hill, 'Development as empowerment'.

122 Sen, *Development as freedom,* 11. In *The idea of justice* Sen argues that the emphasis in theories of justice should be on the actual lives of people and not on the question how just institutions can be established (Sen, *The idea of justice,* xi, 15 ff). Capabilities are 'characteristics of individual advantages' (Sen, *The idea of justice,* 296).

123 Sen, 'Response to commentaries', 79, 85. 'Individual human beings with their various plural identities, multiple affiliations and diverse associations are quintessentially social creatures with different types of societal interactions' (Sen, *The idea of justice,* 247). Sen strongly criticizes the misuse of the views of Adam Smith in neoliberalism. Smith in fact never considered self-interest as the overriding and exclusive motivation of human behaviour but extensively discussed the need for 'non-self-interested behaviour' (Sen, *The idea of justice,* 185). Similarly, Sen argues that there are many more reasons for human cooperation than mutual benefit; asymmetry of power can also create unilateral obligations (Sen, *The idea of justice,* 206-7).

124 Older people for example should be approached as agents rather than victims, argues Peter Lloyd-Sherlock. The approach should also not be merely individualistic since ageing implies the deterioration of internal capabilities (this is curiously not discussed by Nussbaum) but there is also a significant impact of external conditions. Addressing vulnerability requires the removal of structural constraints (Lloyd-Sherlock, 'Nussbaum, capabilities, and older people', 1165, 1167).

125 Nussbaum, *Creating capabilities,* 127 ff.

126 Nussbaum, *Creating capabilities,* 113.

127 Nussbaum, *Creating capabilities,* 65 ff. See also Nussbaum, 'Capabilities as fundamental entitlements: Sen and social justice'. Here she argues: 'To secure a capability to a citizen it is not enough to create a sphere of non-interference: the public conception must design the material and institutional environment so that it provides the requisite affirmative support for all the relevant capabilities'. (Nussbaum, 'Capabilities as fundamental entitlements', 55). Empirical studies in human development have shown that it is not so much the material conditions that are important for adaptive capacity but rather 'access to common property resources' (Bohle, *Living with vulnerability. Livelihoods and human security in risky environments,* 23). Social networks, family, communities, self-help groups and networks of friends are important to help vulnerable subjects to cope. Policies therefore need to go beyond securing basic needs (food, shelter, health) and should promote participation in the social domain so that vulnerable subjects are recognized and empowered. Another difference with neoliberal ideology is pointed out by Burchardt: the capabilities framework rejects income as the sole object of value (Burchardt, 'Capabilities and disabilities: the capabilities framework and the social model of disability', 740).

128 As Nussbaum puts it: 'The capabilities approach rejects utterly the misleading notion of negative liberty: people, especially women, are not free if they are left alone by a lazy state'. (Nussbaum, 'Women's bodies: Violence, security, capabilities', 176).

129 See Robeyns, 'The capability approach in practice'.

130 Nussbaum, *Creating capabilities,* 117–8.

131 Nussbaum, *Creating capabilities,* 166 ff.

132 London, 'Justice and the human development approach to international research', 32 ff.

133 Health is regarded as a *meta-capability*: 'an overarching capability to achieve a cluster of basic capabilities to be and do things that make up a minimally good human life in the contemporary world' (Venkatapuram, *Health justice,* 20). Sen has made a distinction between health achievement and the capability to achieve good health (Sen, 'Why health equity?', 660).

134 Venkatapuram, *Health justice,* 24. The capability to be healthy is determined by the interaction of individual biology, external social and physical environments and behaviour (Venkatapuram, *Health justice,* 74 ff; 143 ff).

135 Venkatapuram, *Health justice,* 147 ff.

136 This approach also has a more restricted view of health capabilities, making a distinction between central health capabilities as 'the capabilities to avoid premature death and escapable morbidity' and non-central health capabilities (Prah Ruger, *Health and social justice,* 4, 43, 76). The first are 'necessary conditions for humanity, regardless of social context'. They should therefore be addressed prior to other health capabilities (Prah Ruger, *Health and social justice,* 76). Health policy is defined as 'public health, pharmaceuticals, health-related research (including biomedical and genetic research) and treatments, individual health care services, and outreach and community-based health services' (Prah Ruger, *Health and social justice,* 101).

137 According to Jennifer Prah Ruger we are far from understanding the societal mechanism influencing health (*Health and social justice,* 6; see also 99–100). In regard to health, she believes, contrary to many other scholars that health policy 'continues to be one of its most influential determinants' (Prah Ruger, *Health and social justice,* 99).

138 Powers and Faden, *Social justice,* 37-41. Sen defends the distinction among others with the argument that it has important policy implications in regard to deprived people. If there is freedom to achieve and people choose not to realize the achievement then it is their personal responsibility to face the consequences. The distinction between capabilities and functionings therefore also demarcates the responsibilities and obligations of societies and individuals. Sen gives the example of the provision of basic healthcare: if the governments guarantees basic healthcare and a person decides not to use that opportunity, then the resulting deprivation is 'not as much of a burning social concern as would be the failure to provide the person with the opportunity for healthcare'.(Sen, *The idea of justice,* 238). In other words, if the health capability is there but not exercised, bad health is the responsibility of the individual. It is clear that this view can easily lead to attitudes of blaming the victim; but it also assumes that health capabilities can be guaranteed.

139 'Our claim is that justice is concerned with securing and maintaining the social conditions necessary for a sufficient level of well-being in all of its essential dimensions for everyone'. (Powers and Faden, *Social justice,* 50).

140 Pettersen, *Comprehending care,* 171–184. See also Hamington and Miller (eds.) *Socializing care. Feminist ethics and public issues,* Hankivsky, *Social policy and the ethic of care.*

141 O'Neill, 'Justice, gender, and international boundaries', 448.

142 See Tronto, *Moral boundaries;* Robinson, *Globalizing care;* Held, *The ethics of care. Personal, political, and global;* Mahon and Robinson (eds.). *Feminist ethics and social policy.*

143 Yeates, 'A global political economy of care', 227.

144 Hochschild, 'Global care chains and emotional surplus value', 130 ff.; Weir, 'Global care chains: Freedom, responsibility, and solidarity', 166. The term 'care drain' is used by Kittay (Kittay, Jennings, and Wasunna, 'Dependency, difference, and the global

ethic of longterm care', 450). See also Mahon and Robinson (eds.). *Feminist ethics and social policy*; Yeates, 'A global political economy of care'.

145   Tronto, 'Vicious circles of privatized caring', 7–8. As Walker explains: 'Care of the dependent and vulnerable remains commonly done by women as an unpaid full-time job or as an unpaid double or triple shift in the home in addition to paid labor, or is done as low-paid "unskilled" work performed by a disproportionately non-white labor force in the workplace (where that workplace is also sometime someone else's home, in the case of home care attendants, babysitters, housecleaners)'. (Walker, 'The curious case of care and restorative justice in the U.S. context', 156–7).

146   Robinson, 'Ethical globalization? States, corporations, and the ethics of care', 171.

147   Koggel and Orme, 'Care ethics: New theories and applications', 109–111. The need for a theory of cosmopolitan care is particularly emphasized by Sarah Miller (Miller, 'Cosmopolitan care').

148   '…the current boundaries of moral and political life are drawn such that the concerns and activities of the relatively powerless are omitted from the central concerns of society' (Tronto, *Moral boundaries*, 20). It is interesting that Tronto's book has been translated into French with the title *Un monde vulnerable. Pour une politique du care.* (A vulnerable world. For a politics of care).

149   '…current moral boundaries preserve and perpetuate the positions of the powerful and privileged…' (Tronto, *Moral boundaries*, 92). Also Robinson argues that relations of care within a global context are constructed by relations of power (Robinson, *The ethics of care*, 5).

150   At both levels, care has four ethical elements: attentiveness, responsibility, competence, and responsiveness (Tronto, *Moral boundaries*, 127–137). See also White and Tronto, 'Political practices of care: Needs and rights'.

151   Weir, 'Global care chains: Freedom, responsibility, and solidarity', 171. The same idea is elaborated by Selma Sevenhuijsen who emphasizes 'selves in networks of care and responsibility' (Sevenhuijsen, 'Caring in the third way', 27). Robinson has put it as follows: 'Care ethics starts from the position that all people exist in webs of relations with others, relying on those others for their care and, hence, their security'. (Robinson, 'Care ethics and the transnationalization of care', 128).

152   '…individuals can only exist because they are members of various networks of care and responsibility, for good or for bad' (Sevenhuijsen, 'Caring in the third way', 9). In the words of Robinson: care ethics is a 'relational ethics' just as care is a relational practice according to Tronto (Robinson, *Globalizing care*, 46; Tronto, 'Creating caring institutions', 161). Robinson is often referring to the 'relational ontology' of care ethics (Robinson, 'Ethical globalization'? 172; Robinson, *The ethics of care*, 4, 29; Robinson, 'Care ethics and the transnationalization of care',131; Robinson, *The ethics of care*, 101). See also Verkerk, 'The care perspective and autonomy'.

153   Dodds, 'Depending on care: Recognition of vulnerability and the social contribution of care provision'. 504. See also: Dodds, 'Dependence, care, and vulnerability'.

154   'A focus on care as the cornerstone of human existence is particularly important in an era of globalization where neo-liberal macroeconomic policies have meant the scaling back of state services…' (Robinson, 'Care, gender, and global social justice: Rethinking "ethical globalization"', 15). Care should be regarded as 'maintenance', the foundation of the human condition that is safeguarding the continuity of society (Robinson, *The ethics of care*, 15). About care as the moral quality of life, see Kleinman and Van der Geest, '"Care" in health care. Remaking the moral world of medicine'.

155 Tronto has pointed out the relative imbalances in individual capacity to care as well as in the amount of care people receive (Tronto, 'Vicious circles of privatized caring', 7). The problem that human dependency as mostly invisible is discussed by Kittay *et al.* (Kittay, Jennings and Wasunna, 'Dependency, difference, and the global ethic of longterm care', 445).

156 Weir, 'Global care chains', 168. Freedom according to Weir is social freedom; it is 'the capacity to be in relationships that one desires'. (Weir, 'Global care chains', 168). She argues that real freedom does not exist if people can only choose among impossible choices (Weir, 'The Global Universal Caregiver: Imagining women's liberation in the new millennium', 322).

157 Tronto, *Moral boundaries,* 168.

158 Tronto, *Moral boundaries,* 102.

159 Weir, 'Global care chains', 170.

160 'The ethics of care is more suited than the ethics of justice for understanding the particularities of different situations, groups, and cultures, to see what really will improve the lives of children, women, and men'. (Held, *The ethics of care,* 164). According to Tronto there are at least three crucial differences with justice approaches: care ethics focuses on responsibility and relationship (rather than rights and rules), it is tied to concrete circumstances (rather than being formal and abstract) and it is an activity (rather than theory of principles) (Tronto, *Moral boundaries,* 79).

161 Hankivsky, *Social policy and the ethics of care,* 2.

162 Kittay, Jennings, and Wasunna, 'Dependency, difference, and the global ethic of long-term care', 456.

163 '...the conception of society as an association of equals masks the inevitable dependencies and asymmetries that form part of the human conditions ...'(Kittay, *Love's labor,* 14).

164 Tronto, *Moral boundaries,* 161–2.

165 Pettersen, 'The ethics of care: Normative structures and empirical implications', 52–3.

166 See in particular, Robinson, *Globalizing care;* Robinson, *The ethics of care.*

167 Robinson, 'Ethical globalization?', 167.

168 Robinson, *The ethics of care,* 70 ff.; Robinson, 'Care, gender, and global social justice: Rethinking "ethical globalization"', 16–17. See also, Kittay, 'The global heart transplant and caring across national boundaries'.

169 Tronto, 'Vicious circles of privatized caring', 3. See also Lynch, Baker, and Lyons, *Affective equality,* 48 ff.

170 Tronto, 'Vicious circles of privatized caring', 17; also Tronto, 'Creating caring institutions', 164.

171 'The language of care in its most characteristic applications reminds people of a largely uncontrollable vulnerability...' (Walker, 'The curious case of care and restorative justice in the U.S. context', 157). The reason, according to Walker, why care language is not accepted and even marginalized or ignored in public policy in the U.S. is not only that it is contrary to the dominant discourse centered on free and self-determining individuals but also that it is in fact threatening to existing practices and institutions.

172 Robinson, *Globalizing care: Ethics, feminist theory, and international affairs,* 137–168; Robinson, 'Ethical globalization?' 171–3; Robinson, 'Care, gender, and global social justice', 17 ff.

173 Held, *The ethics of care,* 124, 154–168. See also Lawson, 'Geographies of care and responsibility'.

174 Robinson, 'Ethical globalization?', 178.

175 Tronto, 'Creating caring institutions', 163 ff.

9

PRACTICAL APPLICATIONS
OF VULNERABILITY

The increasing interest in vulnerability does not only challenge bioethics to expand its theoretical frameworks beyond the principles and approaches established from the 1970s. It also will have practical consequences when the notion is taken seriously and applied within bioethics. As with the theoretical approaches discussed in the previous chapter, much is work in progress and clear-cut deliverables are not available. Nonetheless, the notion of vulnerability is a pointer, orientating bioethical discourse into a new stage of development. This chapter explores what will be the practical characteristics of bioethics inspired by vulnerability. The first sections will outline that one of the consequences will be an expansion of the domain of bioethics. Not only will it develop into a broader discipline but it will also have a richer agenda and a wider focus. Subsequent sections will argue that the notion of vulnerability will motivate bioethics to expand its set of activities, emphasizing advocacy, inclusive and participatory work modes, and cooperative networking. The final section of the chapter will conclude that vulnerability is an inspirational source for a social turn in bioethics.

The impact of vulnerability on bioethics

Global scope

Although bioethics as a new theoretical and practical discourse has emerged in developed countries it is inescapably global nowadays. The notion of vulnerability and the challenges it poses for mainstream bioethics symbolizes this change. It is critical for bioethics, as stated in a recent publication to take into account global changes, 'to incorporate the realities of a globalised world, one with increasing disparities and power differences' [1]. The 'global bioethics' advocated by Van

Rensselaer Potter finally comes into existence [2]. This is not just a matter of scale. While bioethical concerns have undeniably spread across the globe and are now shared by health professionals, researchers, policy-makers, and the general public in all countries, globalization of bioethics is also changing bioethics itself. Even those scholars, like Albert Jonsen, who regard bioethics as a 'native grown American product' observe that the product is not simply exported but thoroughly transformed in the process [3]. Within a global scope, existing problems as poverty, corruption, and inequality are affecting the whole of humankind; but also new concerns such as organ trade, medical tourism, and bioterrorism have emerged. The search for solutions has a global nature since bioethical problems are interconnected. They can only be addressed through cooperation, sharing long-term interests with the goal of safeguarding the survival of humanity. It will be necessary to better define national and global responsibilities and duties of states, transnational corporations, and civil society organizations [4]. The widening of scale and scope is influencing bioethics to develop into a more comprehensive and nuanced system of ethics that can combine traditional professional concerns with ecological issues and the larger problems of society [5]. Examining and analyzing vulnerability therefore calls for a different kind of bioethics.

Broader agenda

That bioethics often has an agenda that is too restricted to properly address contemporary problems is argued for instance by Dan Brock. He criticizes the focus on healthcare, and specifically on inequalities in access to care since it ignores that health is more significantly impacted by social determinants [6]. His call for a broader agenda has been followed by many others. Norman Daniels has pointed out that the traditional concentration of bioethics on relationships between doctor and patient, and researcher and subject has generated a 'myopic view' that does not detect the importance of the institutional context in which these relationships are operating [7]. The awareness of vulnerability shows that there is a need to go farther than the concerns of Brock and Daniels. It is not merely global justice or global health inequality that call for a broader agenda. It is the discourse of bioethics itself that needs expansion. The focus on human vulnerability widens bioethical vision beyond the individualistic perspective of the autonomous decision-maker. It introduces values that are indispensable for a really global approach, such as solidarity, responsibility, and care. Understanding global interdependence and recognition of mutual vulnerability will foster a new mindset reflected in new approaches in global bioethics [8]. Examining and applying the notion of vulnerability can therefore help to overcome the impotence of mainstream bioethics deplored by Paul Farmer. He underlines what is experienced by many others: bioethics is mainly concerned with the quandaries of the fortunate; it is in fact management of inequality because it does not pay attention to the underlying mechanisms and does not introduce a different set of values [9]. The notion of vulnerability, however, encourages bioethics to increase its relevancy for people across the globe.

Critical focus on social context

Emphasizing vulnerability in bioethical discourse entails fundamental criticism of what Judith Butler, following Levinas, has called the 'ontology of individualism' [10]. The recognition that vulnerability is constitutive of being human acknowledges that social and institutional circumstances are not merely the context in which autonomous individuals operate but rather the conditions of possibility for the development of autonomy. Without the context of social relations, and therefore the likelihood of vulnerability, the capability of being autonomous would not develop. This is the reason that Alastair Campbell called the autonomous individual 'a mere philosopher's abstraction' [11]. Vulnerability as anthropological predicament and as produced by social circumstances necessarily directs bioethical attention to these circumstances. Recognizing that vulnerability nowadays is increasingly produced through processes of globalization obliges bioethical discourse to be critical of neoliberal ideology. The advancement of bioethics as a new discipline in the 1970s was initially driven by concerns over the potential impacts of new scientific and technological developments [12]. In this phase, the main challenge was how to empower patients and citizens in light of the new diagnostic and therapeutic powers of medical practice and the sheer endless technological possibilities to improve health, eliminate disease, and extend life. However, the major questions facing global bioethics today are no longer related to the power of science and technology. Nowadays the most important bioethical questions are related to socio-economic conditions and the power of neoliberal market approaches [13]. In the nineteenth century socially concerned physicians such as Rudolph Virchow in Germany and Louis-René Villermé in France expanded the scope of medicine in response to the negative health effects of industrialization [14]. But with the demise of social medicine in the 1990s such critical analysis of social developments is lacking. The notion of vulnerability invites bioethical discourse to resume this critical analysis. The main target should be neoliberal ideology since this is, in the words of Scholte, 'an ideology of the powerful that obscures – and in this way helps to sustain – the sufferings of the vulnerable' [15]. The challenge is to create a constructive, alternative discourse [16]. Arguing that market dynamics should be restricted, especially in healthcare, research, and education, and that alternative public policies must be developed, will not be enough. More specific arguments and practices must be applied, following the reasoning of Sen that policies should provide human beings with the capability to live a dignified life through providing the resources to exist as they wish. It means a reversal of priorities: economic and financial considerations should serve the principles of human dignity and social justice, and no longer be ends in themselves [17]. This implies specific strategies for social inclusion but also institutional support. It means leaving behind the convenient stereotype in neoliberalism that vulnerability and poverty are individual deficiencies that need compensation. When the appropriate institutional supports are provided, and an enhancing context is accomplished, poor and vulnerable subjects will use all chances to improve their condition. It furthermore affirms that the neoliberal claim that processes of globalization are inevitable, irreversible and uncontrollable are mistaken; they are sustained by policies that can be changed. Changes do not require the overhaul of the entire system or the destruction of the ideology; minor interventions can make a difference. Many relatively small changes at the local level are often effective in improving the conditions of the vulnerable [18].

Advocacy and activism

In its orientation towards global problems, bioethics should not only search for new theoretical tools but also practical instruments [19]. What is required is social and political agency complementary to academic enquiry. Volnei Garrafa and Dora Porto have argued in favor of an activist, 'hard' or 'intervention bioethics' that is vigorously focused on power, poverty, violence, and social inequality, and that tries to make alliances with vulnerable populations [20]. In this point of view, global bioethics will do more than reflecting about the world; it will aim for social change. This is one of the challenges of the concept of vulnerability: it is not merely a descriptive or functional notion but it has normative force. Applying this concept means an appeal to action. This implies sustained efforts to identify and analyze patterns of vulnerability with asymmetric impacts of threats but also continual and coordinated activities to influence and change these patterns as well as the encompassing, globalized processes that generate them.

There is no reason to be pessimistic. Jonathan Mann aptly pointed out that empowerment of the vulnerable rests on two pillars: knowledge and understanding, and confidence that the world can change, that there is the possibility of change [21]. Social inequalities and conditions that produce vulnerability are not beyond social and political control. At the same time, as the human rights approach demonstrates, the problem often is not the formulation of a new vision but the implementation of it. Even when the framework of human rights is adopted and the need for social justice affirmed, there are limited possibilities to address transnational corporations and international organizations that drive the globalization processes. Many states are so weakened that they no longer have the ability to address the social and economic environment. The implementation of human rights is therefore more and more dependent on promotion of public awareness of injustices and inequalities. This is in accordance with the views of Amartya Sen. Human rights are ethical claims; advancing them does not only mean making new laws, but it implies a wide range of social activities: monitoring, media exposure and criticism, public debate and agitation, as well as 'naming and shaming' [22]. Implementation of human rights and bioethical principles therefore does not depend on states or political organizations. They motivate citizens, patient organizations, and communities to criticize practices, to target unacceptable professional or corporate conduct, and to demand changes in health and research policies.

This broader range of activities will revitalize the notion of advocacy in the context of bioethics. Although it can take place at many different levels, the use of advocacy is especially interesting at the global level. Using current information and communication technologies, global networks of bioethicists can be created to provide support. Bioethical activities can better liaise with the programs and activities of NGOs that operate in the field of healthcare. Advocacy can take place at the level of international organizations. An example is the appointment of Special Rapporteurs on the right to health at the United Nations Commission on Human Rights. As independent experts they examine country situations and report back to the Human Rights Council. Although their activity is temporary and the effectiveness is often questioned, they have addressed and analyzed many issues [23]. The important point is that at least in the area of human rights monitoring mechanisms exist that can oversee performances and can publicly report on human rights implementation. Similar mechanisms as well as transnational advocacy networks should be established in the field of global bioethics [24].

Promoting participation

In his recent work, Amartya Sen has particularly emphasized the importance of open public discussion. Democracy is regarded as 'government by discussion' [25]. Rather than trying to envision perfect justice and right institutions, one should focus on 'social realizations' [26]. Democratic society is characterized not so much by institutions as well as political participation, dialogue, public interaction, and deliberation. These mechanisms can articulate the voice of the disadvantaged, the deprived, and the vulnerable [27].

This view is refreshing because the voices and views of vulnerable subjects and populations are almost never heard in public discourse. It seems that the vulnerable are disempowered in public and scholarly debates. Somebody else assesses the relevant threats and takes care for protecting them against harm [28]. Better understanding of the experiences of vulnerability will demand participatory approaches including vulnerable subjects, so that 'testimonies of the marginalized' are collected [29]. It also requires the involvement of vulnerable groups in policy development and implementation [30].

Interestingly, several projects have been published about experiences with participation. One concerns monthly Swiss radio programs giving voice to vulnerable groups such as seriously ill persons, people with disabilities, chronic diseases, and addiction so that they are no longer marginalized in public debate. The programs not only gave an equal share of responsibility and power to vulnerable subjects but also brought about an awareness among other people of their own vulnerability; being vulnerable became a mutual experience [31]. Another project reports on a phenomenological study on the experiences of elderly patients in hospital. It shows how observing over a relatively long time ('shadowing') can give voice to experiences of vulnerable subjects who are often not included in research projects because they have difficulties in articulating their experiences [32]. The argument in favor of participation is extended to other types of research. Vulnerability can be reduced through participatory research. It is not sufficient to protect vulnerable subjects or take additional precautions, but they themselves should participate in all phases of the research process and have control over the trials in order to avoid exploitation [33]. This argument is connected with the appeal to recognize the agency of vulnerable subjects and populations in the ethical review process of research, particularly in developing countries [34]. Rather than having review boards representing vulnerable subjects and speaking for them, those subjects need to be included so that they can speak for themselves. Review boards will then not merely be procedural mechanisms but platforms for empowerment of vulnerable groups. Encouraging and extending participation will turn institutional review into 'a mechanism for redress of inequality' [35]. Participation and inclusion are therefore ways to counteract disempowering practices.

Cooperation and sustaining networks

Various responses to vulnerability are required: empowering the individual through an emphasis on rights and capabilities, creating a more just society through changing the social, economic, cultural, and political context, and fostering social and political participation through including the vulnerable as equals in processes of public reasoning. All these responses will be enhanced by international cooperation, forging global alliances, and new networks of solidarity. Present-day societies are risk societies. Many threats are global. They are faced by individuals, groups, and populations and transnational cooperation is the only way to effectively reply to these threats. Vulnerability as a global phenomenon transforms the significance of cooperation. In the words of Ulrich Beck: 'Interdependence is not a *scourge* of humanity but the precondition for its survival. Cooperation is no longer a means but the end' [36].

The unavoidable need for cooperation and solidarity is in fact voiced in the discourse of vulnerability since it questions the current emphasis on personal autonomy. Regarding human beings as self-interested, self-determining subjects disregards the basic importance of cooperation. It does not notice that homo economicus is an 'anthropological monster' [37]. An individualistic ontology makes it impossible to address the root causes of vulnerability, not because they are considered as not relevant from the perspective of individual decision-making but because they are out of the reach of such decision-making capacity. Influencing and changing social conditions requires shared agency and collective capacities to act [38].

In practice, the need for cooperation has two implications. One is pragmatic solidarity [39]. Concrete assistance should be provided and collaboration established.

Global bioethics must be translated into particular and concrete activities in specific settings and locations. In the area of research it implies an emphasis on institutional cooperation to build ethical infrastructure in developing countries. It also refers to the establishment of independent national bioethics committees, creating platforms of stakeholders who can identify what will benefit communities in these countries on the basis of current health needs [40]. In the area of healthcare, collaborative activities should focus on setting up and reinforcing education and training programs in bioethics with a long-term perspective and integrated within institutional curricula for healthcare professionals. This is exactly what international organizations such as UNESCO have been doing for a number of years in developing countries: identifying what teaching programs exists, who will need what kind of support, how collaborative networks can be created and experiences exchanged so that the quality, scope and intensity of bioethics education can be enhanced [41]. The second implication is targeted capacity-building. Many initiatives are taken for international collaboration and a multitude of international projects exists. But most of them are short-term and focused on specific topics and issues. What is needed is long-term cooperation aimed at bioethical infrastructure. The creation of more stable networks and institutions is the only way to resist the spirit of neoliberalism and to foster solidarity arrangements that can reduce vulnerability. Different types of social action are necessary because neoliberal ideology assumes that individuals are free and can engage in voluntary associations. At the same time this ideology assumes that individuals cannot construct strong collective institutions [42]. A broader notion of cooperation will not just emphasize individual initiatives and networking but will aim at establishing structures and institutions that will guarantee the flourishing of all human beings. However, such a notion demands a sense of international solidarity and a social ethics perspective that is currently not well developed in bioethics.

Vulnerability's power of inspiration

Vulnerability is a theoretical notion. It reflects the precariousness of the human condition and the fragility of the human species. It is also a reflection of radical changes in contemporary human existence due to processes of globalization. From a philosophical perspective, vulnerability implies the basic equality of human beings; they are all susceptible to damage. For this reason, humans necessarily build institutions and social arrangements that can protect them from being hurt; vulnerability is the basis for solidarity and mutual care; it encourages cooperation. From a political perspective, it is acknowledged that some people are more susceptible to threats and harms than other people because they are more exposed, have more sensitivity, or have less adaptive capacities, or a combination of all three factors. Specific arrangements are therefore necessary to deal with the special vulnerabilities of individuals, groups, communities, and populations. Philosophical as well as political perspectives demonstrate how contemporary bioethics can move beyond the framing of vulnerability as an individual attribute. Both perspectives make clear that vulnerability is more than a theoretical construct. As a normative notion, it has practical implications, not least for bioethical discourse itself. Three aspects are important to consider since they show how vulnerability can inspire a broader, more social approach in bioethics.

One aspect is that vulnerability is always manifested in individuals, but its sources are located elsewhere. Humanitarian concerns with individuals therefore require critical examination of conditions that produce vulnerability. The notion demonstrates that the emphasis on individual autonomy is inadequate; autonomy itself demands appropriate conditions to arise, to develop, and to flourish. The discourse of vulnerability directs attention precisely towards these underlying conditions for human flourishing.

Another aspect is that vulnerability is not a negative and temporary stage that must be overcome. Since there is the constant possibility of harm, human beings need each other and must cooperate. This furthermore means that the world cannot be divided into agents and passive subjects, protectors and protected. The possibility exists that the discourse of vulnerability could be misused to rearticulate power differences, reinforcing the opposition between the weak and the strong. This danger certainly exists when the notion of anthropological vulnerability is rejected.

The third aspect is that vulnerability is not merely inability or deficiency but it is most of all ability or opportunity. Vulnerable subjects are not victims in need of protection or dependent on the benevolence of the strong. Human capabilities will develop when inequality and structural violence have been removed, and the appropriate social, cultural, political, and economic conditions for human flourishing have been created. Ethics itself has emerged through reflection on the experiences of vulnerability.

Taking vulnerability seriously highlights our common humanity and our interest in other people. Vulnerability is the inspiration for a truly global and different kind of bioethics.

Notes

1 Ganguli Mitra and Biller-Andorno, 'Vulnerability and exploitation in a globalized world', 101.
2 See Potter, *Global bioethics.* Also, Ten Have, 'Potter's notion of bioethics'. For the current development of global bioethics, see Ten Have, *Global Bioethics. An introduction.*
3 Jonsen, *The birth of bioethics,* 377.
4 This is recently advocated by the Joint Action and Learning Initiative on National and Global Responsibility for Health, arguing for a Framework Convention on Global Health. See Gostin, Friedman, Ooms, *et al.,* 'The Joint Action and Learning Initiative: Towards a global agreement on national and global responsibilities for health', and Gostin, 'A framework convention on global health: Health for all, justice for all'.

5   See, Ten Have and Gordijn (eds.). *Handbook of global bioethics*. See also, Rennie and Mupenda, 'Living apart together: reflections on bioethics, global inequality, and social justice'; Benatar and Brock, *Global health and global health ethics*; Pinto and Upshur, *An introduction to global health ethics*.

6   Brock, 'Broadening the bioethics agenda', 22, 31–3.

7   Daniels, 'Equity and population health. Toward a broader bioethics agenda', 23.

8   'That mindset requires recognition that health, human rights, economic opportunities, good governance, peace, and development are all intimately linked within a complex, interdependent world' (Benatar, Daar, and Singer, 'Global health challenges: The need for an expanded discourse on bioethics', 588).

9   Farmer, *Pathologies of power,* 175, 201. According to Farmer, there are four reasons why bioethics is insufficient: it is too far removed from practical experiences; it is a phenomenon of industrialized nations; experts are dominating; and it is 'quandary ethics' for individual patients (Farmer, *Pathologies of power,* 204–5). See also Solbakk, 'Bioethics on the couch'.

10 Butler, *Frames of war,* 33.

11 Campbell, 'Dependency revisited', 105.

12 See Jonsen, *The birth of bioethics* and Rothman, *Strangers at the bedside.*

13 Benatar, 'Global health, vulnerable populations, and law', 43 ff.; Bertomeu, 'Bioethics, globalization, and politics', 36 ff; Downie, 'Glass houses: the power of money in bioethics research', 110. See also Nnaemeka, 'Feminist bioethics and global responsibility: Exploring health care delivery in Kenya'; Benatar and Brock, *Global health and global health ethics.*

14 See Rosen, *From medical police to social medicine* and Léonard, *La medicine entre les pouvoirs et les savoirs.*

15 Scholte, *Globalization. A critical introduction,* 287.

16 See, for example, Steger, Goodman and Wilson, *Justice globalism. Ideology, crises, policy;* Delmas-Marty, *Résister, responsabiliser, anticiper. Ou comment humanizer la mondialisation.*

17 The problem is not, argues Alain Supiot, to 'réguler' the markets but to 'réglementer' them, which means political and legal action to re-establish the order of ends and means according to human needs. (Supiot, *L'esprit de Philadelphie,* 94).

18 Banerjee and Duflo provide many examples of the 'surprising power of small changes'—changes in specific circumstances that substantially improved the living conditions of people. It is necessary to focus on local arrangements that create improved conditions for human flourishing and reducing vulnerability: '...what is needed is a shift in perspective from INSTITUTIONS in capital letters to institutions in lower case – the "view from below"' (Banerjee and Duflo, *Poor economics,* 247, 243).

19 Luna, 'Poverty and inequality', 453.

20 Garrafa and Porto, 'Intervention bioethics: A proposal for peripheral countries in a context of power and injustice', 400–1.

21 Mann, 'Human rights and AIDS: The future of the pandemic', 224–5.

22 Sen, *The idea of justice,* 387, 364–5.

23 Special Rapporteurs Paul Hunt and Anand Grover have issued many reports since 2002 www.ohchr.org/EN/Issues/Health/Pages/SRRightHealthIndex.aspx?pages? SRRightHealthIndex.aspx (Accessed 2 January 2016).

24 See Keck and Sikkink, *Activists beyond borders. Advocacy networks in international politics.* For an application in the field of bioethics, see Ten Have, 'Bioética sem fronteiras' (Bioethics without borders).

25 Sen, *The idea of justice*, ix, 324.
26 Sen, *The idea of justice*, 7; see also 5–6, 15 ff.
27 Sen, *The idea of justice*, 348. See also Schicktanz, Schweda, and Wynne, 'The ethics of "public understanding of ethics" – why and how bioethics expertise should include public and patients' voices'.
28 Dunn, Clare and Holland, 'To empower or to protect? Constructing the "vulnerable adult" in English law and public policy', 243.
29 Koopman, 'Vulnerable church in a vulnerable world'? 252.
30 As argued by Frohlich and Potvin, 'Transcending the known in public health practice. The inequality paradox: The population approach and vulnerable populations', 220.
31 Stutzki, Weber and Reiter-Theil, 'Finding their voices again: a media project offers a floor for vulnerable patients, clients, and the socially deprived'.
32 Van der Meide, Leget, and Olthuis, 'Giving voice to vulnerable people: the value of shadowing for phenomenological healthcare research'.
33 Justo, 'Participatory research: A way to reduce vulnerability', 67.
34 London, 'Ethical oversight of public health research: Can rules and IRBs make a difference in developing countries?', 1082.
35 London, 'Ethical oversight of public health research: Can rules and IRBs make a difference in developing countries?', 1081.
36 Beck, *World risk society,* 208. The example of the SARS epidemic shows according to Beck that the Chinese authorities have been necessitated to engage in transnational risk management. Only international cooperation could provide an appropriate response to the epidemic threat. Transnational cooperation therefore is a 'precondition for successful national and local risk management'. (Beck, *World risk society,* 175). Beck even puts it stronger: 'it is transnational cooperation that makes national sovereignty possible'. (Beck, *World risk society,* 233).
37 Cohen, *Homo economicus, prophète (égaré) des temps nouveaux,* 34. Cohen also explains that Darwin himself has argued that the human being is characterized by sociability. The novelty of the human species is the capacity of sympathy, the ability to attend to and care for other persons. It is the importance of cooperation that makes social life possible. Cohen thus tries to correct the misinterpretation of Darwin in neoliberal ideology, just like Sen has attempted to correct the one-sided image of Adam Smith (Sen, *The idea of justice*, 185). In the words of Charles Taylor: 'A human being alone is an impossibility… As organisms we are separable from society…but as humans this separation is unthinkable' (Taylor, *Human agency and language. Philosophical papers I*, 8).
38 Robinson, *The ethics of care*, 60.
39 Farmer, 'Challenging orthodoxies: The road ahead for health and human rights', 10.
40 Garrafa, Solbakk, Vidal, and Lorenzo, 'Between the needy and the greedy: the quest for a just and fair ethics of clinical research', 504.
41 See Ten Have, 'The activities of UNESCO in the area of ethics' and Ten Have, 'UNESCO's Ethics Education Programme'.
42 Harvey, *A brief history of neoliberalism*, 69.

BIBLIOGRAPHY

Acharya, Amitav. 'Human security: East versus West'. *International Journal* 56, no. 3 (2001): 442–460.

Aday, Lu Ann. *At risk in America: The health and health care needs of vulnerable populations in the United States.* San Francisco: Jossey-Bass, 2003 (Second Ed.).

Adger, W. Neil. 'Vulnerability'. *Global Environmental Change* 16 (2006): 268–281.

Afolabi, Michael O.S. 'Researching the vulnerables: Issues of consent and ethical approval'. In proceeding of the Second Unibadan Biomedical Conference, at Ibadan, Oyo State, Nigeria. Published in: *African Journal of Medicine and Medical Sciences* 12 (2011): 1–8.

African Charter on Human and Peoples' Rights. Banjul, 1981 (http://www1.umn.edu/humanrts/instree/z1afchar.htm) (Accessed 7 April 2015).

Aginam, Obijiofor. 'Between isolationism and mutual vulnerability: A South-North perspective on global governance of epidemics in an age of globalization'. *Temple Law Review* 77 (2004): 297–312.

Aldwin, Carolyn M, and Tracey A. Revenson. 'Vulnerability to economic stress'. *American Journal of Community Psychology* 14, no. 2 (1986): 161–175.

Alesina, Alberto, Edward Glaeser and Bruce Sacerdote. *Why doesn't the US have a European-style welfare system?* Cambridge, MA: National Bureau of Economic Research, working paper 8524, 2001 (http://www.nber.org/papers/w8524) (Accessed 12 October 2015).

Allen, Leslie. 'Will Tuvalu disappear beneath the sea?', *Smithsonian* 35, no. 5 (2004): 44–46, 49–52.

Andanda, Pamela. 'Vulnerability: Sex workers in Nairobi's Majengo slum'. *Cambridge Quarterly of Healthcare Ethics* 18 (2009): 138–146.

Andorno, Roberto. 'Human dignity and human rights as a common ground for a global bioethics'. *Journal of Medicine and Philosophy* 34, no. 3 (2009): 223–240.

Ankrah, E. Maxine. 'AIDS and the social side of health'. *Social Science & Medicine* 32, no. 9 (1991): 967–980.

Annan, Kofi. 'Secretary-General proposes global compact on human rights, labour, environment, in address to world economic forum in Davos', UN Press Release, 1 February 1999, (http://www.un.org/News/Press/docs/1999/19990201.sgsm6881.html) (Accessed 4 April 2015).

Annas, George J. 'Globalized clinical trials and informed consent'. *New England Journal of Medicine* 360, no. 20 (2009): 2050–2053.

Announcements, 'Zbigniew Bankowski lecture fund'. *Pharmacoepidemiology and Drug Safety* 9, no. 4 (2000): 357.

Anthony, E.J. 'A risk-vulnerability intervention model for children of psychotic parents'. In E.J. Anthony, C. Koupernik, and C. Chilands (eds.): *The child in his family: Children at psychiatric risk*. New York: John Wiley and Sons, 1974: 99–121.

Appiah, Kwame Anthony. *Cosmopolitanism. Ethics in a world of strangers*. New York and London: W.W. Norton and Company, 2006.

Arnold, Denis G. 'Exploitation and the sweatshop quandary'. *Business Ethics Quarterly* 13, no. 2 (2003): 243–256.

Ashcroft, Richard E. 'Money, consent, and exploitation in research'. *American Journal of Bioethics* 1, no. 2 (2001): 62–63.

Ashcroft, Richard E. 'The troubled relationship between bioethics and human rights'. In Michael Freeman (ed.). *Law and bioethics: Current legal issues*. Volume II. Oxford: Oxford University Press, 2008: 33–51.

Ashcroft, Richard E. 'Could human rights supersede bioethics?', *Human Rights Law Review* 10, no. 4 (2010): 639–660.

Aulakh, Raveena. 'New Zealand decision created world's first climate refugees'. *The Star*, Sunday 17 August 2014, (http://www.thestar.com/news/world/2014/08/06/new_zealand_decision_created_worlds_first_climate_refugees.html) (Accessed 4 April 2015).

Bagheri, Alireza and Francis L. Delmonico. 'Guest editorial: organ trafficking and transplant tourism: a call for international collaboration'. *Medicine, Health Care, and Philosophy* 16, no. 4 (2013): 885–886.

Baker, Robert. 'Bioethics and human rights: A historical perspective'. *Cambridge Quarterly of Healthcare Ethics* 10 (2001): 241–252.

Baldwin, Moyra A. 'Patient advocacy: a concept analysis'. *Nursing Standard* 17, no. 21 (2002): 33–39.

Ballantyne, Angela. 'HIV international clinical research: exploitation and risk'.'' *Bioethics* 19, no. 5–6 (2005): 476–491.

Ballantyne, Angela. 'How to do research fairly in an unjust world'. *American Journal of Bioethics* 10, no. 6 (2010): 26–35.

Banerjee, Anhijit V. and Esther Duflo. *Poor economics*. London: Penguin Books, 2011.

Bankoff, Gregory. 'Rendering the world unsafe: "Vulnerability" as Western discourse'. *Disasters* 25, no. 1 (2001): 19–35.

Bankowski, Zbigniew, and John F. Dunne. 'History of the WHO/CIOMS project for the development of guidelines for the establishment of ethical review procedures for research involving human subjects'. In Z. Bankowski and N. Howard-Jones (eds.): *Human experimentation and medical ethics. Proceedings of the XVth CIOMS Round Table Conference Manila, 13-16 September 1981*. Geneva: CIOMS, 1982: 441–452.

Bankowski, Zbigniew, Jack J. Bryant and John Last (eds). *Ethics and epidemiology: International guidelines. Proceedings of the XXVth CIOMS Conference*. Geneva: CIOMS, 1991.

Bankowski, Zbigniew. 'Ethics and health'. *World Health Forum* 16 (1995): 115–125.

Basnyat, Iccha. 'Beyond biomedicine: health through social and cultural understanding'. *Nursing Inquiry* 18, no. 2 (2011): 123–134.

Bauman, Zygmunt. *Globalization. The human consequences*. New York: Columbia University Press, 1998.

Bayer, Ronald. 'In memoriam, Jonathan Mann, 1947–1998'. *American Journal of Public Health* 88, no. 11 (1998): 1608–9.

Beauchamp, Dan E. *The health of the republic. Epidemics, medicine, and moralism as challenges to democracy*. Philadelphia: Temple University Press, 1988.

Beauchamp, Tom L. 'The Belmont report'. In Ezekiel J. Emanuel, Christine Grady. Robert A. Crouch, Reidar K. Lie, Franklin G. Miller and David Wendler (eds.). *The Oxford Textbook on Clinical Research Ethics*. Oxford and New York: Oxford University Press, 2008: 149–155.

Beauchamp, Tom L. and James F. Childress. *Principles of biomedical ethics*. New York and Oxford: Oxford University Press, 1979.

Beauchamp, Tom L., Bruce Jennings, Eleanor D. Kinney and Robert J. Levine. 'Pharmaceutical research involving the homeless'. *Journal of Medicine and Philosophy* 27, no. 5 (2002): 547–564.

Beck, Ulrich. *Risk society: Towards a new modernity*. London: Sage, 1992.

Beck, Ulrich. *World risk society*. Cambridge, UK: Polity Press, 2009.

Benaroyo, Lazare. 'The notion of vulnerability in the philosophy of Emmanuel Levinas and its significance for medical ethics and aesthetics'. 2007. (http://www.api.or.at/aebm/download/docs/web_levinas.pdf) (Accessed 7 August 2015).

Benatar, David. 'Bioethics and health and human rights: a critical view'. *Journal of Medical Ethics* 32 (2006): 17–20.

Benatar, Solomon R, Abdallah S. Daar and Peter A. Singer. 'Global health challenges: The need for an expanded discourse on bioethics'. *PLoS Medicine* 2, no. 7 (2005): e143.

Benatar, Solomon R. 'Global health, vulnerable populations, and law'. *Journal of Law, Medicine & Ethics* 41, no. 1 (2013): 42–47.

Benatar, Solomon and Gillian Brock (eds). *Global health and global health ethics*. Cambridge, UK and New York: Cambridge University Press, 2011.

Bennett, Belinda and George F. Tomossy (eds.). *Globalization and health. Challenges for health law and bioethics*. Dordrecht: Springer, 2006.

Bergoffen, Debra. 'February 22, 2001: Toward a politics of the vulnerable body'. *Hypatia* 18, no. 1 (2003): 116–134.

Bertomeu, M. Julia. 'Bioethics, globalization, and politics'. *International Journal of Feminist Approaches to Bioethics* 2, no. 1 (2009): 33–51.

Bezner Kerr, Rachel and Paul Mkandawire. 'Imaginative geographies of gender and HIV/AIDS: moving beyond neoliberalism'. *GeoJournal* 77 (2012): 459–473.

Bielby, Phil. *Competence and vulnerability in biomedical research*. Dordrecht: Springer Publishers, 2008.

Birch, Kean. 'Neoliberalising bioethics: Bias, enhancement and economistic ethics'. *Genomics, Society and Policy* 4 (2008): 1–10.

Blacksher, Erika and John Stone. 'Introduction to "vulnerability" issues'. *Theoretical Medicine and Bioethics* 23, no. 6 (2002): 421–424.

Bluhm, Robyn. 'Vulnerability, health, and illness'. *International Journal of Feminist Approaches to Bioethics* 5, no. 2 (2012): 147–161.

Bohle, Hans-Georg. *Living with vulnerability. Livelihoods and human security in risky environments*. Bonn: UNU Institute for Environment and Human Security, 2007 (http://www.microinsuranceconference2005.com/dms/MRS/Documents/InterSection2007_Bohle_Vulnerability.pdf) (Accessed 4 October 2015).

Botbol-Baum, Mylène. 'The necessary articulation of autonomy and vulnerability'. In Rendtorff, Jacob Dahl and Peter Kemp (eds). *Basic ethical principles in European bioethics and biolaw. Volume II. Partners' research*. Copenhagen and Barcelona: Centre for Ethics and Law, and Institut Borja de Bioetica, 2000: 57–64.

Boukus, Ellyn, Alwyn Cassil, Ann O'Malley, 'A snapshot of U.S. physicians: Key findings from the 2008 health tracking physician survey', http://www.hschange.com/CONTENT/1078/) (Accessed 10 October 2013).

Braedley, Susan and Meg Luxton (eds.). *Neoliberalism and everyday life*. Montreal and Kingston: McGill-Queen's University Press, 2010.

Braybrooke, David. *Meeting needs*. Princeton, NJ: Princeton University Press, 1987.

Brenner, Joel. *America the vulnerable. Inside the new threat matrix of digital espionage, crime, and warfare*. New York: The Penguin Press, 2011.

Bresson, Maryse. *Sociologie de la précarité*. (Second Edition) Paris: Armand Colin, 2012.

Briguglio, Lino. 'Small island developing states and their economic vulnerabilities'. *World Development* 23, no. 9 (1995): 1615–1632.

Briones, Leah. 'Rights with capabilities: A new paradigm for social justice in migrant activism'. *Studies in Social Justice* 5, no. 1 (2011): 127–143.

Brock, Dan W. 'Broadening the bioethics agenda'. *Kennedy Institute of Ethics Journal* 10, no. 1 (2000): 21–38.

Brock, Dan W. 'Health resource allocation for vulnerable populations'. In Marion Danis, Carolyn Clancy, and Larry Churchill (eds.). *Ethical Dimensions of Health Policy*. New York: Oxford University Press, 2002: 283–309.

Brock, Gillian (ed). *Necessary goods. Our responsibilities to meet others' needs*. Lanham, Maryland: Rowman and Littlefield Publishers, 1998.

Brodie, Janine. 'Reforming social justice in neoliberal times'. *Studies in Social Justice* 1, no. 2 (2007): 93–107.

Brody, Baruch A. *The ethics of biomedical research. An international perspective*. New York and Oxford: Oxford University Press, 1998.

Brooks, Tom. 'Capabilities'. In Hugh LaFollette (ed). *The International Encyclopedia of Ethics*. Oxford: Wiley-Blackwell Publishing, 2013: 692–698.

Brown, Kate. '"Vulnerability": Handle with care'. *Ethics and Social Welfare* 5, no. 3 (2011): 313–321.

Brown, Neville. 'Climate change and human history, some indications from Europe, AD 400–1400'. *Environmental Pollution* 83 (1994): 37–43.

Brown, Tim. '"Vulnerability is universal": Considering the place of "security" and "vulnerability" within contemporary global health discourse'. *Social Science & Medicine* 72 (2011): 319–326.

Brundtland, Gro Harlem. 'Global health and international security'. *Global Governance* 9 (2003): 417–423.

Buchanan, Allen. 'Exploitation, alienation, and injustice'. *Canadian Journal of Philosophy* 9, no. 1 (1979): 121–139.

Budiani-Saberi, Debra A. and Frank L. Delmonico. 'Organ trafficking and transplant tourism: A commentary on the global realities'. *American Journal of Transplantation* 8 (2008): 925–929.

Buncombe, Andrew. '"A heaven for clinical trials, a hell for India": Court orders government to regulate drugs testing by international pharmaceutical companies.' *The Independent* 30 September 2013 (http://www.independent.co.uk/news/world/asia/a-heaven-for-clinical-trials-a-hell-for-india-court-orders-government-to-regulate-drugs-testing-by-international-pharmaceutical-companies-8849461.html) (Accessed 6 October 2015).

Buncombe, Andrew and Nina Lakhani, 'Without consent: how drugs companies exploit Indian "guinea pigs",' *The Independent*, 14 November 2011, http://www.independent.co.uk/news/world/asia/without-consent-how-drugs-companies-exploit-indian-guinea-pigs-6261919.html (Accessed 4 April 2015).

Burchardt, Tania. 'Capabilities and disabilities: the capabilities framework and the social model of disability'. *Disability & Society* 19, no. 7 (2004): 735–751.

Burggraeve, Roger. 'Violence and the vulnerable face of the Other: The vision of Emmanuel Levinas on moral evil and our responsibility'. *Journal of Social Philosophy* 30, no. 1 (1999): 29–45.

Butler, Judith. *Precarious life. The powers of mourning and violence*. London and New York: Verso, 2006.

Butler, Judith. 'Reply from Judith Butler to Mills and Jenkins'. *Differences: A Journal of Feminist Cultural Studies* 18, no. 2 (2007): 180–185.

Butler, Judith. *Frames of war. When is life grievable?* London and New York: Verso, 2010.

Buytendijk, F.J.J. *Prolegomena to an anthropological physiology*. Pittsburgh: Duquesne University Press, 1975 (original in Dutch, 1965).

Cadwallader, Jessica Robyn. '(Un)expected suffering: The corporeal specificity of vulnerability'. *International Journal of Feminist Approaches to Bioethics* 5, no. 2 (2012): 105–125.

Callahan, Daniel. 'The vulnerability of the human condition'. In Peter Kemp, Jacob Rendtorff and Niels Mattson Johansen (eds.). *Bioethics and Biolaw. Vol. II: Four ethical principles*. Copenhagen: Rhodos International Science and Art Publishers & Centre for Ethics and Law, 2000: 115–122.

Cambridge Dictionaries Online, http://dictionary.cambridge.org/dictionary/american-english/vulnerable?q=vulnerability (Accessed 4 April 2015).

Campbell, Alastair. 'Dependency revisited: The limits of autonomy in medical ethics'. In Margaret Brazier and Mary Lobjoit (eds). *Protecting the vulnerable. Autonomy and consent in health care*. London: Routledge, 1991: 101–112.

Campbell, Amy T. '"Vulnerability" in context: Recognizing the sociopolitical influences'. *American Journal of Bioethics* 4, no. 3 (2004): 58–59.

Campbell, Courtney S. 'Body, self, and the property paradigm'. *Hastings Center Report* 22, no. 5 (1992): 34–42.

Carel, Havi. 'Phenomenology and its application in medicine'. *Theoretical Medicine and Bioethics* 32 (2011): 33–46.

Carpenter, Mick. 'The capabilities approach and critical social policy: Lessons from the majority world?' *Critical Social Policy* 29, no. 3 (2009): 351–373.

Carse, Alisa L. and Margaret Olivia Little. 'Exploitation and the enterprise of medical research'. In Jennifer S. Hawkins and Ezekiel J. Emanuel (eds). *Exploitation and developing countries. The ethics of clinical research*, Princeton, NJ and Oxford: Princeton University Press 2008: 206–245.

Chamayou, Grégoire. *Théorie du drone*. Paris: La Fabrique éditions, 2013.

Chambers, Robert. 'Editorial introduction: vulnerability, coping and policy'. *IDS Bulletin* 20, no. 2 (1989): 1–7.

Chambers, Anne, and Keith Chambers. 'Five takes on climate and cultural change in Tuvalu'. *The Contemporary Pacific* 19, no. 1(2007): 294–306.

Chapman, Audrey R. 'Globalization, human rights, and the social determinants of health'. *Bioethics* 23, no. 2 (2009): 97–111.

Charter of Fundamental Rights of the European Union, *Official Journal of the European Communities*, 2000, C/ 364/01 (http://www.europarl.europa.eu/charter/pdf/text_en.pdf) (Accessed 7 April 2015).

Chennells, Roger. 'Vulnerability and indigenous communities: Are the San of South Africa a vulnerable people?', *Cambridge Quarterly of Healthcare Ethics* 18 (2009): 147–154.

Church of Norway. *Vulnerability and security. Current challenges in security policy from an ethical and theological perspective*. Commission on International Affairs in Church of Norway Council on Ecumenical and International Relations. 2002 (http://www.kirken.no/english/doc/Kisp_vulnerab_00.pdf) (Accessed 6 April 2015).

Christakis, Nicholas A., and Morris J. Panner. 'Existing international ethical guidelines for human subjects research: Some open questions'. *Law, Medicine & Health Care* 19, no. 3–4 (1991): 214–220.

CIOMS. 'Proposed International Guidelines for Biomedical Research Involving Human Subjects'. In Z. Bankowski and N. Howard-Jones (eds): *Human experimentation and medical ethics. Proceedings of the XVth CIOMS Round Table Conference Manila, 13–16 September 1981*. Geneva: CIOMS, 1982: 389–440.

CIOMS. *International Guidelines for Ethical Review of Epidemiological Studies*. Geneva: CIOMS, 1991 (http://www.cioms.ch/publications/guidelines/1991_texts_of_guidelines.htm) (Accessed 25 January 2015).

CIOMS. *International Ethical Guidelines for Biomedical Research Involving Human Subjects*. Geneva: CIOMS, 1993 (http://www.codex.uu.se/texts/international.html) (Accessed 26 January 2015).

CIOMS. *International Ethical Guidelines for Biomedical Research Involving Human Subjects*. Geneva: CIOMS, 2002 (http://www.cioms.ch/publications/guidelines/guidelines_nov_2002_blurb.htm) (Accessed 26 January 2015).

Clark, Ian. *The vulnerable in international society*. Oxford: Oxford University Press, 2013.

Cockerham, Geoffrey B. and William C. Cockerham. *Health and globalization*. Cambridge, UK and Malden, MA: Polity Press, 2010.

Cohen, Daniel. *Homo economicus, prophète (égaré) des temps nouveaux*. Paris: Albin Michel, 2012.

Coleman, Carl H. 'Vulnerability as a regulatory category in human subject research'. *Journal of Law, Medicine & Ethics* 37, no. 1 (2009): 12–18.

Cooper, Melinda. *Life as surplus. Biotechnology and capitalism in the neoliberal era*. Seattle and London: University of Washington Press, 2008.

Cox, Robert W. *The political economy of a plural world: Critical reflections on power, morals, and civilization*. London: Routledge, 2002.

Crocker, David A. 'Functioning and capability: The foundations of Sen's and Nussbaum's development ethic'. *Political Theory* 20, no. 4 (1992): 584–612.

Culp, Kristine A. *Vulnerability and glory. A theological account*. Louisville, Kentucky: Westminster John Knox Press, 2010.

Cutler, Stephen J. 'Safety on the streets: Cohort changes in fear'. *International Journal of Aging and Human Development* 10, no. 4 (1979–80): 373–384.

Daley, Suzanne. 'Spain's jobless rely on family, a frail crutch', *The New York Times*, 28 July 2012, http://www.nytimes.com/2012/07/29/world/europe/spains-elders-bearing-burden-of-recession.html?pagewanted=all&_r=0. (Accessed 4 April 2015).

Daniel, Linda E. 'Vulnerability as a key to authenticity'. *Image – The Journal of Nursing Scholarship* 30, no. 2 (1998): 191–192.

Daniels, Norman. 'Equity and population health. Toward a broader bioethics agenda'. *Hastings Center Report* 36, no. 4 (2006): 22–35.

Danis, Marion and Donald Patrick. 'Health policy, vulnerability, and vulnerable populations'. In Marion Danis, Carolyn Clancy, and Larry Churchill (eds.). *Ethical Dimensions of Health Policy*. New York: Oxford University Press, 2002: 310–334.

Dean, Hartley. 'Critiquing capabilities: The distractions of a beguiling concept'. *Critical Social Policy* 29, no. 2 (2009): 261–273.

DeBruin, Debra. 'Reflections on "vulnerability"'. *Bioethics Examiner* 5, no. 2 (2001): 1, 4 and 7.

DeBruin, Debra A. 'Looking beyond the limitations of "vulnerability": Reforming safeguards in research'. *American Journal of Bioethics* 4, no. 3 (2004): 76–78.

De Chesnay, Mary. 'Vulnerable populations: vulnerable people'. In Mary de Chesnay (ed.). *Caring for the vulnerable. Perspectives in nursing theory, practice, and research*. Sudbury, MA: Jones and Bartlett Publishers, 2005: 3–12.

De Chesnay, Mary, Rebecca Wharton, and Christopher Pamp, 'Cultural competence, resilience, and advocacy'. In Mary de Chesnay (ed.): *Caring for the vulnerable. Perspectives in nursing theory, practice, and research*. Sudbury, MA: Jones and Bartlett Publishers, 2005: 31–41.

Declaration of Istanbul on organ trafficking and transplant tourism. *Transplantation* 86, no. 8 (2008): 1013–1018.

Dekkers, Wim J.M. 'F.J.J. Buytendijk's concept of an anthropological physiology'. *Theoretical Medicine* 16, no. 1 (1995): 15–39.

Delmas-Marty, Mireille. *Résister, responsabiliser, anticiper. Ou comment humanizer la mondialisation*. Paris: Éditions du Seuil, 2013.

Delor, Francois and Michel Hubert. 'Revisiting the concept of "vulnerability"'. *Social Science and Medicine* 50 (2000): 1557–1570.

DeMarco, Joseph P. 'Vulnerability: A needed moral safeguard'. *American Journal of Bioethics* 4, no. 3 (2004): 82–84.

Deneulin, Séverine. '"Necessary thickening". Ricoeur's ethic of justice as a complement to Sen's capability approach'. In Séverine Deneulin, Mathias Nebel, and Nicholas Sagovsky (eds). *Transforming unjust structures. The capability approach*. Dordrecht: Springer, 2006: 27–45.

Deneulin, Séverine, Mathias Nebel, and Nicholas Sagovsky (eds). *Transforming unjust structures. The capability approach*. Dordrecht: Springer, 2006.

Deneulin, Séverine and J. Allister McGregor. 'The capability approach and the politics of a social conception of wellbeing'. *European Journal of Social Theory* 13, no. 4 (2010): 501–519.

Derrida, Jacques. 'The animal that there I am (more to follow)'. *Critical Inquiry* 28. no. 2 (2002): 369–418.

DHEW (Department of Health, Education, and Welfare) Guidelines. 'Protection of human subjects'. *Federal Register* 39, no. 105, Part II (1974): 18913–18920.

Diedrich, W. Wolf, Roger Burggraeve and Chris Gastmans. 'Towards a Levinasian care ethic: A dialogue between the thoughts of Joan Tronto and Emmanuel Levinas'. *Ethical Perspectives: Journal of the European Ethics Network* 13, no. 1 (2003): 33–61.

Diniz, Debora, Dirce Bellezi Guilhem and Volnei Garrafa. 'Bioethics in Brazil'. *Bioethics* 13, no. 3/4 (1999): 244–248;

Dodds, Susan. 'Depending on care: Recognition of vulnerability and the social contribution of care provision'. *Bioethics* 21, no. 9 (2007): 500–510.

Dodds, Susan. 'Dependence, care, and vulnerability'. In Cathriona Mackenzie, Wendy Rogers, and Susan Dodds (eds.). *Vulnerability. New essays in ethics and feminist philosophy*. Oxford and New York: Oxford University Press, 2014: 181–203.

Dong, Doris and Beverley Temple. 'Oppression: A concept analysis and implications for nurses and nursing'. *Nursing Forum* 46, no. 3 (2011): 169–176.

D'Oronzio, Joseph C. 'The integration of health and human rights: An appreciation of Jonathan M. Mann'. *Cambridge Quarterly of Healthcare Ethics* 10 (2001): 231–240.

Dos Anjos, Marcio Fabri. 'Medical ethics in the developing world: A liberation theology perspective'. *Journal of Medicine and Philosophy* 21, no. 6 (1996): 629–637.

Dower, Nigel. *World ethics. The new agenda*. (Second Edition). Edinburgh: Edinburgh University Press, 2007.

Dower, Nigel. 'Globalization'. In Hugh LaFolette (ed.). *The International Encyclopedia of Ethics*. Oxford: Blackwell Publishing, 2013: 2174–2186.

Downie, Jocelyn. 'Glass houses: the power of money in bioethics research'. *International Journal of Feminist Approaches to Bioethics* 2, no. 2 (2009): 97–115.

Dubois, Jean-Luc and Sophie Rousseau. 'Reinforcing households' capabilities as a way to reduce vulnerability and prevent poverty in equitable terms'. In Flavio Comim, Mazaffar Qizilbash and Sabina Alkire (eds.). *The capability approach. Concepts, measures and applications*. Cambridge, UK: Cambridge University Press, 2008: 421–436.

Dujardin, Bruno. 'Health and human rights: The challenge for developing countries'. *Social Science and Medicine* 39, no. 9 (1994): 1261–1274.

Dunn, Michael C, Isabel C.H. Clare and Anthony J. Holland. 'To empower or to protect? Constructing the "vulnerable adult" in English law and public policy'. *Legal Studies* 28, no. 2 (2008): 234–253.

Dwyer, James. 'Global health and justice'. *Bioethics* 19, no. 5–6 (2005): 460–475.

Dyer, Janyce G. and Teena Minton McGuinness. 'Resilience: Analysis of the concept'. *Archives of Psychiatric Nursing* 10, no. 5 (1996): 276–282.

Earvolino-Ramirez, Marie. 'Resilience: a concept analysis'. *Nursing Forum* 42, no. 2 (2007): 73–82.

Eckenwiler, Lisa. 'Moral reasoning and the review of research involving human subjects'. *Kennedy Institute of Ethics Journal* 11, no. 1 (2001): 37–69.

Eckenwiler, Lisa, Carolyn Ellis, Dafna Feinholz and Toby Schonfeld, 'Hopes for Helsinki: reconsidering "vulnerability"'. *Journal of Medical Ethics* 34, no. 10 (2008): 765–6.

Edgar, Bill, Joe Doherty, and Henk Meert. *Review of Statistics on Homelessness in Europe*. Brussels: European Federation of National Associations working with the Homeless, 2003 (http://www.feantsa.org/files/transnational_reports/EN_StatisticsReview_2003 .pdf) (Accessed 20 March 2015).

Edgar, Bill, and Henk Meert. *Fourth review of statistics on homelessness in Europe*. Brussels: European Federation of National Associations working with the Homeless, 2005 (http://www.feantsa.org/files/transnational_reports/EN_Stats_2005.pdf) (Accessed 20 March 2015).

Eggertsen, Laura. 'Helsinki doctrine under review'. *Canadian Medical Association Journal* 184, no. 16 (2012): E827–828.

Ehrenreich, Nancy. 'Conceptualism by any other name'. *Denver University Law Review* 74 (1997): 1281–1305.

Eichner, Maxine. 'Dependency and the liberal polity: On Martha Fineman's The Autonomy Myth'. *California Law Review* 93, no. 4 (2005): 1285–1321.

Ells, Carolyn. 'Respect for people in situations of vulnerability: A new principle for health-care professionals and health-care organizations'. *International Journal of Feminist Approaches to Bioethics* 5, no. 2 (2012): 180–185.

Elster, Jon. 'Exploring exploitation'. *Journal of Peace Research* 15, no. 3 (1978): 3–17.

Epstein, Paul R. 'Framework for an integrated assessment of health, climate change, and ecosystem vulnerability'. *Annals New York Academy of Sciences* 740 (1994): 423–435.

Erlen, Judith A. 'Who speaks for the vulnerable?' *Orthopaedic Nursing* 25, no. 2 (2006): 133–136.

Evans, Donald. 'Commentary on the UNESCO IBC report on respect for vulnerability and personal integrity'. *International Journal of Feminist Approaches to Bioethics* 5, no. 2 (2012): 170–173.

Evans, Peter. 'Collective capabilities, culture, and Amartya Sen's *Development as freedom*'. *Studies in Comparative International Development* 37, no. 2 (2002): 54–60.

FAO (Food and Agriculture Organization of the United Nations). *World Food Summit. Rome Declaration on World Food Security.* Rome (Italy): FAO, 1996 (http://www.fao.org/docrep/003/w3613e/w3613e00.htm) (Accessed 22 March 2015).

FAO (Food and Agriculture Organization of the United Nations). *Report on the development of food insecurity and vulnerability information and mapping systems (FIVIMS).* Rome (Italy): FAO, Committee on World Food Security, 1998 (http://www.fao.org/docrep/meeting/W8497e.htm) (Accessed 22 March 2015).

Farbotko, Carol. 'Tuvalu and climate change: Constructions of environmental displacement in the *Sydney Morning Herald*'. *Geografiska Annaler. Series B, Human Geography* 87, no. 4 (2005): 279–293.

Farmer, Paul. 'On suffering and structural violence: A view from below'. *Daedalus* 125, no. 1 (1996): 261–283.

Farmer, Paul. *Pathologies of power. Health, human rights, and the new war on the poor.* Berkeley, Los Angeles, London: University of California Press, 2005.

Farmer, Paul. 'Challenging orthodoxies: The road ahead for health and human rights'. *Health and Human Rights* 10, no. 1 (2008): 5–19.

Faunce, Thomas A. 'Will international human rights subsume medical ethics? Intersections in the UNESCO *Universal Bioethics Declaration*'. *Journal of Medical Ethics* 31 (2005): 173–178.

Feito, Lydia. 'Vulnerabilidad'. *Anales del Sistema Sanitario de Navarra* 30, Supl. 3 (2007): 7–22.

Fillenbaum, Gerda G. 'An examination of the vulnerability hypothesis'. *International Journal of Aging and Human Development* 8, no. 2 (1977–78): 155–160.

Fineman, Martha Albertson. *The autonomy myth. A theory of dependency.* New York and London: The New Press, 2004.

Fineman, Martha Albertson. 'The vulnerable subject: Anchoring equality in the human condition'. *Yale Journal of Law and Feminism* 20, no. 1 (2008): 1–23.

Fineman, Martha Albertson. 'The vulnerable subject and the responsive state'. *Emory Law Journal* 60 (2010): 251–275.

Fisher, Jill A. *Medical research for hire. The political economy of pharmaceutical clinical trials.* New Brunswick, New Jersey and London: Rutgers University Press, 2009.

Flanigan, Rosemary. 'Vulnerability and the bioethics movement'. *Bioethics Forum* 16, no. 2 (2000): 13–18.

Flaskerud, Jacquelyn H and Betty W. Winslow. 'Conceptualizing vulnerable populations health-related research'. *Nursing Research* 47, no. 2 (1998): 69–78.

Flaskerud, Jacquelyn H. and Betty W. Winslow. 'Vulnerable populations and ultimate responsibility'. *Issues in Mental Health Nursing* 31 (2010): 298–299.

Folke, Carl. 'Resilience: the emergence of a perspective for social-ecological systems analyses'. *Global Environmental Change* 16 (2006): 253–267.

Forster, Heidi P., Ezekiel Emanuel and Christine Grady. 'The 2000 revision of the Declaration of Helsinki: a step forward or more confusion?', *The Lancet* 358 (2001): 1449–1453.

Foucault, Michel. *The birth of biopolitics. Lectures at the Collège de France 1978–1979.* New York: Palgrave Macmillan, 2008.

Fox, Ken. 'Hotep's story: Exploring the wounds of health vulnerability in the US'. *Theoretical Medicine* 23 (2002): 471–497.

Fraser, Nancy. *Justice interruptus. Critical reflections on the 'postsocialist' condition.* New York and London: Routledge, 1997.

Fraser, Nancy. 'Reframing justice in a globalizing world'. *New Left Review* 36 (2005): 1–19.

Fraser, Nancy. *Scales of justice. Reimagining political space in a globalizing world.* New York: Columbia University Press, 2010.

Frodeman, Robert, Adam Briggle and J. Britt Holbrook. 'Philosophy in the age of neoliberalism'. *Social Epistemology* 26, no. 3–4 (2012): 311–330.

Frohlich, Katherine L. and Louise Potvin. 'Transcending the known in public health practice. The inequality paradox: The population approach and vulnerable populations'. *American Journal of Public Health* 98, no. 2 (2008): 216–221.

Füssel, Hans-Martin. 'Vulnerability: A generally applicable conceptual framework for climate change research'. *Global Environmental Change* 17 (2007): 155–167.

Gaille, Marie (ed). *Philosophie de la Médecine. Frontière, savoir, clinique.* Paris: Vrin, 2011.

Gallopín, Gilberto C. 'Linkages between vulnerability, resilience, and adaptive capacity'. *Global Environmental Change* 16 (2006): 293–303.

Galtung, Johan. 'Violence, peace, and peace research'. *Journal of Peace Research* 6, no. 3 (1969): 167–191.

Ganguli Mitra, Agomoni and Nikola Biller-Andorno. 'Vulnerability in healthcare and research ethics'. In Ruth Chadwick, Henk ten Have and Eric M Meslin (eds.). *The SAGE Handbook of health care ethics: Core and emerging issues.* Los Angeles: Sage, 2011: 239–250.

Ganguli Mitra, Agomoni. 'Off-shoring clinical research: Exploitation and the reciprocity constraint'. *Developing World Bioethics* 13, no. 3 (2013): 111–118.

Ganguli Mitra, Agomoni and Nikola Biller-Andorno. 'Vulnerability and exploitation in a globalized world'. *International Journal of Feminist Approaches to Bioethics* 6, no. 1 (2013): 91–102.

GAO (United States Government Accountability Office). *Human subject research. Undercover tests show the Institutional Review Board system is vulnerable to unethical manipulation.* GAO-09-448T, 2009 (http://www.gao.gov/new.items/d09448t.pdf) (Accessed 8 October 2015).

Garrafa, Volnei and Dora Porto. 'Intervention bioethics: A proposal for peripheral countries in a context of power and injustice'. *Bioethics* 17, no. 5–6 (2003): 399–416.

Garrafa, Volnei, Jan Helge Solbakk, Susan Vidal and Claudio Lorenzo. 'Between the needy and the greedy: the quest for a just and fair ethics of clinical research'. *Journal of Medical Ethics* 36 (2010): 500–504.

Gasper, Des and Irene van Staveren. 'Development as freedom – and as what else?', *Feminist Economics* 9, no. 2–3 (2003): 137–161.

Gbadegesin, Segun and David Wendler. 'Protecting communities in health research from exploitation'. *Bioethics* 20, no. 5 (2006): 248–253.

Gehlen, Arnold. *Man. His nature and place in the world.* New York, Columbia University Press: 1988 (original German edition, 1974).

Giddens, Anthony. *The politics of climate change*. Cambridge, UK: Polity Press, 2009.

Gilson, Erinn. 'Vulnerability, ignorance, and oppression'. *Hypatia* 26, no. 2 (2011): 308–332.

Goldstein, Pierre. *Vulnérabilité et autonomie dans la pensée de Martha C. Nussbaum*. Paris: Presses Universitaires de France, 2011.

Goodin, Robert E. *Protecting the vulnerable. A reanalysis of our social responsibilities*. Chicago and London: The University of Chicago Press, 1985.

Goodin, Robert E. 'Exploiting a situation and exploiting a person'. In Andrew Reeve (ed.). *Modern theories of exploitation*. London: Sage Publications, 1987: 166–200.

Goodin, Robert E. 'Relative needs'. In Alan Ware and Robert E. Goodin (eds.). *Needs and welfare*. London: Sage Publications, 1990: 12–33.

Goodin, Robert E. 'Exploitation'. *American Political Science Review* 91, no. 3 (1997): 733–734.

Goodin, Robert E. 'Vulnerabilities and responsibilities: An ethical defense of the welfare state'. In Gillian Brock (ed.). *Necessary goods. Our responsibilities to meet others' needs*. Lanham, Maryland: Rowman & Littlefield Publishers, 1998: 73–94.

Goodman, David M. *The demanded self. Levinasian ethics and identity in psychology*. Pittsburgh: Duquesne University Press, 2012.

Gordijn, Bert and Henk ten Have. 'Future perspectives'. In Henk ten Have and Bert Gordijn (eds.). *Handbook of global bioethics*. Dordrecht: Springer Publishers, 2013: 829–844.

Gordon, John-Stewart. 'Human rights in bioethics – Theoretical and applied'. *Ethical Theory and Moral Practice* 15 (2012): 283–294.

Gorovitz, Sam. 'Reflections on the vulnerable'. In Z. Bankowski and J.H. Bryant (eds.). *Poverty, vulnerability, the value of human life, and the emergence of bioethics. Highlights and papers of the XXVIIIth CIOMS Conference, Iztapa, Guerrero State, Mexico, 17–20 April 1994*. Geneva: CIOMS, 1994: 63–65.

Gostin, Lawrence. 'Macro-ethical principles for the conduct of research on human subjects: Population-based research and ethics'. In Zbigniew Bankowski, Jack J. Bryant and John Last, (eds.). *Ethics and epidemiology: International guidelines. Proceedings of the XXVth CIOMS Conference*. Geneva: CIOMS, 1991: 29–46.

Gostin, Lawrence. 'Ethical principles for the conduct of human subject research: Population-based research and ethics'. *Law, Medicine & Health Care* 19, no. 3–4 (1991): 191–201.

Gostin, Lawrence O. (ed.). *Public health law and ethics. A reader*. Revised and updated Second Edition. Berkeley and New York: University of California Press and The Milbank Memorial Fund, 2010.

Gostin, Lawrence O., Eric A. Friedman, Gorik Ooms, Thomas Gebauer, Narendra Gupta, Devi Sridhar, Wang Chenguang, John-Arne Røttingen, and David Sanders. 'The Joint Action and Learning Initiative: Towards a global agreement on national and global responsibilities for health'. *PLoS Medicine* 8, no 5 (2011): e1001031. DOI:10.1371/journal.pmed.1001031.

Gostin, Lawrence O. 'A framework convention on global health. Health for all, justice for all'. *JAMA* 307, no. 19 (2012): 2087–92.

Gracia, Diego. 'Ownership of the human body: Some historical remarks'. In Henk A.M.J. ten Have and Jos V.M. Welie (eds.). *Ownership of the human body. Philosophical considerations on the use of the human body and its parts in healthcare*. Dordrecht/Boston/London: Kluwer Academic Press, 1998: 67–79.

Gulp, Kristine A. *Vulnerability and glory. A theological account*. Louisville, KY: Westminster John Knox Press, 2010.

Gunaratnam, Yasmin. 'From competence to vulnerability: Care, ethics, and elders from racialized minorities'. *Mortality* 13, no. 1 (2008): 24–41.

Gutlove, Paula, and Gordon Thompson. 'Human security: Expanding the scope of public health'. *Medicine, Conflict & Survival* 19, no. 1 (2003): 17–34.

Hamington, Maurice and Dorothy C. Miller (eds.). *Socializing care. Feminist ethics and public issues*. Lanham: Rowman and Littlefield Publishers, 2006.

Hand, Sean (ed.). *The Levinas reader*. Oxford, UK and Malden, MA: Blackwell Publishers, 1989.

Hankivsky, Olena. *Social policy and the ethic of care*. Vancouver and Toronto: UBC Press, 2004.

Harrison, Paul. 'Corporeal remains: vulnerability, proximity, and living on after the end of the world'. *Environment and Planning* 40, no. 2 (2008): 423–445.

Harvey, David. *A brief history of neoliberalism*. Oxford and New York: Oxford University Press, 2005.

Haugen, Hans Morten. 'Inclusive and relevant language: the use of the concepts of autonomy, dignity and vulnerability in different contexts'. *Medicine, Health Care and Philosophy* 13, no. 3 (2010): 203–213.

Hawkins, Jennifer S. and Ezekiel J. Emanuel (eds.). *Exploitation and developing countries. The ethics of clinical research*. Princeton, NJ and Oxford: Princeton University Press, 2008.

Held, Virginia. *The ethics of care. Personal, political, and global*. Oxford and New York: Oxford University Press, 2006.

Henderson, Virginia A. *The nature of nursing. A definition and its implications for practice, research, and education. Reflections after 25 years*. New York: National League for Nursing Press, 1991.

Henderson, Gail E., Arlene M. Davis, and Nancy M.P. King. 'Vulnerability to influence: A two-way street'. *American Journal of Bioethics* 4, no. 3 (2004): 50–52.

Hibou, Béatrice. *La bureaucratisation du monde à l'ère néolibérale*. Paris: Éditions La Découverte, 2012.

Higgins, Jenny A, Susie Hoffman, and Shari L. Dworkin. 'Rethinking gender, heterosexual men, and women's vulnerability to HIV/AIDS'. *American Journal of Public Health* 100, no. 3 (2010): 435–445.

Higgins, Michael. 'Toward an ethical economy'. Dublin City University, 11 September 2013 (http://www.president.ie/speeches/toward-an-ethical-economy-michael-d-higgins-dublin-city-university-11th-september-2013/). (Accessed 18 September 2015).

Hill, John Lawrence. 'Exploitation'. *Cornell Law Review* 79 (1994): 631–699.

Hill, Marianne T. 'Development as empowerment'. *Feminist Economics* 9, no. 2–3 (2003): 117–135.

Hinkelammert, Franz. 'Globalization as cover-up: An ideology to disguise and justify current wrongs'. In Jon Sobrino and Felix Wilfred (eds.). *Globalization and its victims*. London: SCM Press, 2001: 25–34.

Hochschild, Arlie Russell. 'Global care chains and emotional surplus value'. In Will Hutton and Anthony Giddens (eds.). *Global capitalism*. New York: The New Press, 2000: 130–146.

Hoffmaster, C. Barry. 'What does vulnerability mean?'. *Hastings Center Report* 36, no. 2 (2006): 38–45.

Holmstrom, Nancy. 'Exploitation'. *Canadian Journal of Philosophy* 7, no. 2 (1977): 353–369.

Holmstrom, Nancy. 'Exploitation'. In Hugh LaFollette (ed.). *The international encyclopedia of ethics*. Oxford: Blackwell Publishing, 2013: 1860–1870.

Homeland Security Council. *National strategy of pandemic influenza implementation plan: one year summary*. July 2007. (http://www.flu.gov/planning-preparedness/federal/pandemic-influenza-oneyear.pdf). (Accessed 12 March 2015).

Homer, *The Iliad*, Book IV: 606–607, p. 103 (E-book edition 2005) (http://www.gutenberg.org/files/16452/16452-h/16452-h.htm#page_083) (Accessed 4 April 2015).

Howard-Jones, Norman. 'Human experimentation in historical and ethical perspectives'. *Social Science and Medicine* 16 (1982): 1429–1448.

Hughes, Cheryl. 'The primacy of ethics: Hobbes and Levinas'. *Continental Philosophy Review* 31 (1998): 79–94.

Hurst, Samia. 'Vulnerability in research and health care; describing the elephant in the room?' *Bioethics* 22, no. 4 (2008): 191–202.

Hutchins, Sonja S., Benedict I. Truman, Toby L. Merlin and Stephen C. Redd. 'Protecting vulnerable populations from pandemic influenza in the United States: A strategic imperative'. *American Journal of Public Health* 99, no. S2 (2009): S243–S248.

Hyder, A.A., S.A. Wali, A.N. Khan, N.B. Teoh, N.E. Kass and L. Dawson. 'Ethical review of health research: A perspective from developing country researchers'. *Journal of Medical Ethics* 30, no. 1 (2004): 68–72.

IBC (International Bioethics Committee). *Report of the IBC on the principle of respect for human vulnerability and personal integrity*. Paris: UNESCO, 2011. (http://unesdoc.unesco.org/images/0018/001895/189591e.pdf). (Accessed 21 February 2015).

Ignatieff, Michael. *Human rights as politics and idolatry*. Princeton, NJ: Princeton University Press, 2001.

Ikuenobe, Polycarp. *Philosophical perspectives on communalism and morality in African traditions*. Lanham: Lexington Books, 2006.

Iltis, Ana S. 'Introduction: Vulnerability in biomedical research'. *Journal of Law, Medicine & Ethics* 37, no. 1 (2009): 6–11.

Iltis, Ana S., Anji Wall, Jason Lesandrini, Erica K. Rangel and John T. Chibnall. 'Federal interpretation and enforcement of protections for vulnerable participants in human research'. *Journal of Empirical Research on Human Research Ethics* 4, no. 1 (2009): 37–41.

INSERM. *Improving public health responses to extreme weather/heat waves – EuroHEAT*. Meeting report. Bonn, Germany, March 2007. (http://www.euro.who.int/__data/assets/pdf_file/0018/112473/E91350.pdf). (Accessed 16 March 2015).

Jackson, Ben. 'The conceptual history of social justice'. *Political Studies Review* 3 (2005): 356–373.

Jaggar, Alison M. 'Vulnerable women and neo-liberal globalization: debt burdens undermine women's health in the global South'. *Theoretical Medicine* 23, no. 6 (2002): 425–440.

Jaggar, Alison M. 'Reasoning about well-being: Nussbaum's methods of justifying the capabilities'. *Journal of Political Philosophy* 14, no. 3 (2006): 301–322.

Janssen, Marco A., Michael L. Schoon, Weimao Ke, and Katy Börner. 'Scholarly networks on resilience, vulnerability, and adaptation within the human dimensions of global environmental change'. *Global Environmental Change* 16 (2006): 240–252.

Janssen, Marco A. and Elinor Ostrom. 'Resilience, vulnerability, and adaptation: a cross-cutting theme of the International Human Dimensions Programme on Global Environmental Change'. *Global Environmental Change* 16 (2006): 237–239.

Jecker, Nancy S. 'Exploiting subjects in placebo-controlled trials'. *American Journal of Bioethics* 2, no. 2 (2002): 19–20.

Jecker, Nancy S. 'Protecting the vulnerable'. *American Journal of Bioethics* 4, no. 3 (2004): 60–62.

Jecker, Nancy S. 'A broader view of justice'. *American Journal of Bioethics* 8, no. 10 (2008): 2–10.

Jegede, Ayodele S. 'Understanding informed consent for participation in international health research'. *Developing World Bioethics* 9, no. 2 (2009): 81–87.

Jenkins, Fiona. 'Towards a nonviolent ethics: Response to Catherine Mills'. *Differences: A Journal of Feminist Cultural Studies* 18, no. 2 (2007): 157–179.

Johnson, Jeannette L. and Shelly A. Wiechelt. 'Introduction to the special issue on resilience'. *Substance Use & Misuse* 39, no. 5 (2004): 657–670.

Johnstone, Megan-Jane. 'Ethics and human vulnerability'. *Australian Nursing Journal* 16, no. 10 (2009): 25.

Jonsen, Albert R. *The birth of bioethics*. New York and Oxford: Oxford University Press, 1998.

Jonsen, Albert R. 'Foreword'. In Edwin R. DuBose, Ronald P. Hamel, Laurence J. O'Connell (eds): *A. matter of principles? Ferment in U.S. bioethics*. Valley Forge, PA: Trinity Press International, 1994: ix–xvii.

Jotkowitz, Alan. 'Vulnerability from a global medicine perspective'. *American Journal of Bioethics* 4, no. 3 (2004): 62–63.

Justo, Luis. 'Participatory research: A way to reduce vulnerability'. *American Journal of Bioethics* 4, no. 3 (2004): 67–68.

Khan, Kausar S. 'Epidemiology and ethics: The perspective of the Third World'. In Zbigniew Bankowski, Jack J. Bryant and John Last (eds.). *Ethics and epidemiology: International guidelines. Proceedings of the XXVth CIOMS Conference*. Geneva: CIOMS, 1991: 70–75.

Kahn, Kausar and Bryant, John H. 'The vulnerable in developed and developing countries – A conceptual approach'. In Zbigniew Bankowski and John H. Bryant (eds.). *Poverty,*

Vulnerability, the value of human life, and the emergence of bioethics. Geneva: CIOMS, 1994: 57–63.

Khan, Ali S, David L. Swerdlow, and Dennis D. Juranek. 'Precautions against biological and chemical terrorism directed at food and water supplies'. *Public Health Reports* 116 (2001): 3–14.

Kaldor, Mary. *Human security. Reflections on globalization and intervention.* Cambridge, UK: Polity Press, 2007.

Karlsen, Jan Reinert, Jan Helge Solbakk and Søren Holm. 'Ethical endgames: Broad consent for narrow interests; open consent for closed mind'. *Cambridge Quarterly of Healthcare Ethics* 20 (2011): 572–583.

Keck, Margaret E. and Kathryn Sikkink. *Activists beyond borders. Advocacy networks in international politics.* Ithaca, NY and London: Cornell University Press, 1998.

Kemp, Peter. 'Four ethical principles in biolaw'. In Peter Kemp, Jacob Rendtorff and Niels Mattson Johansen (eds.). *Bioethics and biolaw. Vol. II: Four ethical principles.* Copenhagen: Rhodos International Science and Art Publishers & Centre for Ethics and Law, 2000: 13–22.

Kemp, Peter, Jacob Rendtorff and Niels Mattson Johansen (eds.). *Bioethics and biolaw. Vol. II: Four ethical principles.* Copenhagen: Rhodos International Science and Art Publishers & Centre for Ethics and Law, 2000.

King, Gary, and Christopher Murray. 'Rethinking human security'. *Political Science Quarterly* 116, no. 4 (2001–2002): 585–610.

Kipnis, Kenneth. 'Vulnerability in research subjects: A bioethical taxonomy'. In *Ethical and Policy Issues in Research Involving Human Research Participants.* Bethesda, MD: National Bioethics Advisory Commission, 2001: G1–G13.

Kipnis, Kenneth. 'Seven vulnerabilities in the pediatric research subject'. *Theoretical Medicine and Bioethics* 24 (2003): 107–120.

Kipnis, Kenneth. 'Vulnerability in research subjects. An analytic approach'. In David C. Thomasma and David N. Weisstub (eds.). *The variables of moral capacity.* Dordrecht: Kluwer Academic Publishers, 2004: 217–231.

Kipnis, Kenneth. 'The limitations of "limitations"'. *American Journal of Bioethics* 4, no. 3 (2004): 70–72.

Kirby, Peadar. *Vulnerability and violence. The impact of globalisation.* London and Ann Arbor, MI: Pluto Press, 2006.

Kirby, Peadar. 'Vulnerability and globalization. The social impact of vulnerability'. In Bryan S. Turner (ed.). *Handbook of globalization studies.* London: Routledge, 2010: 113–134.

Kittay, Eva Feder. *Love's labor. Essays on women, equality, and dependency.* New York and London: Routledge, 1999.

Kittay, Eva Feder, Bruce Jennings and Angela A. Wasunna. 'Dependency, difference and the global ethic of longterm care'. *Journal of Political Philosophy* 13, no. 4 (2005): 443–469.

Kittay, Eva Feder. 'The global heart transplant and caring across national boundaries'. *Southern Journal of Philosophy* 46 (2008): 138–165.

Kittay, Eva Feder. 'The ethics of care, dependence, and disability'. *Ratio Juris. An International Journal of Jurisprudence and Philosophy of Law* 24, no. 1 (2011): 49–58.

Kleinman, Arthur and Sjaak van der Geest. '"Care" in health care. Remaking the moral world of medicine'. *Medische Antropologie* 21, no. 1 (2009): 159–168.

Knowles, Lori P. 'The lingua franca of human rights and the rise of a global bioethic'. *Cambridge Quarterly of Healthcare Ethics* 10 (2001): 253–263.

Koggel, Christine and Joan Orme. 'Care ethics: New theories and applications'. *Ethics and Social Welfare* 4, no. 2 (2010): 109–114.

Kompanje, Erwin J.O. 'Strong words, but still a step back for researchers in emergency and critical care research? The proposed revision of the Declaration of Helsinki'. *Intensive Care Medicine* 39 (2013): 1469–1470.

Koopman, Nico. 'Vulnerable Church in a vulnerable world? Towards an ecclesiology of vulnerability'. *Journal of Reformed Theology* 2 (2008): 240–254.

Kopelman, Loretta M. 'Research policy: II. Risk and vulnerable groups'. In Stephen G. Post (ed.). *Encyclopedia of Bioethics*. Third Edition, volume 4. New York: Macmillan Reference USA, 2004: 2365–2372.

Korsgaard, Christine M. *The sources of normativity*. Cambridge, UK: Cambridge University Press, 1996.

Kottow, Michael H. 'The vulnerable and the susceptible'. *Bioethics* 17, no. 5–6 (2003): 460–471.

Kottow, Michael H. 'Vulnerability: What kind of principle is it?' *Medicine, Health Care and Philosophy* 7 (2004): 281–287.

Kukathas, Chandran. 'The mirage of global justice'. *Social Philosophy and Policy* 23, no. 1 (2006): 1–28.

Kuntz, J.R. 'A litmus test for exploitation: James Stacey Taylor's *Stakes and Kidneys*'. *Journal of Medicine and Philosophy* 34 (2009): 552–572.

Labonté, Ronald and Ted Schrecker (eds.). *Towards health-equitable globalization: Rights, regulation, and redistribution. Final report to the Commission on Social Determinants of Health*. Ottawa: Institute of Population Health, 2007 (http://www.who.int/social_determinants/resources/gkn_report_06_2007.pdf) (Accessed 2 August 2015).

Lawson, Victoria. 'Geographies of care and responsibility'. *Annals of the Association of American Geographers* 97, no. 1 (2007): 1–11.

Lear, Jonathan. *Radical hope. Ethics in the face of cultural devastation*. Cambridge, MA and London, UK: Harvard University Press, 2006.

Leavitt, Frank J. 'Is any medical research population not vulnerable?' *Cambridge Quarterly of Healthcare Ethics* 15 (2006): 81–88.

Leder, Drew. *The absent body*. Chicago and London: The University of Chicago Press, 1990.

Leder, Drew (ed.). *The body in medical thought and practice*. Dordrecht, Boston, London: Kluwer Academic Publishers, 1992.

Lemke, Thomas. *Biopolitics. An advanced introduction*. New York and London: New York University Press, 2011.

Léonard, Jacques. *La medicine entre les pouvoirs et les savoirs. Histoire intellectuelle et politique de la médecine française aux XIXe siècle*. Paris: Aubier Montaigne, 1981.

Levinas, Emmanuel. *Ethics and infinity. Conversations with Philippe Nemo*. Pittsburgh, PA: Duquesne University Press, 1985.

Levinas, Emmanuel. *Otherwise than being or beyond essence*. Pittsburgh, PA: Duquesne University Press, 1998 (2009, Eighth printing).

Levinas, Emmanuel. *Humanism of the Other*. Urbana and Chicago: University of Illinois Press, 2003.

Levinas, Emmanuel. 'Ethics of the Infinite'. In Richard Kearney (ed.). *Debates in continental philosophy. Conversations with contemporary thinkers*. New York: Fordham University Press, 2004: 65–84.

Levine, Carol. 'The concept of vulnerability in disaster research'. *Journal of Traumatic Stress* 17, no. 5 (2004): 395–402.

Levine, Carol, Ruth Faden, Christine Grady, Dale Hammerschmidt, Lisa Eckenwiler and Jeremy Sugarman. '"Special scrutiny": A targeted form of research protocol review'. *Annals of Internal Medicine* 140 (2004): 220–223.

Levine, Carol, Ruth Faden, Christine Grady, Dale Hammerschmidt, Lisa Eckenwiler and Jeremy Sugarman, 'The limitations of "vulnerability" as a protection for human research participants'. *American Journal of Bioethics* 4, no. 3 (2004): 44–49.

Levine, Robert J and Karen Lebacqz. 'Some ethical considerations in clinical trials'. *Clinical Pharmacology and Therapeutics* 25, no. 5, part 2 (1979): 728–741.

Levine, Robert J. *Ethics and regulation of clinical research*. Baltimore-Münich: Urban & Schwarzenberg, 1981.

Levine, Robert J. 'Informed consent: Some challenges to the universal validity of the Western model'. In Zbigniew Bankowski, Jack J. Bryant and John Last (eds.). *Ethics and epidemiology: International guidelines. Proceedings of the XXVth CIOMS Conference*. Geneva: CIOMS, 1991: 47–58.

Levine, Robert J. 'International codes and guidelines for research ethics: A critical appraisal'. In Harold Y. Vanderpool (ed.). *The ethics of research involving human subjects: Facing the twenty-first century*. Frederick, MD: University Publishing Group, 1996: 235–259.

Lewis, James. 'Some perspectives on natural disaster vulnerability in Tonga'. *Pacific Viewpoint* 22, no. 2 (1981): 145–162.

Lewis, James. 'The vulnerability of small island states to sea level rise: The need for holistic strategies'. *Disasters* 14, no. 3 (1990): 241–249.

Lloyd-Sherlock, Peter. 'Nussbaum, capabilities and older people'. *Journal of International Development* 14 (2002): 1163–1173.

Logar, Tea. 'Exploitation as wrongful use: Beyond taking advantage of vulnerabilities'. *Acta Analytica* 25 (2010): 329–346.

London, Alex John. 'Justice and the human development approach to international research'. *Hastings Center Report* 35, no. 1 (2005): 24–37.

London, Leslie. 'Ethical oversight of public health research: Can rules and IRBs make a difference in developing countries?', *American Journal of Public Health* 92, no. 7 (2002): 1079–1084.

Lott, Jason P. 'Module three: Vulnerable/special participant populations'. *Developing World Bioethics* 5, no. 1 (2005): 30–54.

Luna, Florencia. 'Poverty and inequality: Challenges for the IAB: IAB Presidential Address'. *Bioethics* 19, no. 5–6 (2005): 451–459.

Luna, Florencia. *Bioethics and vulnerability. A Latin American view*. Amsterdam and New York: Rodopi, 2006.

Luna, Florencia. 'Elucidating the concept of vulnerability: Layers not labels.' *International Journal of Feminist Approaches to Bioethics* 2, no. 1 (2009): 121–139.

Luthar, Suniya S., Dante Cicchetti, and Bronwyn Becker. 'The construct of resilience: a critical evaluation and guidelines for future work'. *Child Development* 71, no. 3 (2000): 543–562.

Luxton, Meg. 'Doing neoliberalism: Perverse individualism in personal life'. In Susan Braedley and Meg Luxton (eds.). *Neoliberalism and everyday life*. Montreal and Kingston: McGill-Queen's University Press, 2010: 163–183.

Lynch, Kathleen, John Baker and Maureen Lyons. *Affective equality. Love, care, and injustice*. Houndmills, UK: Palgrave Macmillan, 2009.

Macfarlane, Sarah, Mary Racelis and Florence Muli-Musiime. 'Public health in developing countries'. *Lancet* 356 (2000): 841–846.

MacIntyre, Alasdair. *Dependent rational animals. Why human beings need the virtues*. Chicago and La Salle, IL: Open Court Publishing Company, 1999.

Mackenzie, Cathriona, Wendy Rogers and Susan Dodds (eds.). *Vulnerability. New essays in ethics and feminist philosophy*. Oxford and New York: Oxford University Press, 2014.

Macklin, Ruth. 'Bioethics, vulnerability, and protection'. *Bioethics* 17, no. 5–6 (2003): 472–486.

Macklin, Ruth. 'A global ethics approach to vulnerability'. *International Journal of Feminist Approaches to Bioethics* 5, no. 2 (2012): 64–81.

Macrae, Duncan J. 'The Council for International Organizations and Medical Sciences (CIOMS) Guidelines on ethics of clinical trials'. *Proceedings of the American Thoracic Society* 4 (2007): 176–179.

Mahon, Rianne and Fiona Robinson (eds.). *Feminist ethics and social policy. Towards a new global political economy of care*. Vancouver and Toronto: UBC Press, 2011.

Maillard, Nathalie. *La vulnérabilité. Une nouvelle catégorie morale?* Geneva: Editions Labor et Fides, 2011.

Malterud, Kirsti and Hanne Hollnagel. 'The doctor who cried: A qualitative study about the doctor's vulnerability'. *Annals of Family Medicine* 3, no. 4 (2005): 348–352.

Malterud, Kirsti, Lise Fredriksen, and Mette Haukaas Gjerde. 'When doctors experience their vulnerability as beneficial for the patients'. *Scandinavian Journal of Primary Health Care* 27 (2009): 85–90.

Mann, Jonathan M., Lawrence Gostin, Sofia Gruskin, Troyen Brennan, Zita Lazzarini and Harvey V. Fineberg. 'Health and human rights'. *Health and Human Rights* 1, no. 1 (1994): 6–23.

Mann, Jonathan M. and Daniel Tarantola. 'From vulnerability to human rights'. In Jonathan M. Mann and Daniel Tarantola (eds.). *AIDS in the World II.* New York: Oxford University Press, 1996: 463–467.

Mann, Jonathan M. 'Medicine and public health, ethics, and human rights'. *Hastings Center Report* 27, no. 3 (1997): 6–13.

Mann, Jonathan, M. Sofia Gruskin, Michael A. Grodin and George J. Annas (eds.). *Health and human rights.* New York and London: Routledge, 1999.

Mann, Jonathan M. 'Human rights and AIDS: The future of the pandemic'. In Mann, Jonathan, M. Sofia Gruskin, Michael A. Grodin and George J. Annas (eds.). *Health and human rights.* New York and London: Routledge, 1999: 216–226.

Marcum, James A. *An introductory philosophy of medicine. Humanizing modern medicine.* Dordrecht: Springer Science + Business Media, 2008.

Marmot, Michael G. 'Social differentials in health within and between populations'. *Daedalus* 123, no. 4 (1994): 197–216.

Marmot, Michael and Richard G. Wilkinson (eds.). *Social determinants of health.* Second Edition. Oxford and New York: Oxford University Press, 2006.

Martin, Adrienne M. 'Hope and exploitation'. *Hastings Center Report* 38, no. 5 (2008): 49–55.

Mayer, Robert. 'A Walzerian theory of exploitation'. *Polity* 34, no. 3 (2002): 337–354.

Mayer, Robert. 'Sweatshops, exploitation, and moral responsibility'. *Journal of Social Philosophy* 38, no. 4 (2007): 605–619.

Mayer, Robert. 'What's wrong with exploitation?' *Journal of Applied Philosophy* 24, no. 2 (2007): 137–150.

McCall, Grant. 'Clearing confusion in a disembedded world: The case for nissology'. *Geographische Zeitschrift* 84, no. 2 (1996): 74–85.

McEntire, David. 'Understanding and reducing vulnerability: from the approach of liabilities and capabilities'. *Disaster Prevention and Management* 20, no. 3 (2011): 294–313.

McGilloway, F.A. 'Dependency and vulnerability in the nurse/patient situation'. *Journal of Advanced Nursing* 1 (1976): 229–236.

McInnis, Colin and Kelley Lee. *Global health and international relations.* Cambridge, UK: Polity, 2012.

McLaughlin, Paul. 'The ethics of exploitation'. *Studia Philosophica Estonica* 1, no. 3 (2008): 5–16.

McRobbie, Angela. 'Vulnerability, violence and (cosmopolitan) ethics: Butler's *Precarious Life*'. *British Journal of Sociology* 57, no. 1 (2006): 69–86.

Mechanic, David and Jennifer Tanner, 'Vulnerable people, groups, and populations: Societal view'. *Health Affairs* 26, no. 5 (2007): 1220–1230.

Memorandum. 'Concerns around the human papilloma virus (HPV) vaccine'. *Indian Journal of Medical Ethics* 7, no. 1 (2010): 38–41.

Merleau-Ponty, Maurice. *Phenomenology of perception.* London and Henley: Routledge and Kegan Paul, 1962.

Miller, David. *Social justice.* Oxford: Clarendon Press, 1976.

Miller, David. *Principles of social justice.* Cambridge, MA and London, UK: Harvard University Press, 2001.

Miller, Sarah Clark. 'Cosmopolitan care.' *Ethics and Social Welfare* 4, no. 2 (2010): 145–157.

Mills, Catherine. 'Normative violence, vulnerability, and responsibility'. *Differences: A Journal of Feminist Cultural Studies* 18, no. 2 (2007): 133–156.

Mimura, Nobua. 'Vulnerability of island countries in the South Pacific to sea level rise and climate change'. *Climate Research* 12 (1999): 137–143.

Mitra, Sophie. 'The capability approach and disability'. *Journal of Disability Policy Studies* 16, no. 4 (2006): 236–247.

Moazzam, Farhat, Riffat Moazam Zaman, and Aamir M. Jafarey. 'Conversations with kidney vendors in Pakistan. An ethnographic study'. *Hastings Center Report* 39, no. 3 (2009): 29–44.

Morawa, Alexander. 'Vulnerability as a concept in international human rights law'. *Journal of International Relations and Development* 6, no. 2 (2003): 139–155.

Morin, Edgar. *Introduction à la pensée complexe*. Paris: Éditions du Seuil, 2005.

Morris, Kelly. 'Revising the Declaration of Helsinki'. *The Lancet* 381 (2013): 1532–1533.

Mortreux, Colette and Jon Barnett. 'Climate change, migration and adaptation in Funafuti, Tuvalu'. *Global Environmental Change* 19 (2009): 105–112.

Moulin, Madeleine. 'La vulnérabilité entre sciences et solidarité'. In Jacob Dahl Rendtorff and Peter Kemp (eds). *Basic ethical principles in European bioethics and biolaw. Volume II. Partners' research*. Copenhagen and Barcelona: Centre for Ethics and Law, and Institut Borja de Bioetica, 2000: 195–201.

Muller, Jerry Z. 'Capitalism and inequality: What the Right and the Left get wrong'. *Foreign Affairs* 92, no. 2 (2013): 30–40, 42–51.

Murphy, Ann V. 'Corporeal vulnerability and the new humanism'. *Hypatia* 26, no. 3 (2011): 575–590.

Murray, Karen Bridget. 'Do not disturb: "vulnerable populations" in federal government policy discourses and practices'. *Canadian Journal of Urban Research* 13, no. 1 (2004): 50–69.

NaRanong, Anchana and Viroj NaRanong. 'The effects of medical tourism: Thailand's experience'. *Bulletin of the World Health Organization* 89, no. 5 (2011): 336–344.

Nathanson, Vivienne. 'Revising the Declaration of Helsinki. Your chance to influence research governance'. *British Medical Journal* 346 (2013): f2837.

National Bioethics Advisory Commission, *Ethical and policy issues in research involving human participants. Volume I: Report and recommendations of the National Bioethics Advisory Commission*. Bethesda, MD, August 2001.

National Coalition for the Homeless. *How many people experience homelessness?* Washington, D.C.: National Coalition for the Homeless, 2009 (http://www.nationalhomeless.org/factsheets/How_Many.html) (Accessed 22 March 2015).

National Commission for the Protection of Human Subjects of Biomedical and Behavioral Research. 'The Belmont Report: Ethical principles and guidelines for the protection of human subjects of research'. *Federal Register* 44(76) (1979): 23191–7.

National Commission for the Protection of Human Subjects of Biomedical and Behavioral Research. *The Belmont Report. Appendix Volume I & II*. Washington, D.C.: U.S. Government Printing Office, DHEW Publication No.(OS) 78-0013, 1978.

Navarro, Vicente. 'Development and quality of life: A critique of Amartya's Sen's *Development as freedom*'. In Vicente Navarro (ed.). *The political economy of social inequalities. Consequences for health and quality of life*. New York: Baywood Publishing Company, 2002: 461–474.

Nef, Jorge. *Human security and mutual vulnerability. The global political economy of development and underdevelopment*. (Second Edition) Ottawa: IDRC Books, 1999.

Newman, Edward. 'Human security and constructivism'. *International Studies Perspectives* 2 (2001): 239–251.

Nichiata, Lucia, Maria Bertolozzi, Renata Takahashi, and Lislaine Fracolli. 'The use of the "vulnerability" concept in the nursing area'. *Revista Latino-Americana de Enfermagem* 16, no. 5 (2008): 923–8 (http://www.scielo.br/pdf/rlae/v16n5/20.pdf) (Accessed 8 April 2015).

Nicholson, Richard. 'Who is vulnerable in clinical research?' *Bulletin of Medical Ethics* no. 181 (2002): 19–24.

Nickel, Philip J. 'Vulnerable populations in research: The case of the seriously ill'. *Theoretical Medicine and Bioethics* 27 (2006): 245–264.

Nnaemeka, Obioma. 'Feminist bioethics and global responsibility: Exploring health care delivery in Kenya'. *International Journal of Feminist Approaches to Bioethics* 2, no. 1 (2009): 71–76.

Nortvedt, Per. 'Subjectivity and vulnerability: reflections on the foundation of ethical sensibility'. *Nursing Philosophy* 4 (2003): 222–230.

Nussbaum, Martha C. *Women and human development. The capabilities approach.* New York: Cambridge University Press, 2000.

Nussbaum, Martha C. 'Capabilities as fundamental entitlements: Sen and social justice'. *Feminist Economics* 9, no. 2–3 (2003): 33–59.

Nussbaum, Martha C. 'Women's bodies: Violence, security, capabilities'. *Journal of Human Development* 6, no. 2 (2005): 168–183.

Nussbaum, Martha C. *Frontiers of justice. Disability, nationality, species membership.* Cambridge, MA and London, UK: The Belknap Press of Harvard University Press, 2006.

Nussbaum, Martha C. *Creating capabilities. The human development approach.* Cambridge, MA and London, UK: The Belknap Press of Harvard University Press, 2011.

O'Keefe, Phil, Ken Westgate, and Ben Wisner. 'Taking the naturalness out of natural disasters'. *Nature* 260, no. 5552 (1976): 566–567.

O'Neill, Onora. 'Justice, gender and international boundaries'. *British Journal of Political Science* 20, no. 4 (1990): 439–459.

O'Neill, Onora. *Towards justice and virtue. A constructive account of practical reasoning.* Cambridge, UK: Cambridge University Press, 1996.

O'Neill, Onora. *Bounds of justice.* Cambridge, UK: Cambridge University Press, 2000.

Outka, Gene. 'Social justice and equal access to health care'. *Journal of Religious Ethics* 2, no. 1 (1974): 11–32.

Parker, Richard. 'Sexuality, culture, and power in HIV/AIDS research'. *Annual Review of Anthropology* 30 (2001): 163–179.

Parks, Bradley and J. Timmons Roberts. 'Globalization, vulnerability to climate change, and perceived injustice'. *Society and Natural Resources* 19 (2006): 337–355.

Patel, Samir S. 'A sinking feeling'. *Nature* 440, no. 7085 (2006): 734–736.

Paton, Kathryn, and Peggy Fairbairn-Dunlop. 'Listening to local voices: Tuvaluans respond to climate change'. *Local Environment* 15, no. 7 (2010): 687–698.

Patrão Neves, Maria. 'The new vulnerabilities raised by biomedical research'. In M. Hayri, T. Takala and P. Herrisone-Kelly (eds.). *Ethics in biomedical research.* Amsterdam: Rodopi, 2007: 181–192.

Patrão Neves, Maria. 'Article 8: Respect for human vulnerability and personal integrity'. In Henk ten Have and Michèle S. Jean (eds.). *The UNESCO Universal Declaration on Bioethics and Human Rights. Background, principles and application.* Paris: UNESCO Publishing, 2009: 155–164.

Peck, Jamie. *Constructions of neoliberal reason.* Oxford: Oxford University Press, 2010.

Pellegrino, Edmund D. *Humanism and the physician.* Knoxville, TN: The University of Tennessee Press, 1979.

Pellegrino, Edmund D. and David C. Thomasma. *A philosophical basis of medical practice. Toward a philosophy and ethics of the healing professions.* New York and Oxford: Oxford University Press, 1981.

Pellegrino, Edmund D and David C. Thomasma. *The virtues in medical practice.* New York and Oxford: Oxford University Press, 1993.

Pelluchon, Corine. *La raison du sensible. Entretiens autour de la bioéthique.* Perpignan: Artège, 2009.

Pelluchon, Corine. *L'autonomie brisée. Bioéthique et philosophie.* Paris: PUF, 2009.

Pelluchon, Corine. *Éléments pour une éthique de la vulnérabilité. Les hommes, les animaux, la nature.* Paris: Les Éditions du Cerf, 2011.

Peperzak, Adriaan. *To the other. An introduction to the philosophy of Emmanuel Levinas.* West Lafayette, IN: Purdue University Press, 1993.

Perez, Martine. 'Canicule 2003: 70 000 morts en Europe', *Le Figaro*, 15 October 2007. http://www.lefigaro.fr/sciences/2007/03/23/01008-20070323ARTFIG90028-canicule_morts_en_europe.php. (Accessed 4 April 2015)

Pettersen, Tove. *Comprehending care. Problems and possibilities in the ethics of care.* Lanham, MD: Lexington Books, 2008.

Pettersen, Tove. 'The ethics of care: Normative structures and empirical implications'. *Health Care Analysis* 19 (2011): 51–64.

Petryna, Adriana. *When experiments travel. Clinical trials and the global search for human subjects.* Princeton and Oxford: Princeton University Press, 2009.

Phillips, Trisha. 'Exploitation in payments to research subjects'. *Bioethics* 25, no. 4 (2011): 209–219.

Pinto, Andrew D. and Ross E.G. Upshur (eds). *An introduction to global health ethics.* London and New York: Routledge, 2013.

Pinto, Paula. 'Beyond the state: The making of disability and gender under neoliberalism in Portugal'. In Susan Braedley and Meg Luxton (eds.). *Neoliberalism and everyday life.* Montreal and Kingston: McGill-Queen's University Press, 2010: 113–135.

Placher, William C. *Narratives of a vulnerable God. Christ, theology, and scripture.* Louisville, KY: Westminster John Knox Press, 1994.

Plügge, Herbert. *Wohlbefinden und Missbefinden. Beiträge zu einer Medizinischen Anthropologie.* Tübingen: Max Niemeyer Verlag, 1962.

Plügge, Herbert. *Der Mensch und sein Leib.* Tübingen: Max Niemeyer Verlag, 1967.

Plügge, Herbert. 'Man and his body'. In Stuart F. Spicker (ed). *The philosophy of the body. Rejections of Cartesian dualism.* Chicago: Quadrangle Books, 1970: 293–311.

Pogge, Thomas. 'The international significance of human rights'. *The Journal of Ethics* 4, no. 1–2 (2000): 45–69.

Pogge, Thomas. 'Testing our drugs on the poor abroad'. In Jennifer S. Hawkins and Ezekiel J. Emanuel (eds.). *Exploitation and developing countries. The ethics of clinical research.* Princeton and Oxford: Princeton University Press, 2008: 105–141.

Pogge, Thomas. *World poverty and human rights. Cosmopolitan responsibilities and reforms.* Second Edition. Cambridge, UK and Malden, MA: Polity Press, 2008.

Pollock, Elizabeth. *Tuvalu: That sinking feeling,* PBS Frontline, 6 December 2005. http://www.pbs.org/frontlineworld/rough/2005/12/tuvalu_that_sin_1.html.

Potter, Van Rensselaer. *Global bioethics. Building on the Leopold legacy.* East Lansing: Michigan State University Press, 1988.

Powers, Madison and Ruth Faden. *Social justice. The moral foundations of public health and health policy.* Oxford and New York: Oxford University Press, 2006.

Prah Ruger, Jennifer. *Health and social justice.* Oxford and New York: Oxford University Press, 2010.

Pratt, Bridget, Deborah Zion and Bebe Loff. 'Evaluating the capacity of theories of justice to serve as a justice framework for international clinical research'. *American Journal of Bioethics* 12, no. 11 (2012): 30–41.

Ramanathan, Mala, and Amar Jesani. 'The legacy of scandals and non-scandals in research and its lessons for bioethics in India'. *Indian Journal of Medical Ethics* 9, no. 1 (2012): 4–6.

Ramsay, L.E., Tidd, M.J., Butler, J.K., and Venning, G.R. 'Ethical review in the pharmaceutical industry'. *British Journal of Clinical Pharmacology* 4 (1977): 73–76.

Raphael, D. D. *Concepts of justice.* Oxford: Clarendon Press, 2001.

Rawlinson, Mary C. 'Women and special vulnerability: Commentary on the principle of respect for human vulnerability and personal integrity, Unesco, International Bioethics Committee report'. *International Journal of Feminist Approaches to Bioethics* 5, no. 2 (2012): 174–179.

Rawls, John. *A theory of justice.* Cambridge, MA: The Belknap Press of Harvard University Press, 1971.

Reich, Warren. 'The power of a single idea'. In Maria Patrão Neves and Manuela Lima (eds.). *Bioetica ou bioeticas na evolução das sociedades*. Coimbra: Grafica de Coimbra, 2005: 380–382.

Rendtorff, Jacob Dahl and Peter Kemp. *Basic ethical principles in European bioethics and biolaw. Volume I. Autonomy, dignity, integrity and vulnerability*. Copenhagen and Barcelona: Centre for Ethics and Law, and Institut Borja de Bioetica, 2000.

Rendtorff, Jacob. 'The second international conference about bioethics and biolaw: European principles in bioethics and biolaw. A report from the conference'. In Peter Kemp, Jacob Rendtorff and Niels Mattsson Johansen (eds.). *Bioethics and Biolaw. Volume II. Four ethical principles*. Copenhagen: Rhodos International Science and Art Publishers and Centre for Ethics and Law, 2000: 157–166.

Rendtorff, Jacob and Peter Kemp. 'Vulnérabilité (Principe de'). In Gilbert Hottois and Jean-Noël Missa (eds). *Nouvelle Encyclopédie de bioéthique, médecine, environnement, biotechnologie*. Bruxelles: De Boeck Université, 2001: 869–876.

Rendtorff, Jacob and Peter Kemp. 'Vulnérabilité (Personne)'. In Gilbert Hottois and Jean-Noël Missa (eds). *Nouvelle Encyclopédie de bioéthique, médecine, environnement, biotechnologie*. Bruxelles: De Boeck Université, 2001: 876–881.

Rendtorff, Jacob D. 'Basic ethical principles in European bioethics and biolaw: Autonomy, dignity, integrity and vulnerability – Towards a foundation of bioethics and biolaw'. *Medicine, Health Care and Philosophy* 5 (2002): 235–244.

Rendtorff, Jacob D. 'European perspectives'. In Henk ten Have and Bert Gordijn (eds.). *Handbook of Global Bioethics*. Dordrecht: Springer Publishers, 2014: 293–310.

Rennie, Stuart and Bavon Mupenda. 'Living apart together: reflections on bioethics, global inequality and social justice'. *Philosophy, Ethics, and Humanities in Medicine* 3:25 (2008): DOI: 10.1186/1747-5341-3-15.

Resnik, David B. 'Exploitation and the ethics of clinical trials'. *American Journal of Bioethics* 2, no. 2 (2002): 28–30.

Resnik, David B. 'Exploitation in biomedical research'. *Theoretical Medicine* 24 (2003): 233–259.

Resnik, David B. 'Research subjects in developing nations and vulnerability'. *American Journal of Bioethics* 4, no. 3 (2004): 63–64.

Ricoeur, Paul. *Reflections on the just*. Chicago and London: The University of Chicago Press, 2007. (Original French edition, 2001).

Robeyns, Ingrid. 'The capability approach: a theoretical survey'. *Journal of Human Development* 6, no. 1 (2005): 93–114.

Robeyns, Ingrid. 'The capability approach in practice'. *Journal of Political Philosophy* 14, no. 3 (2006): 351–376.

Robinson, Fiona. *Globalizing care: Ethics, feminist theory, and international affairs*. Boulder, CO: Westview Press, 1999.

Robinson, Fiona. 'Ethical globalization? States, corporations, and the ethics of care'. In Maurice Hamington and Dorothy C. Miller (eds.). *Socializing care. Feminist ethics and public issues*. Lanham (Md): Rowman and Littlefield Publishers, 2006: 163–181.

Robinson, Fiona. 'Care, gender and global social justice: Rethinking "ethical globalization"'. *Journal of Global Ethics* 2, no. 1 (2006): 5–25.

Robinson, Fiona. 'Care ethics and the transnationalization of care. Reflections on autonomy, hegemonic masculinities, and globalization'. In Rianne Mahon and Fiona Robinson (eds.). *Feminist ethics and social policy. Towards a new global political economy of care*. Vancouver and Toronto: UBC Press, 2011: 127–144.

Robinson, Fiona. *The ethics of care. A feminist approach to human security*. Philadelphia: Temple University Press, 2011.

Rogers, Ada C. 'Vulnerability, health and health care'. *Journal of Advanced Nursing* 26 (1997): 65–72.

Rogers, Wendy and Angela Ballantyne. 'Gender and trust in medicine: Vulnerabilities, abuses, and remedies'. *International Journal of Feminist Approaches to Bioethics* 1, no. 1 (2008): 48–66.

Rogers, Wendy, Catriona Mackenzie and Susan Dodds. 'Why bioethics needs a concept of vulnerability'. *International Journal of Feminist Approaches to Bioethics* 5, no. 2 (2012): 11–38.

Rogers, Wendy. 'Vulnerability and bioethics'. In Cathriona Mackenzie, Wendy Rogers and Susan Dodds (eds.). *Vulnerability. New essays in ethics and feminist philosophy*. Oxford and New York: Oxford University Press, 2014: 60–87.

Rolfsen, Raag. 'Vulnerability and the role of churches'. Evangelical Lutheran Church in America. *Journal of Lutheran Ethics* 4, no. 4 (2004): (http://www.elca.org/What-We-Believe/Social-Issues/Journal-of-Lutheran-Ethics/Issues/April-2004/Vulnerability-and-the-Role-of-the-Churches.aspx) (Accessed 2 August 2015).

Rose, Marion H. and Marcia Killien. 'Risk and vulnerability: a case for differentiation'. *Advances in Nursing Science* 5, no. 3 (1983): 60–73.

Rosen, George. *From medical police to social medicine: Essays on the history of health care*. New York: Science History Publications, 1974.

Rothman, David J. *Strangers at the bedside. A history of how law and bioethics transformed medical decision making*. New York: Basic Books, 1991.

Rothschild, Emma. 'What is security?', *Daedalus* 124, no. 3 (1995): 53–98.

Rougeau, Vincent D. 'Enter the poor. American welfare reform, solidarity, and the capability of human flourishing'. In Séverine Deneulin, Mathias Nebel and Nicholas Sagovsky (eds.). *Transforming unjust structures. The capability approach*. Dordrecht: Springer, 2006: 161–176.

Rudowski, Withold. 'World Health Organisation biomedical research guidelines and the conduct of clinical trials'. *Journal of Medical Ethics* 6 (1980): 58–60.

Ruof, Mary C. 'Vulnerability, vulnerable populations, and policy'. *Kennedy Institute of Ethics Journal* 14, no. 4 (2004): 411–425.

Sample, Ruth J. *Exploitation: What it is and why it's wrong*. Lanham, MD: Rowman and Littlefield, 2003.

Sandel, Michael J. *Public philosophy. Essays on morality in politics*. Cambridge, MA and London, UK: Harvard University Press, 2005.

Sartre, Jean-Paul. 'The body'. In Stuart F. Spicker (ed.). *The philosophy of the body. Rejections of Cartesian dualism*. Chicago: Quadrangle Books, 1970: 218–240.

Savard, Jacqueline. 'Personalised medicine: A critique on the future of health care'. *Bioethical Inquiry* 10 (2013): 197–203.

Schachter, Oscar. 'Human dignity as a normative concept'. *American Journal of International Law* 77, no. 4 (1983): 848–854.

Scheper-Hughes, Nancy. *Death without weeping. The violence of everyday life in Brazil*. Berkeley, Los Angeles, London: University of California Press, 1992.

Scheper-Hughes, Nancy. 'The global traffic in human organs'. *Current Anthropology* 41, no. 2 (2000): 191–224.

Scheper-Hughes, Nancy. 'Rotten trade: millennial capitalism, human values and global justice in organs trafficking'. *Journal of Human Rights* 2, no. 2 (2003): 197–226.

Schicktanz, Silke, Mark Schweda, and Brian Wynne. 'The ethics of "public understanding of ethics" – why and how bioethics expertise should include public and patients' voices'. *Medicine, Health Care and Philosophy* 15 (2012): 129–139.

Schipper, I., and F. Weyzig. *Briefing paper on ethics in clinical trials; #1: Examples of unethical trials*. Amsterdam: SOMO, 2008 (http://somo.nl/publications-en/Publication_2534?set_language=en) (Accessed 4 November 2014).

Scholle Connor, S. and H. L. Fuenzalida-Puelma (eds). *Bioethics. Issues and Perspectives*. Washington, D.C.: Pan American Health Organization, 1990: 220–226.

Scholte, Jan Aart. *Globalization. A critical introduction*. Houndmills and New York: Palgrave, 2000.

Schonfeld, Toby. 'The perils of protection: vulnerability and women in clinical research'. *Theoretical Medicine and Bioethics* 34 (2013): 189–206.

Schotsmans, Paul. 'Ownership of the body: A personalist perspective'. In Henk A.M.J. ten Have and Jos V.M. Welie (eds.). *Ownership of the human body. Philosophical considerations on the use of the human body and its parts in healthcare*. Dordrecht/Boston/London: Kluwer Academic Press, 1998: 159–172.

Schramm, Fermin and Marlène Braz. 'Bioethics of protection: A proposal for the moral problems of developing countries'? *Journal International de Bioéthique* 19, no. 1–2 (2008): 73–86.

Schroeder, Doris, and Eugenijus Gefenas. 'Vulnerability: Too vague and too broad?'. *Cambridge Quarterly of Healthcare Ethics* 18 (2009): 113–121.

Schwartz, Justin. 'What's wrong with exploitation?'. *Noûs* 29, no. 2 (1995): 158–188.

Sebastian, Juliann G. 'Homelessness: a state of vulnerability'. *Family & Community Health* 8, no. 3 (1985): 11–24.

Sellman, Derek. 'Towards an understanding of nursing as a response to human vulnerability'. *Nursing Philosophy* 6 (2005): 2–10.

Sen, Amartya. *Commodities and capabilities*. New Delhi: Oxford University Press, 2012 (1987).

Sen, Amartya. *Inequality examined*. New York and Cambridge, MA: Russell Sage Foundation and Harvard University Press, 1992.

Sen, Amartya. *Development as freedom*. New York: Anchor Books, 1999.

Sen, Amartya. 'Why health equity'? *Health Economics* 11 (2002): 659–666.

Sen, Amartya. 'Response to commentaries'. *Studies in Comparative International Development* 37, no. 2 (2002): 78–86.

Sen, Amartya. 'Human rights and capabilities'. *Journal of Human Development* 6, no. 2 (2005): 151–166.

Sen, Amartya. *The idea of justice*. Cambridge, MA: The Belknap Press of Harvard University Press, 2009.

Sen, Amartya. 'The global reach of human rights'. *Journal of Applied Philosophy* 29, no. 2 (2012): 91–100.

Sengupta, Amit. 'Fatal trials: clinical trials are killing people'. *Indian Journal of Medical Ethics* 6, no. 3 (2009): 118–119.

Sermons, M. William, and Meghan Henry. *Demographics of homelessness series: The rising elderly population.* Washington DC: Homelessness Research Institute, National Alliance to End Homelessness, 2010 (http://b.3cdn.net/naeh/9c130dfb64e7ddbdf7_88m6bnd7g.pdf). (Accessed 3 March 2015).

Sevenhuijsen, Selma. 'Caring in the third way: The relation between obligation, responsibility and care on *Third Way* discourse'. *Critical Social Policy* 20, no. 1 (2000): 5–37.

Shainess, Natalie. 'Vulnerability to violence: Masochism as process'. *American Journal of Psychotherapy* 33, no. 2 (1979): 174–189.

Shi, Leiyu. 'The convergence of vulnerable characteristics and health insurance in the US'. *Social Science & Medicine* 53 (2001): 519–529.

Shi, Leiyu and Stevens, Gregory D. *Vulnerable populations in the United States*. San Francisco: Jossey-Bass, 2005.

Shildrick, Margrit. 'Becoming vulnerable: Contagious encounters and the ethics of risk'. *Journal of Medical Humanities* 21, no. 4 (2000): 215–227.

Shildrick, Margrit. *Embodying the monster. Encounters with the vulnerable self*. London, Thousands Oaks, New Delhi: Sage Publications, 2002.

Shivas, Tricha. 'Contextualizing the vulnerability standard'. *American Journal of Bioethics* 4, no. 3 (2004): 84–86.

Silvers, Anita. 'Historical vulnerability and special scrutiny: Precautions against discrimination in medical research'. *American Journal of Bioethics* 4, no. 3 (2004): 56–57.

Singer, Peter. *One world. The ethics of globalization*. Second Edition. New Haven and London: Yale University Press, 2004.

Smit, Barry, and Johanna Wandel. 'Adaptation, adaptive capacity and vulnerability'. *Global Environmental Change* 16 (2006): 282–292.

Smith, Kristen. 'The problematization of medical tourism: A critique of neoliberalism'. *Developing World Bioethics* 12, no. 1 (2012): 1–8.

Snarr, D. Neil, and E. Leonard Brown. 'Permanent post-disaster housing in Honduras: Aspects of vulnerability to future disasters'. *Disasters* 3, no. 3 (1979): 287–292.

Snyder, Jeremy C. 'Needs exploitation'. *Ethical Theory and Moral Practice* 11 (2008): 389–405.

Snyder, Jeremy. 'Exploitation and sweatshop labor: Perspectives and issues'. *Business Ethics Quarterly* 20, no. 2 (2010): 187–213.

Snyder, Jeremy. 'Multiple forms of exploitation in international research: The need for multiple standards of fairness'. *American Journal of Bioethics* 10, no. 6 (2010): 40–41.

Sobrino, Jon and Felix Wilfred (eds.). *Globalization and its victims.* London: SCM Press, 2001.

Solbakk, Jan Helge. 'Vulnerability: A futile or useful principle in healthcare ethics?'. In Ruth Chadwick, Henk ten Have and Eric M. Meslin (eds.). *The SAGE Handbook of Health Care Ethics: Core and Emerging Issues.* London: Sage, 2011: 228–238.

Solbakk, Jan Helge and Susana Vidal. 2012. 'Research ethics, clinical'. In Ruth Chadwick (ed.). *Encyclopedia of Applied Ethics*, vol. 3 (Second Edition). San Diego, CA: Academic Press, 2012: 775–785.

Solbakk, Jan Helge. 'Bioethics on the couch'. *Cambridge Quarterly of Healthcare Ethics* 22 (2013): 319–326.

Solomon, Stephanie R. 'Protecting and respecting the vulnerable: existing regulations or further protections?' *Theoretical Medicine and Bioethics* 34 (2013): 17–28.

Spiers, Judith. 'New perspectives on vulnerability using emic and etic approaches'. *Journal of Advanced Nursing* 31, no. 3 (2000): 715–721.

Srinivasan, Sandhya. 'The clinical trial scenario in India'. *Economic and Political Weekly*, August 29-September 4 (2009): 29–33.

Srinivasan, Sandhya. 'HPV vaccine trials and sleeping watchdogs'. *Indian Journal of Medical Ethics* 8, no. 2 (2011): 73–74;

Staudigl, Michael. 'The vulnerable body: Towards a phenomenological theory of violence'. In Anna-Teresa Tymienicka (ed.): *Logos of phenomenology and phenomenology of the logos. Book two. The human condition in-the-unity-of-everything-there-is-alive. Individuation, self, person, self-determination, freedom, necessity.* Dordrecht: Springer, 2006: 259–272.

Steady, Filomena Chiona. 'African women, industrialization and another development. A global perspective'. *Development Dialogue* 1–2 (1982): 51–64.

Steger, Manfred B. *Globalism: The new market ideology.* First Edition. Lanham, MD: Rowman and Littlefield Publishers, 2002.

Steger, Manfred B. *Globalism. Market ideology meets terrorism.* Second Edition. Lanham, MD: Rowman and Littlefield Publishers, 2005.

Steger, Manfred B. *Globalisms. The great ideological struggle of the twenty-first century.* Third Edition. Lanham, MD: Rowman and Littlefield Publishers, 2009.

Steger, Manfred B. *Globalization. A very short introduction.* Oxford and New York: Oxford University Press, 2003.

Steger, Manfred B., James Goodman and Erin K. Wilson. *Justice globalism. Ideology, crises, policy.* Los Angeles, London, New Delhi, Singapore, Washington, D.C.: Sage Publications, 2013.

Stenbock-Hult, Bettina and Anneli Sarvimäki. 'The meaning of vulnerability to nurses caring for older people'. *Nursing Ethics* 18, no. 1 (2011): 31–41.

Stewart, Frances and Severine Deneulin. 'Amartya Sen's contribution to development thinking'. *Studies in Comparative International Development* 37, no. 2 (2002): 61–70.

Stewart, Frances. 'Groups and capabilities'. *Journal of Human Development* 6, no. 2 (2005): 185–204.

Stiglitz, Joseph E. *The price of inequality.* New York and London: W.W. Norton & Company, 2012.

Stone, John. 'Race and healthcare disparities: Overcoming vulnerability'. *Theoretical Medicine* 23, no. 6 (2002): 499–518.

Stone, T. Howard. 'The invisible vulnerable: The economically and educationally disadvantaged subjects of clinical research'. *Journal of Law, Medicine & Ethics* 31, no. 1 (2003): 149–153.

Stutzki, Ralf, Markus Weber and Stella Reiter-Theil. 'Finding their voices again: a media project offers a floor for vulnerable patients, clients and the socially deprived'. *Medicine, Health Care and Philosophy* 16, no. 4 (2013): 739–750.

Suhrke, Astri. 'Human security and the interests of states'. *Security Dialogue* 30, no. 3 (1999): 265–276.

Sunder Rajan, Kaushik. *Biocapital. The construction of postgenomic life*. Durham and London: Duke University Press, 2006.

Supiot, Alain. *L'esprit de Philadelphie. La justice sociale face au marché total*. Paris: Seuil, 2010.

Svenaeus, Fredrik. 'Illness as unhomelike being-in-the-world: Heidegger and the phenomenology of medicine'. *Medicine, Health Care and Philosophy* 14 (2011): 333–343.

Taylor, Charles. *Human agency and language. Philosophical papers I*. Cambridge: Cambridge University Press, 1985.

Taylor, James Stacey. *Stakes and Kidneys: Why markets in human body parts are morally imperative*. Aldershot: Ashgate, 2005.

Ten Have, Henk A.M.J. and Jos V.M. Welie. 'Medicine, ownership, and the human body'. In Henk A.M.J. ten Have and Jos V.M. Welie (eds.). *Ownership of the human body. Philosophical considerations on the use of the human body and its parts in healthcare*. Dordrecht, Boston, London: Kluwer Academic Press, 1998: 1–15.

Ten Have, Henk A.M.J. 'Bodies of knowledge, philosophical anthropology, and philosophy of medicine'. In H. Tristram Engelhardt (ed.). *The philosophy of medicine. Framing the field*. Dordrecht, Boston, London: Kluwer Academic Publishers, 2000: 19–36.

Ten Have, Henk. 'The activities of UNESCO in the area of ethics'. *Kennedy Institute of Ethics Journal* 16, no. 4 (2006): 333–351.

Ten Have, Henk. 'UNESCO's Ethics Education Programme'. *Journal of Medical Ethics* 34, no. 1 (2008): 57–59.

Ten Have, Henk and Michèle S. Jean (eds.). *The UNESCO Universal Declaration on Bioethics and Human Rights. Background, principles and application*. Paris: UNESCO Publishing, 2009.

Ten Have, Henk. 'Global bioethics and communitarianism'. *Theoretical Medicine and Bioethics* 32 (2011): 315–326.

Ten Have, Henk. 'Potter's notion of bioethics'. *Kennedy Institute of Ethics Journal* 22, no. 1 (2012): 59–82.

Ten Have, Henk. 'Bioética sem fronteiras' (Bioethics without borders). In Dora Porto, Volnei Garrafa, Gerson Zalafon Martins and Swedenberger do Nacimento Barbosa (eds.). *Bioéticas, poderes e injustiças – 10 anos depois*. Conselhos Federal de Medicina, Cátedra Unesco de Bioética, Sociedade Brasileira de Bioética: Brasilia 2012: 43–61.

Ten Have, Henk. 'Bioethics and Human Rights – Wherever the twain shall meet'. In Silja Vöneky, Britta Beylage-Haarmann, Anja Höfelmeier and Anna-Katharina Hübler (eds.). *Ethik und Recht – Die Ethisierung des Rechts / Ethics and Law - The ethicalization of law*. Heidelberg, New York, Dordrecht, London: Springer Publishers, 2013: 149–175.

Ten Have, Henk and Bert Gordijn (eds.). *Handbook of global bioethics*. Dordrecht: Springer Publishers, 2013.

Ten Have, Henk. 'The principle of vulnerability in the UNESCO Declaration on Bioethics and Human Rights'. In Joseph Tham, Alberto Garcia and Gonzalo Miranda (eds.). *Religious perspectives on human vulnerability in bioethics*. Dordrecht: Springer Publishers, 2014: 15–28.

Ten Have, Henk. *Global bioethics. An introduction*. London and New York: Routledge, 2016.

Thomas, Caroline. 'Globalization and human security'. In Anthony McGrew and Nana K. Poku (eds.). *Globalization, development and human security*. Cambridge, UK and Malden, MA: Polity Press, 2007: 107–131.

Thomasma, David C. 'Bioethics and international human rights'. *Journal of Law, Medicine and Ethics* 25 (1997): 295–306.

Thomasma, David. 'The vulnerability of the sick'. *Bioethics Forum* 16 (2000): 5–12.

Thomasma, David C. 'Proposing a new agenda: Bioethics and international human rights'. *Cambridge Quarterly of Healthcare Ethics* 10 (2001): 299–310.

Titmuss, Richard. *The gift relationship: from human blood to social policy*. New York: Pantheon Books, 1971.

Tomm-Bonde, Laura. 'The naïve nurse; revisiting vulnerability for nursing'. *BMC Nursing* 155:5 (2012); DOI: 10.1186/1472-6955-11-5.

Toombs, S. Kay. *Handbook of phenomenology and medicine*. Dordrecht, Boston, London: Kluwer Academic Publishers, 2001.

Traquair, H.M. 'The special vulnerability of the macular fibres and "sparing of the macula"'. *British Journal of Opthalmology* 9, no. 2 (1925): 53–57.

Tronto, Joan C. *Moral boundaries. A political argument for an ethic of care*. New York and London: Routledge, 1993.

Tronto, Joan. 'Vicious circles of privatized caring'. In Maurice Hamington and Dorothy C. Miller (eds.). *Socializing care. Feminist ethics and public issues*. Lanham, MD: Rowman and Littlefield Publishers, 2006: 3–25.

Tronto, Joan. 'Creating caring institutions: Politics, plurality, and purpose'. *Ethics and Social Welfare* 4, no. 2 (2010): 158–171.

Turner II, B.L., Roger E. Kasperson, Pamela A. Matson, James J. McCarthy, Robert W Corell, Kindsey Christensen, Noelle Eckley, Jeanne X. Kasperson, Amy Luers, Marybeth L. Martello, Colin Polsky, Alexander Pulsipher, and Andrew Schiller. 'A framework for vulnerability analysis in sustainability science'. *Proceedings of the National Academy of Science of the United States of America* 100, no. 14 (2003): 8074–8079.

Turner, Bryan S. *The body and society. Explorations in social theory*. (Second Edition). London: Sage Publications, 1996.

Turner, Bryan S. *Vulnerability and human rights*. University Park, PA: The Pennsylvania State University Press, 2006.

Turner, Bryan S. and Alex Dumas. 'Vulnerability, diversity and scarcity: on universal rights'. *Medicine, Health Care and Philosophy* 16, no. 4 (2013): 663–670.

Tusaie, Kathleen and Janyce Dyer. 'Resilience: a historical review of the construct'. *Holistic Nursing Practice* 18, no. 1 (2004): 3–8.

UN Conference on Environment and Development. *Rio Declaration on Environment and Development*. Rio de Janeiro, June 1992 (http://www.unesco.org/pv_obj_cache/pv_obj_id_5153A80E5D000D9118833F33BE125378F1050100/filename/RIO_E.PDF) (Accessed 18 June 2015).

UN (United Nations). *Human security in theory and practice. Application of the human security concept and the United Nations Trust Fund for Human Security*. New York: Human Security Unit, Office for the Coordination of Humanitarian Affairs, 2009 (http://hdr.undp.org/en/media/HS_Handbook_2009.pdf) (Accessed 18 June 2015).

UN (United Nations). Committee on Economic, Social and Cultural Rights. *General comment No. 14: The right to the highest attainable standard of health*. Geneva: Economic and Social Council, UN Doc.E/C.12/2000/4, 2000 (http://www.unhchr.ch/tbs/doc.nsf/(symbol)/E.C.12.2000.4.En) (Accessed 18 June 2015).

UNDP (United Nations Development Programme). *Human Development Report 1990. Concept and measurement of human development*. New York and Oxford: Oxford University Press, 1990.

UNDP (United Nations Development Programme). *Human Development Report*. New York: UNDP, 1994.

UNDP (United Nations Development Programme). *Human Development Report 1999*. New York: Oxford University Press, 1999.

UNEP (United Nations Environment Programme). *Assessing human vulnerability to environmental change. Concepts, issues, methods and case studies*. Nairobi: UNEP, 2003.

UNESCO (United Nations Educational Scientific and Cultural Organisation). *Universal Declaration on the Human Genome and Human Rights*. Paris: UNESCO, 1997. (http://unesdoc.unesco.org/images/0012/001229/122990eo.pdf) (Accessed 25 January 2015).

UNESCO (United Nations Educational Scientific and Cultural Organisation). *International Declaration on Human Genetic Data*. Paris: UNESCO, 2003. (http://portal.unesco.org/en/ev.php-URL_ID=17720&URL_DO=DO_TOPIC&URL_SECTION=201.html) (Accessed 25 January 2015).

UNESCO (United Nations Educational Scientific and Cultural Organisation). *Universal Declaration on Bioethics and Human Rights*. Paris: UNESCO, 2005. (http://unesdoc .unesco.org/images/0014/001461/146180e.pdf) (Accessed 12 October 2015).

UNFPA and HelpAge International. *Ageing in the twenty-first century. A celebration and a challenge*. New York and London: United Nations Population Fund & HelpAge International; New York and London, 2012. (http://www.unfpa.org/webdav/site/ global/shared/documents/publications/2012/UNFPA-Exec-Summary.pdf). (Accessed 12 March 2015).

UNICEF. *Investing in vulnerable children*. Dhaka: Unicef Bangladesh, 2010. (http://www.unicef .org/socialpolicy/files/Investing_in_children_Web.pdf).(Accessed 12 March 2015).

United Nations Department of Economic and Social Affairs. *Report on the World Social Situation, 2003. Social vulnerability: Sources and Challenges*. New York: United Nations, 2003.

United Nations General Assembly, 23rd Session. Resolution 2450 (XXIII). *Human rights and scientific and technological developments*. 18 December 1968, 54. (http://www.un .org/ga/search/view_doc.asp?symbol=A/RES/2450(XXIII)&Lang=E&Area= RESOLUTION) (Accessed 12 April 2015).

United States Senate, ninety-third Congress. *Quality of Health Care – Human experimentation, 1973*. Hearings before the Subcommittee on Health of the Committee on Labor and Public Welfare. Washington, D.C.: U.S. Government Printing Office, 1973.

Uscher-Pines, Lori, Patrick S. Duggan, Joshua P. Garoon, Ruth A Karron, and Ruth R. Faden. 'Planning for an influenza pandemic. Social justice and disadvantaged groups'. *Hastings Center Report* 37, no. 4 (2007): 32–39.

Valdman, Mikhail. 'Exploitation and injustice'. *Social Theory and Practice* 34, no. 4 (2008): 551–572.

Valdman, Mikhail. 'A theory of wrongful exploitation'. *Philosophers' Imprint* 9, no. 6 (2009): 1–14.

Vallentyne, Peter. 'Debate: Capabilities versus opportunities for well-being'. *Journal of Political Philosophy* 13, no. 3 (2005): 359–371.

Van der Meide, Hanneke, Carlo Leget and Gert Olthuis. 'Giving voice to vulnerable people: the value of shadowing for phenomenological healthcare research'. *Medicine, Health Care and Philosophy* 16, no. 4 (2013): 731–737.

Van Wolputte, Steven and Patrick Meurs. 'Lichaam, zorg, kwetsuur en verzorging. Een antropologie van de menselijke kwetsbaarheid'. *Medische Antropologie* 14, no. 1 (2002): 170–189.

Venkatapuram, Sridhar. *Health justice. An argument from the capabilities approach*. Cambridge, UK: Polity Press, 2011.

Verkerk, Marian A. 'The care perspective and autonomy'. *Medicine, Health Care and Philosophy* 4 (2001): 289–294.

Vladeck, Bruce C. 'How useful is "vulnerable" as a concept?' *Health Affairs* 26, no. 5 (2007): 1231–1234.

Vlieghe, Joris. 'Judith Butler and the public dimension of the body: Education, critique and corporeal vulnerability'. *Journal of Philosophy of Education* 44, no. 1 (2010): 153–170.

Vrousalis, Nicholas. 'Exploitation, vulnerability, and social domination'. *Philosophy and Public Affairs* 41, no. 2 (2013): 131–157.

Wadhwa, Vandana, Jayati Ghosh, and Ezekiel Kalipeni. 'Factors affecting the vulnerability of female slum youth to HIV/AIDS in Delhi and Hyderabad, India'. *GeoJournal* 77 (2012): 475–488.

Waiton, Stuart. *The politics of antisocial behaviour. Amoral panics*. New York and London: Routledge, 2008.

Walker, Margaret Urban. 'The curious case of care and restorative justice in the U.S. context'. In Maurice Hamington and Dorothy C. Miller (eds.). *Socializing care. Feminist ethics and public issues*. Lanham, MD: Rowman and Littlefield Publishers, 2006: 145–162.

Watts, Michael J. and Hans G. Bohle, 'The space of vulnerability: The causal structure of hunger and famine'. *Progress in Human Geography* 17, no. 1 (1993)): 43–67.

Weir, Alison. 'The Global Universal Caregiver: Imagining women's liberation in the new millennium'. *Constellations* 12, no. 3 (2005): 308–330.

Weir, Alison. 'Global care chains: Freedom, responsibility, and solidarity'. *Southern Journal of Philosophy* 46 (2008): 166–175.

Werner, Emmy E. 'Risk, resilience, and recovery: Perspectives from the Kauai longitudinal study'. *Development and Psychopathology* 5 (1993): 503–515.

Wertheimer, Alan. *Exploitation*. Princeton, NJ: Princeton University Press, 1996.

Wertheimer, Alan. 'Exploitation in clinical research'. In Jennifer S. Hawkins and Ezekiel J. Emanuel (eds.). *Exploitation and developing countries. The ethics of clinical research.* Princeton and Oxford: Princeton University Press, 2008: 63–104.

WFP (World Food Programme). *Sampling guidelines for vulnerability analysis.* Rome: WFP, 2004.

Whelan, Daniel, 'Human rights approaches to an expanded response to address women's vulnerability to HIV/AIDS'. *Health and Human Rights* 3, no. 1 (1998): 20–36.

White, Julie A. and Joan C. Tronto. 'Political practices of care: Needs and rights'. *Ratio Juris* 17, no. 4 (2004): 425–453.

WHO (World Health Organization). *Health aspects of human rights: with special reference to developments in biology and medicine.* Geneva: WHO, 1976.

WHO (World Health Organization). *The world health report 2007 – A safer future: global public health security in the twenty-first century.* Geneva: WHO, 2007.

WHO (World Health Organization). *Everybody's business: Strengthening health systems to improve health outcomes. WHO's framework for action.* Geneva: WHO, 2007. (http://www .who.int/healthsystems/strategy/everybodys_business.pdf) (Accessed 7 April 2015).

WHO. *World Health Statistics 2011.* Geneva: WHO, 2011.

Wild, Verina. 'How are pregnant women vulnerable research participants?'. *International Journal of Feminist Approaches to Bioethics* 5, no. 2 (2012): 82–104.

Wilkinson, Richard. *Unhealthy societies. The afflictions of inequality.* London and New York: Routledge, 1996.

Wilkinson, Stephen. '"The exploitation argument against commercial surrogacy'. *Bioethics* 17, no. 2 (2003): 169–187.

Williams, Bernard. *Ethics and the limits of philosophy.* London: Fontana Press and Collins, 1985.

Wilson, Cecil B. 'An updated Declaration of Helsinki will provide more protection'. *Nature Medicine* 19, no. 6 (2013): 664.

Wilson, John, R. S. 'In one another's power'. *Ethics* 88, no. 4 (1978): 299–315.

WMA (World Medical Association). *Declaration of Helsinki*, Fifth Revision. 2000. (http:// web.archive.org/web/20071027224123/www.wma.net/e/policy/pdf/17c.pdf) (Accessed 26 January 2015).

WMA (World Medical Association). *Declaration of Helsinki*, Sixth Revision. 2008. (http:// www.wma.net/en/30publications/10policies/b3/index.html.pdf?print-media-type&- footer-right=[page]/[toPage] (Accessed 26 January 2015).

WMA (World Medical Association). *Declaration of Helsinki*, Seventh Revision 2013. (http:// www.wma.net/en/30publications/10policies/b3/) (Accessed 20 August 2015).

Wood, Allen W. 'Exploitation'. *Social Philosophy and Policy* 12, no. 2 (1995): 136–158.

Wyn Schofer, Jonathan. *Confronting vulnerability. The body and the divine in rabbinic ethics.* Chicago and London: The University of Chicago Press, 2010.

Yeates, Nicola. 'A global political economy of care'. *Social Policy & Society* 4, no. 2 (2005): 227–234.

Young, Iris Marion. 'Responsibility and global labor justice'. *Journal of Political Philosophy* 12, no. 4 (2004): 365–388.

Young, Iris Marion. 'Responsibility and global justice: A social connection model'. *Social Philosophy and Policy* 23, no. 1 (2006): 102–130.

Young, Iris Marion. 'Five faces of oppression'. In George Henderson and Marvin Waterstone (eds.): *Geographic thought. A praxis perspective*. London and New York: Routledge, 2009: 55–71.

Young, Iris Marion. *Justice and the politics of difference*. Princeton and Oxford: Princeton University Press, 2011 (Second Edition).

Young, Iris Marion. *Responsibility for justice*. Oxford and New York: Oxford University Press, 2013.

Zaidi, Akbar. 'Poverty and disease: Need for structural change'. *Social Science and Medicine* 27, no. 2 (1988): 119–127.

Zaner, Richard. *Ethics and the clinical encounter*. Englewood Cliffs, NJ: Prentice Hall, 1988.

Zaner, Richard, 'Power and hope in the clinical encounter: A meditation on vulnerability'. *Medicine, Health Care and Philosophy* 3 (2000): 265–275.

Zaner, Richard. 'Physicians and patients in relation: Clinical interpretation and dialogues of trust'. In George Khushf (ed.). *Handbook of Bioethics: Taking stock of the field from a philosophical perspective*. Dordrecht, Boston, London: Kluwer Academic Publishers, 2004: 223–250.

Zion, Deborah, Lynn Gillam and Bebe Loff. 'The Declaration of Helsinki, CIOMS and the ethics of research on vulnerable populations'. *Nature Medicine* 6, no. 6 (2000): 615–617.

Zwolinski, Matt. 'Structural exploitation'. *Social Philosophy and Policy* 29, no. 1 (2012): 154–179.

INDEX